D0944623

Unimolecular Reactions

Unimolecular Reactions

P. J. ROBINSON

Chemistry Department
The University of Manchester Institute of Science and
Technology

K. A. HOLBROOK

Chemistry Department
University of Hull

WILEY–INTERSCIENCE

A division of John Wiley & Sons Ltd

London · New York · Sydney · Toronto

Library of Congress catalog card number
70-161690

ISBN 0 471 72814 4

Printed in Great Britain by John Wright and Sons Ltd. at The Stonebridge Press, Bristol, England

To M. E. R. and A. M. H.

Foreword

The past twenty years have witnessed a high level of experimental activity in the field of unimolecular reactions—stimulated in large part by the near-concurrence of the advent of the special Slater theory, on the one hand, and the Marcus–Rice equilibrium modification (RRKM) of the Rice–Ramsperger–Kassel theory, on the other. Since these treatments maintain different central assumptions with regard to the time scale for intramolecular relaxation of internal energy of excitation, considerable zest was added thereby to critical experimentation. The advances that followed have involved both the study of neutrals and of charged species, although the former came earlier into context with the RRKM theory as a practical, working calculational model. The comparative success of this treatment provides a convenient and hospitable way-station on the path of progress, even as the study of unimolecular reactions continues to advance and refine. Recent experimental work, including the study of ionic species, gives promise of further developments, while on the theoretical side the connections of the equilibrium treatment with phase space theory, resonant scattering theory and other quantum mechanical aspects are being strengthened.

This writer, along with many readers, knows of the difficulties presently encountered by adherents of these subjects in wending their way through the accumulated literature that deals with the RRKM theory, and in finding clear and explicit delineation of the several theoretical and calculational advances. The authors of this book have set for themselves a limited, but very worth-while task. They have undertaken to give a clear and detailed description of the RRKM treatment of unimolecular reactions—with full attention to the mechanics of actual computation. At the same time, they have surveyed the experimental literature on reactions of neutrals and have provided many insights into the findings of the past decades and the connections between these and various statements that may be derived from the RRKM formalism. In all of this they will have won for themselves the grateful thanks of students as well as those of practising kineticists.

B. S. Rabinovitch
Trinity College, Oxford
July 1971

Preface

A unimolecular gas-phase reaction can be regarded as the simplest type of elementary reaction. The theory of such processes is therefore of considerable importance, especially since many aspects of the theory have now been developed to the stage of providing a practical calculation of kinetic behaviour from the fundamental properties of the reacting molecules. Although the foundations of current theories were laid about forty years ago, practising kineticists have been slow to make use of the existing theoretical treatments and apply them to their experimental results. A major reason for the lack of progress in this direction is the substantial complexity of the modern theories, and this book attempts to remedy the situation by discussing in considerable detail the derivations of the basic equations and their numerical applications to illustrative cases. The most important theory discussed is the RRKM (Marcus–Rice) theory, which has only been extensively used and tested since about 1960. Its practical application raises the difficult question of the evaluation of the necessary vibrational and rotational energy level densities and sums, and the various procedures for these calculations are described and assessed. The earlier theories of Lindemann, Hinshelwood, Rice, Ramsperger and Kassel, as well as Slater's harmonic theory, are treated in sufficient depth to provide a basis for the treatment of the RRKM theory and to permit comparison of the various theories to be made.

A second major function of the book is to provide an up-to-date compilation of experimental kinetic data for unimolecular reactions, including data for the fall-off regions, and these data are reviewed with particular emphasis on the applications of rate theories in cases where the unimolecular behaviour is well established. Other topics treated include the studies of chemical activation, kinetic isotope effects and collisional energy transfer in unimolecular reaction systems. The very rapid growth in the study of unimolecular reactions in recent years has inevitably meant that the scope of this book has had to be limited. Some important topics such as the unimolecular decompositions of charged species and recent quantum–mechanical theories of unimolecular reactions have been omitted in order to keep the book to a reasonable length.

ix

The book is intended primarily for the postgraduate physical chemist with a reasonable standard of mathematics. The less mathematical parts and the development of the fundamental theoretical equations should, however, prove useful reading for the average honours undergraduate.

P. J. ROBINSON
K. A. HOLBROOK
July 1971

Acknowledgements

We are greatly indebted to Professor B. S. Rabinovitch for reading the manuscript and providing many invaluable comments upon it. Thanks are also due to Dr. J. H. Knox, Professor N. B. Slater and Professor K. J. Laidler for their helpful remarks on various parts of the work, and to Professor P. G. Ashmore for his valuable comments, advice and encouragement.

We also acknowledge the co-operation of the following bodies in giving permission for the reproduction of diagrams from their publications:

The American Chemical Society
Figures 7.5 (*J. Phys. Chem.*, **69**, 436) and 7.10 (*J. Amer. Chem. Soc.*, **84**, 4217).

The American Institute of Physics
Figures 5.3 (*J. Chem. Phys.*, **38**, 2471), 5.6 (**38**, 2473), 8.4 (**38**, 411), 8.5 (**45**, 3234), 10.2 (**38**, 1679), 10.3 (**52**, 2146), 10.4 (**45**, 3726) and 10.5 (**48**, 1291).

The Chemical Society
Figure 8.6 [*J. Chem. Soc. (A)*, **1968**, 3062].

The Faraday Society
Figures 7.3 (*Trans. Faraday Soc.*, **64**, 936), 7.4 (**64**, 934), 7.6 (**64**, 938), 7.7 (**65**, 443) and 7.11 (**64**, 89).

The Royal Society
Figure 2.6 [*Proc. Roy. Soc. (A)*, **217**, 567].

Contents

1 Introduction and Early Theories

GENERAL INTRODUCTION

A unimolecular reaction is in principle the simplest kind of elementary reaction, since it involves the isomerization or decomposition of a single isolated reactant molecule A through an activated complex A^+ which involves no other molecule:

$$A \longrightarrow A^+ \longrightarrow \text{Products}$$

This book is concerned almost exclusively with gas-phase reactions, since the so-called 'unimolecular reactions' in condensed phases must necessarily involve participation of the surrounding molecules, and are therefore not unimolecular in the strict sense of the term.

The experimental study of gas-phase unimolecular reactions is often complicated by the simultaneous occurrence of surface processes or free-radical chain reactions, and Chapter 1 therefore starts (section 1.1) with a discussion of the experimental criteria for unimolecular reactions, emphasizing the need for careful assessment of the data for any particular reaction. Section 1.2 describes some general background of reaction rate theories, and deals in particular with potential energy surfaces and the Absolute Rate Theory. Most modern theories of unimolecular reaction rates are based on the fundamental Lindemann mechanism involving collisional energization of the reactant molecules, and more specifically on Hinshelwood's development of the original treatment. These two theories are described and their application is illustrated in section 1.3, and section 1.4 outlines the bases of the subsequent developments leading to the more detailed theories discussed in the following chapters.

The logical connections between the various chapters are indicated in Figure 1.1. Chapter 2 describes the Slater theory, which is a *dynamical* theory concerned with the detailed treatment of molecular vibrations and the behaviour of particular molecular coordinates as a function of time. Reaction is postulated to occur in Slater theory when a chosen coordinate achieves a critical extension by the phase-coincidence of certain modes of vibration. In the more realistic *statistical* theories there is assumed to be a rapid statistical redistribution of energy among the various degrees of

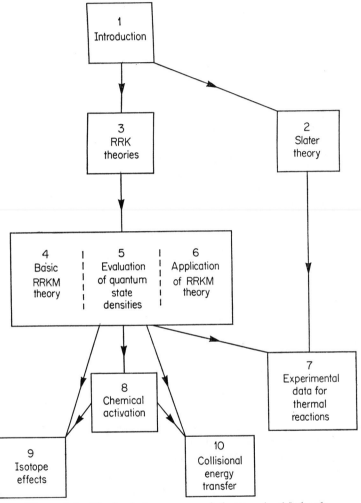

Figure 1.1   The logical sequence of chapters in this book

freedom of the molecule, and reaction may be said loosely to depend on the probability that sufficient energy will be found in the appropriate part of the molecule. The RRK theory (Chapter 3) was an early statistical theory which formed the basis of the later RRKM theory (Chapter 4), currently accepted to be the most accurate and practical treatment of unimolecular reactions. Application of the RRKM theory depends very much on an assessment of the distribution of quantum states of molecular systems, the relevant quantities being the density of quantum states at a given energy and the number of quantum states with energies less than this value. The evaluation of these quantities is a common source of difficulty and is dealt with separately in Chapter 5. Chapter 6 discusses the technique of application of the RRKM theory, including methods for constructing a model of a given reaction, and a detailed numerical example of the application of RRKM theory to a typical unimolecular reaction.

These chapters complete the basic discussion of unimolecular reaction rate theories, and are followed in Chapter 7 by a compilation of experimental data for a large number of unimolecular reactions. Data for both the high-pressure and low-pressure regions are presented, and the applications of various rate theories are discussed.

The remaining chapters discuss both theoretical and experimental aspects of some further topics of great importance to unimolecular reaction studies. Chapter 8 is concerned with chemical activation studies, Chapter 9 with kinetic isotope effects, and Chapter 10 with collisional energy transfer in unimolecular reactions.

The reader's attention is also drawn to Appendix 1, which summarizes the more important nomenclature of this book and compares it with that of other authors, and to Appendix 2, which reviews some aspects of Statistical Mechanics which are of particular importance to the present work. The RRKM theory in particular depends heavily on statistical-mechanical theory, and the aspects which are important here do not always feature prominently in the standard texts.

1.1   EXPERIMENTAL STUDY OF UNIMOLECULAR REACTIONS

1.1.1   Experimental methods

Although there are many excellent books[1] which deal with experimental techniques in the study of gas-phase reactions, some mention must be made here of the general methods used. It is important to be able to recognize a unimolecular reaction experimentally and to be aware of the complications which can occur. Most gas-phase kineticists study reactions either by the *static* or *flow* techniques. In the former, a known pressure of reactant is admitted to a constant-volume reaction vessel kept at a constant temperature and usually constructed of quartz or Pyrex glass. The reaction is then followed by the change in some physical property of the reacting gas with time, e.g. by measuring the total pressure (if a pressure change occurs) or by measuring light absorption in some convenient region of the spectrum. Alternatively, the reacting mixture can be analysed at various time intervals by one of a variety of methods such as infrared analysis, mass spectrometry or gas chromatography. In the flow method, on the other hand, the gas is made to flow at a known rate through a heated reaction vessel of known volume. The average residence-time of the gas in the heated zone can then be calculated and the rate of reaction evaluated from analyses of the inlet and exit gases.

In addition to conventional kinetic studies involving the pyrolysis or isomerization of molecules at room temperatures up to 500–600 °C, valuable information of relevance to the understanding of unimolecular reactions has been provided in recent years by the use of new techniques. For example, the temperature range of conventional studies has been extended by the use of shock tubes, and temperatures of 1000–2000 °C can be achieved with reactant residence-times in the region of 1 to 10^4 microseconds.[2]

Much experimental work recently has also been directed towards the problems of energy transfer between molecules (e.g. translational–vibrational energy transfer occurring upon collision) and of intramolecular vibrational–vibrational energy transfer. The experimental methods used have varied widely from spectroscopic methods[3] to the use of molecular beams[4] and experiments involving the dispersion of sound waves.[5] Although these experiments are of importance in relation to unimolecular reactions, and some of their results will be quoted later, it is not possible within the scope of this book to review all the experimental methods used. The reader is referred to the recent book on this topic by Stevens.[6] The production of vibrationally excited molecules by means of *chemical activation* has become particularly important in recent years, however, and an account of this topic is given in Chapter 8.

1.1.2 Deductions concerning mechanism

The observations of a large number of reaction products and complicated orders of reaction are almost certain indications of a free radical reaction rather than a unimolecular one. Unfortunately the reverse conclusion is not necessarily true. Rice and Herzfeld[7] were the first to show how complicated free radical mechanisms for the pyrolysis of organic compounds can give simple overall orders and often relatively few major products. In principle, free radicals in gaseous systems can be detected by a number of physical methods such as mass spectrometry and e.s.r. spectroscopy. In practice, however, the concentrations of free radicals generated thermally are usually near the limits of detection by most methods, and therefore the presence of free radicals must be inferred by indirect methods.

It should be noted that the presence of free radicals in a system does not in itself preclude the occurrence of unimolecular reactions. Indeed, many free radicals are themselves generated by the unimolecular fission of stable molecules, e.g. for ethane the process (1.1). The problem is that the

$$C_2H_6 \longrightarrow 2\ CH_3 \qquad (1.1)$$

presence of free radicals often indicates that the mechanism of the reaction consists of a series of reaction steps. The overall orders may be simple, but the rate-constants deduced are combinations of rate-constants relating to the individual reaction steps. Sometimes it is still possible to derive information relating to a single unimolecular rate-constant, as for example in the unimolecular decomposition of methyl chloride. In this reaction the unimolecular process (1.2) is followed by a non-chain sequence in which

$$CH_3Cl \longrightarrow CH_3 + Cl \qquad (1.2)$$

each chlorine atom produced by the above reaction leads ultimately to three molecules of hydrogen chloride (see section 7.3). Information about the unimolecular rate-constant is then obtainable from measurements of the rate of production of hydrogen chloride.

It is often possible to write alternative molecular and free radical chain processes which lead to the same products. For ethyl chloride, for example, we can write mechanisms (A) and (B) which both predict initial rates which

$$C_2H_5Cl \longrightarrow C_2H_4 + HCl \qquad (A)$$

$$C_2H_5Cl \longrightarrow C_2H_5 + Cl \qquad (a)$$

$$C_2H_5 + C_2H_5Cl \longrightarrow C_2H_4Cl + C_2H_6 \quad (b)$$

$$Cl + C_2H_5Cl \longrightarrow C_2H_4Cl + HCl \quad (c) \qquad (B)$$

$$C_2H_4Cl \longrightarrow C_2H_4 + Cl \qquad (d)$$

$$C_2H_4Cl + Cl \longrightarrow C_2H_4Cl_2 \qquad (e)$$

are first-order in ethyl chloride at high pressures of reactant, but whereas the former first-order rate-constant is simply the unimolecular rate-constant k_A, the latter scheme yields the rate-constant k_B which is given by (1.3). If in fact the reaction proceeds by a chain mechanism, any interpreta-

$$k_B = (k_a k_c k_d / k_e)^{\frac{1}{2}} \qquad (1.3)$$

tion of the behaviour of k_B must involve the dependence of k_B upon the other rate-constants involved. One must therefore decide on some grounds other than the overall order of reaction whether the molecular or the chain mechanism is the more likely. In the case of ethyl chloride pyrolysis there are reasons to believe that step (*d*) in the chain mechanism is slow, thus favouring the molecular mechanism.[8] Often in the past the distinction between molecular and free radical chain mechanisms has been based upon the action of inhibitors or radical chain initiators upon the reaction. Reactions proceeding by free radical intermediates are more likely to be influenced by the addition of trace amounts of substances which can start or stop chains than are molecular processes. It must be emphasized, however, that firm conclusions cannot be drawn solely from observations of the rate in the presence of such substances as oxygen, nitric oxide, and propene. In particular, inhibition studies carried out in the presence of nitric oxide have been the subject of much controversy in recent years and their explanation is certainly much more complicated than was originally proposed. The fact that nitric oxide can actually initiate chain reactions under suitable circumstances[9] makes it very unwise to assume that the residual reaction occurring in the presence of excess inhibitor is a purely molecular reaction.

Bauer has recently suggested an 'operational criterion' for the recognition of concerted unimolecular elimination reactions.[10] In the elimination of small molecules such as N_2, CO or CO_2 from ring compounds there is frequently a large change in a particular bond distance (e.g. the N—N distance) on going from reactant to product. If the reaction occurs by a concerted process the transition state has a very short lifetime and the small molecular product is likely to be produced in a geometrically distorted and hence vibrationally excited state. A stepwise process, on the other hand, involves fission of a single bond to give an intermediate with a lifetime of several molecular vibrations before breaking the second bond. The high efficiency of intramolecular energy transfer then permits geometric relaxation of the fragment ejected. Since vibrational excitation of the products can be detected by laser techniques the two cases can in principle be distinguished experimentally.

Whilst no single test for the presence of a purely molecular mechanism rather than a chain mechanism is completely unambiguous, it is neverthe-less possible to be fairly confident about the conclusions if in a particular case several independent lines of reasoning agree with each other. In Chapter 7, the experimental evidence concerning a series of well-known examples of unimolecular reactions will be discussed in more detail.

1.1.3   Heterogeneous processes

One further complication often overlooked by gas kineticists is the question of the role of the surface in reactions presumed to occur in the gas phase. The usual test for surface activity has been to pack the reaction vessel with short lengths of glass tubing and to determine the effect upon the reaction rate of the change in surface/volume ratio. The absence of an effect of surface/volume ratio upon reaction rate does not necessarily imply a homogeneous reaction, however, since it is possible for a radical-chain process which is both surface-initiated and surface-terminated to be unaffected by changes in the surface/volume ratio. Tests for homogeneity and for molecularity are therefore sometimes related to each other. It is often a better test of homogeneity to change the nature of the surface rather than simply to change its extent. The pyrolyses of hydrocarbons involve non-polar transition states and have been shown to be largely unaffected by the nature and extent of the vessel surface.[11, 12] The pyrolyses of alkyl halides, on the other hand, involve some degree of charge separation in the activated complex and the rates and product compositions have sometimes been found to vary with the nature of the vessel surface.[13]

The conditioning of reaction vessel surfaces by the production of carbon films can lead to surfaces which possess high free-electron contents.[14, 15] Such surfaces may be active in promoting *cis-trans* isomerizations and possibly in initiating other free radical reactions. The carbon films derived from allyl bromide and certain other halogenated hydrocarbons are known to be particularly active catalysts. Whether this activity is due to the free-electron content or to some other property of the films (such as their halogen content) is not certain at present. It is clear, however, that such surfaces produced by conditioning must be demonstrated to be inert rather than simply assumed to be so.

1.2   POTENTIAL ENERGY SURFACES, ACTIVATED COMPLEXES AND ABSOLUTE RATE THEORY

In order to undergo unimolecular reaction, molecules must first acquire the necessary internal energy; molecules which possess sufficient energy to react are termed *energized molecules* (or sometimes activated molecules). In thermal systems energization is achieved by intermolecular collisions, although it can also be brought about by other means such as the absorption of radiation. Energized molecules have sufficient energy to react but do not necessarily do so; the energy may need to be redistributed among the internal degrees of freedom of the molecule and even with a suitable energy distribution the molecular vibrations must come correctly into phase before the molecular rearrangement can occur. The molecule then attains a *critical configuration* which is structurally intermediate between reactant and products, and is said to be an activated complex (or transition state). A more detailed discussion of these points will be found in section 4.1.

The processes of energy transfer and molecular motion are conveniently discussed in terms of *potential energy surfaces*. The potential energy for a molecule can be expressed in general as a function of n internal coordinates of the molecule. For a non-linear molecule $n = 3N - 6$, and for a linear molecule $n = 3N - 5$, where N is the number of atoms in the molecule. If the potential energy is plotted against these n coordinates, then a hyper-surface of dimension $(n + 1)$ results.

The simplest case to consider is that of a diatomic molecule which has a single internal coordinate—the internuclear distance. When the potential energy of the system is plotted against this distance a two-dimensional diagram results. For real molecules, the result is of the general form shown as a solid curve in Figure 1.2. Various empirical equations have been

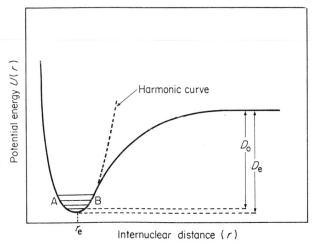

Figure 1.2 The potential energy curve for a diatomic molecule

proposed to represent the shape of this curve, the best known being the Morse equation (1.4) in which D_e is the electronic binding energy (equal

$$\text{Potential energy } U(r) = D_e\{1 - \exp[-a(r - r_e)]\}^2 \qquad (1.4)$$

to the ground-state dissociation energy D_0 plus the zero-point energy) and a is a constant for a given molecule. Near the equilibrium internuclear distance ($r \approx r_e$), the curve is approximately harmonic. Vibrational energy levels are represented by horizontal lines such as AB in Figure 1.2. After the first few vibrational levels the solid curve diverges from the harmonic curve and tends asymptotically to the energy level representing the two separated atoms at rest at infinite internuclear separation.

In order for unimolecular decomposition of a diatomic molecule to occur, the molecule must acquire the necessary energy, D_0. Since there is only one mode of vibration for a diatomic molecule, every energized molecule will achieve a critical extension (say $r = r_c$) and thus be considered to dissociate during the period of its first vibration. The fraction of molecules with energy $\geqslant D_0$ in thermally energized systems is given by the Boltzmann factor $\exp(-D_0/kT)$, and hence the rate-constant for the reaction at high pressure is given by (1.5) where ν is the vibration frequency of the molecule.

$$k_\infty = \nu \exp(-D_0/kT) \qquad (1.5)$$

For a triatomic molecule restricted to linear configurations, the potential energy can be represented in terms of two internuclear distances. The resulting contour diagram obtained by plotting potential energy for the symmetrical molecule Y_1XY_2 against the distances Y_1X and XY_2 is shown in Figure 1.3. This Figure differs from the more familiar contour diagram for the $H + H_2$ system in that the triatomic species exists in a deep potential well and is thus a stable molecule. Normally the molecule vibrates about its mean configuration in its various modes of vibration. The symmetrical

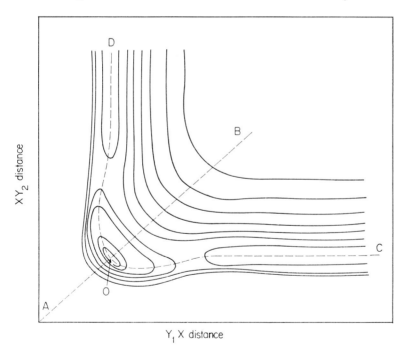

Figure 1.3 Potential energy surface for a symmetrical linear triatomic molecule Y_1XY_2

vibration corresponding to simultaneous increase or decrease of the two XY distances is represented by a motion along the line AB. Unimolecular dissociation of the molecule into $Y_1 + XY_2$ corresponds more to the asymmetric vibration and is represented by motion along OC, and dissociation into $Y_1X + Y_2$ corresponds to motion along OD. The path OC represents the lowest energy path on the surface by which the molecule undergoes the reaction $Y_1XY_2 \rightarrow Y_1 + XY_2$ and is known as the *reaction coordinate* for this reaction. A section through the surface along the reaction coordinate is the potential-energy profile shown in Figure 1.4.

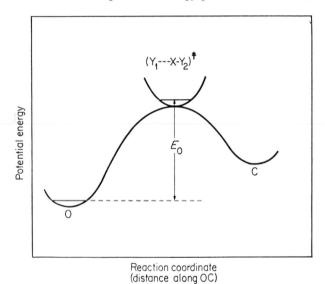

Figure 1.4   Potential energy profile for the reaction
$$Y_1XY_2 \longrightarrow Y_1 + XY_2$$

The potential-energy profile shows the activated complex $(Y_1 \cdots XY_2)^{\ddagger}$ corresponding to the highest energy point on this path. It is customary to regard as activated complexes those species lying within an arbitrary length δ at the top of the energy barrier, and to consider them as normal molecules except that they have one mode of vibration replaced by a free translational motion along the reaction coordinate (see also sections 4.1 and 4.5.1). The potential-energy profile along the reaction coordinate represents the minimum energy path for the reaction, and the *critical energy* (or threshold energy) E_0 is the difference between the zero-point energies of the reactant and the activated complex. This critical energy is not in general exactly equal to the experimental activation energy E_{Arr}, which is defined quite generally by (1.6) in terms of the slope of the

Arrhenius plot. The predicted expression for E_{Arr} from any given

$$E_{Arr} = -k\frac{d\ln k}{d(1/T)} = kT^2\frac{d\ln k}{dT} \tag{1.6}$$

theoretical treatment is similarly obtained by applying (1.6) to the theoretical expression for k. Tolman[16] showed that E_{Arr} for a unimolecular reaction at its high-pressure limit was equal to the average total energy $\overline{E^{\pm}}$ of the molecules which are actually undergoing the reaction (i.e. the activated complexes) minus the average total energy $\overline{E}$ of all the molecules (1.7). Since both molecules and activated complexes have average energies

$$E_{Arr} = \overline{E^{\pm}} - \overline{E} \tag{1.7}$$

in excess of their zero-point energies, due mainly to vibrational and rotational contributions which will not in general be equal, E_{Arr} will not in general be equal to E_0; usually $E_{Arr} > E_0$. The heat or enthalpy of activation, $\Delta H^{\pm}$, is sometimes quoted, and for a unimolecular reaction is equal to $E_{Arr} - kT$. Except for the special case where k is a first-order rate-constant, the value of E_{Arr} from (1.6) depends on whether the rate-constant is expressed in concentration or pressure units. For a second-order rate-constant k_2, for example, it is readily shown that

$$E_{Arr}(k_2 \text{ in conc. units}) = E_{Arr}(k_2 \text{ in press. units}) + kT.$$

For reactions involving more complicated molecules, the potential-energy surfaces are usually too difficult to visualize, since too many dimensions are involved. Sometimes it is possible to simplify matters by considering only two coordinates and keeping all others fixed, thus obtaining a section through the complete hypersurface. This is in fact what has been done here with the linear triatomic molecule, since the bending modes of vibration have been ignored.

Accurate calculations of potential-energy surfaces are difficult for even the simplest chemical systems, and the use of such calculations to predict activation energies and other properties of the activated complex, although possible in principle,[17] is in practice very limited. The concepts of the potential-energy surface and the activated complex are, however, invaluable to an understanding of the way in which chemical reactions occur, and form the basis of the Absolute Rate Theory.

The *Absolute Rate Theory*† (ART) applies thermodynamic and statistical-mechanical arguments to a system in which activated complexes are effectively at equilibrium with the reactant molecules (see also section 4.12.3 and ref. 18). For a general reaction $A + B \rightarrow$ Products, the

† Also known as Transition-State Theory or, perhaps most desirably, Activated Complex Theory.[18]

reaction scheme is (1.8). The equilibrium concentration of activated

$$A + B \underset{\text{eqm}}{\rightleftharpoons} AB^{\ddagger} \xrightarrow{\ k^{\ddagger}\ } \text{Products} \tag{1.8}$$

complexes is evaluated from statistical-mechanical equations, and together with expressions for the rate-constant $k^{\ddagger}$ leads[18, 19] to the equation (1.9)

$$k = \frac{-d[A]/dt}{[A][B]} = \frac{kT}{h} \frac{Q^{\ddagger}}{Q_A Q_B} \exp(-E_0/kT) \tag{1.9}$$

for the overall rate-constant. In this equation E_0 is the critical energy referred to above, Q_A and Q_B are the complete partition functions for the reactants, and $Q^{\ddagger}$ is the partition function for all the degrees of freedom of the activated complex except the reaction coordinate. The motion in the reaction coordinate has been considered separately and its partition function included in the factor kT/h.

For a unimolecular reaction there is only one reactant and the ART expression becomes (1.10); it will be seen later that ART gives the limiting

$$k_{\infty} = \frac{kT}{h} \frac{Q^{\ddagger}}{Q} \exp(-E_0/kT) \tag{1.10}$$

high-pressure rate-constant k_{∞} (see section 4.7). The translational contributions to Q and $Q^{\ddagger}$ are identical since the activated complex has the same mass as the reactant molecule. Assuming that rotational contributions also cancel and that classical harmonic vibrational partition functions may be inserted ($Q = kT/h\nu$), the approximation (1.11) results, in which $\nu_1, \ldots, \nu_n$

$$k_{\infty} = \frac{\nu_1 \nu_2 \nu_3 \ldots \nu_n}{\nu_1^{\ddagger} \nu_2^{\ddagger} \nu_3^{\ddagger} \ldots \nu_{n-1}^{\ddagger}} \exp(-E_0/kT) \tag{1.11}$$

are the n frequencies of the molecule and $\nu_1^{\ddagger}, \ldots, \nu_{n-1}^{\ddagger}$ are the $(n-1)$ frequencies of the activated complex. The pre-exponential factor at high pressures is therefore generally expected to be of the order of a vibration frequency, i.e. about $10^{13}\,\text{s}^{-1}$. Factors of this order of magnitude have in fact been found for a large number of unimolecular reactions (see Chapter 7). If the geometry of the activated complex is very different from that of the reactant molecule, the external rotational contributions to Q and $Q^{\ddagger}$ do not cancel, and the rate-constant is multiplied by a factor

$$(I_A^{\ddagger} I_B^{\ddagger} I_C^{\ddagger} / I_A I_B I_C)$$

involving the principal moments of inertia of the reactant and activated complex. When 'loose' activated complexes are formed, vibrational degrees of freedom in the reactant molecule are replaced by rotational degrees of freedom in the activated complex, and this factor may contribute to the much higher than normal pre-exponential factors found for such unimolecular reactions (see section 6.2.1).

The various theories of unimolecular reactions differ in their detailed descriptions of how the energized molecules become activated complexes. Crucial to this process is the question of whether all configurations with a given total energy are freely interconvertible by redistribution of the energy between the different degrees of freedom. If they are, then every energized molecule may in sufficient time become an activated complex. This is the case with the 'statistical' theories due to Rice, Ramsperger, Kassel and Marcus. The Slater theory, on the other hand, has more stringent requirements concerning the energization of the molecule; in the harmonic form of the theory there is no energy-flow between different modes of vibration, and not all molecules with total energy greater than E_0 are able to react.

These theories are all built on a common mechanism of collisional energization and de-energization, and these basic ideas are dealt with in the following section.

1.3 BASIC THEORIES OF UNIMOLECULAR REACTIONS

At the beginning of the twentieth century, many gas-phase reactions were known to be first-order processes and were assumed to be unimolecular, and unimolecular reactions were thought to be first-order under all conditions. Many reactions studied then, such as the pyrolyses of simple ketones, aldehydes and ethers, have been subsequently found not to be unimolecular processes according to the modern definition, but to involve free radical chains. Despite this complexity, the early studies of these reactions were important in the development of unimolecular reaction theory. They focused attention upon the central problem of how the reacting molecule acquires the activation energy needed for reaction to take place. It was difficult to see how first-order processes could result if molecules were energized by bimolecular collisions which would be expected to be second-order processes. It was argued (quite wrongly) that, even at very low pressures where molecular collisions were rare, unimolecular reactions would continue to occur with the same first-order rate-constant and hence molecular collisions would appear to be unimportant. In 1919, Perrin[20] therefore proposed the radiation hypothesis in which molecules were supposed to acquire energy by the absorption of infrared radiation from the walls of the reaction vessel. The rate-constant k for a first-order reaction would then be given by (1.12) in which ν is the frequency of the radiation absorbed. It was soon shown by Langmuir[21] and others

$$k = \text{constant} \times \exp(-h\nu/kT) \qquad (1.12)$$

that the density of infrared radiation available from the walls at the temperatures concerned was not sufficient to account for the observed reaction rates. In addition, experimental evidence was rapidly accumulated to show that infrared radiation is generally ineffective photochemically, and indeed many molecules do not absorb in the frequency region implied by (1.12) and the observed rate-constant. These facts led to the abandonment of the radiation theory and its replacement by theories in which molecular collisions were involved as the means of providing the activation energy.

In the theory of Christiansen and Kramers[22] an overall first-order rate was achieved, despite second-order collisional energization, by supposing that product molecules were produced with an excess of energy which could be used to re-energize reactant molecules. The process is described by the steps (1.13)–(1.15). Assuming stationary-state concentrations of A*

$$A + A \; \underset{k_2}{\overset{k_1}{\rightleftarrows}} \; A^* + A \qquad \text{Collisional energization/} \qquad (1.13)$$
$$\text{de-energization}$$

$$A^* \; \xrightarrow{\;k_3\;} \; B^* \qquad \text{Production of 'energy-rich'} \qquad (1.14)$$
$$\text{products}$$

$$B^* + A \; \xrightarrow{\;k_4\;} \; A^* + B \qquad \text{Energization of further A} \qquad (1.15)$$
$$\text{molecules}$$

and B* it can easily be shown that the overall rate v is given by (1.16). An equilibrium concentration of energized molecules A* is maintained

$$v = \frac{-d[A]}{dt} = k_3[A^*] = \left(\frac{k_3 k_1}{k_2}\right)[A] \qquad (1.16)$$

irrespective of the total pressure, and the overall rate is their first-order rate of reaction to give products. This theory proved to be unsatisfactory in two major respects. First, most unimolecular reactions are endothermic rather than exothermic processes and so product molecules are not formed with sufficient internal energy to energize more reactant molecules. Secondly, inert gases would be expected to remove the excess energy of the product molecules B* and hence to reduce the overall rate of the reaction. In practice it is found that inert gases often *increase* the rates of unimolecular reactions.

The disadvantages of earlier theories were overcome by the theory of Lindemann, outlined in the next section. The importance of molecular collisions in the energization process was finally established when it was found that the first-order rate-constant for unimolecular reactions is not a true constant but does decline at low pressures. The decline or 'fall-off' in the first-order rate-constant with pressure has since become recognized as an important experimental criterion of unimolecular reactions (see section 7.2).

A consequence of this decline in the rate-constant is that at low pressures the initial rate becomes proportional to the total pressure as well as to the concentration of reactant. The reaction is then second-order overall, although for reasons which will be apparent later the time-development of a given reaction mixture remains first-order. At still lower pressures it is possible that wall effects may become important in energization processes and there is some evidence that the rate-constant then becomes a true first-order constant again.

1.3.1 The Lindemann theory

The theory known as *Lindemann theory* forms the basis for all modern theories of unimolecular reactions and has been developed from ideas published almost simultaneously by Lindemann[23] and Christiansen.[24] The main concepts of the theory can be stated briefly as follows:

(a) By collisions, a certain fraction of the molecules become energized, i.e. gain energy in excess of a critical quantity E_0. The rate of the energization process depends upon the rate of bimolecular collisions. In the most general terms this process can be written as (1.17) where

$$A + M \xrightarrow{k_1} A^* + M \tag{1.17}$$

M can represent a product molecule, an added 'inert' gas molecule, or a second molecule of reactant. In the simple Lindemann theory k_1 is taken to be energy-independent and is calculated from the simple collision theory equation.

(b) Energized molecules are de-energized by collision. This is the reverse of process (1.17) and may be written as (1.18). The rate-

$$A^* + M \xrightarrow{k_2} A + M \tag{1.18}$$

constant k_2 is taken to be energy-independent, and is equated with the collision number Z_2, i.e. it is assumed that every collision of A^* leads to de-energization.

(c) There is a time-lag between the energization and unimolecular dissociation or isomerization of the energized molecule. The unimolecular dissociation process (1.19) also occurs with a rate-constant independent of the energy-content of A^*.

$$A^* \xrightarrow{k_3} B + C \tag{1.19}$$

When Lindemann first proposed his theory, he thought of the critical energy acquired by the molecule as rotational energy and wrote of a 'centrifugal bursting' of the molecule. Although the RRKM theory (to be dealt with later) considers rotational energy sometimes to be important, there is no doubt that the major contribution is usually from the vibrational energy of the molecule, and the energization process is considered largely as one of translational–vibrational energy transfer.

The consequences of the Lindemann mechanism, expressed by reactions (1.17)–(1.19), may be seen by application of the steady-state hypothesis to the concentration of A*. The overall rate of reaction is then given by

$$v = k_3[A^*] = \frac{k_1 k_3[A][M]}{k_2[M]+k_3} = \frac{(k_1 k_3/k_2)[A]}{1+k_3/k_2[M]} \tag{1.20}$$

At high pressures, when $k_2[M] \gg k_3$, (1.20) becomes

$$v_\infty = \left(\frac{k_1 k_3}{k_2}\right)[A] = k_\infty[A] \tag{1.21}$$

At low pressures, $k_2[M] \ll k_3$ and (1.20) becomes

$$v_{\text{bim}} = k_1[A][M] = k_{\text{bim}}[A][M] \tag{1.22}$$

i.e. the rate of reaction is then equal to the second-order rate of energization.

Initially $[M] = [A]$ (in the absence of added gases) and the initial rate is second-order in the concentration of A. However, unless the products and reactants differ substantially in their ability to energize molecules of A, the concentration $[M]$ remains virtually constant with time during a given experiment and the time-behaviour remains first-order.[25] This follows since (1.22) should really be written

$$v = \sum_i k_{1i}[A][M_i] = [A]\{k_{1A}[A]+k_{11}[B_1]+k_{12}[B_2]\dots\}$$

where B_1, B_2, etc. are the various reaction products. If there is only one reaction product B, i.e. for an isomerization reaction, and reactant and product are equally efficient at transferring energy to produce A*, then $k_{1A} = k_{11} = k_{12} = \dots = k_1$, and

$$v = k_1[A]\{[A]+[B]\}$$

Initially $[B] = 0$ and the initial rate of reaction is therefore given by

$$v_{t=0} = k_1[A]_{t=0}^2$$

At some time t, we can replace $[A]_t + [B]_t$ by the total pressure p which is constant, and therefore $v_t = k_1 p[A]_t$, i.e. the rate is first-order in [A] during a given run. For decomposition reactions, the situation is more complicated since the total pressure increases throughout the reaction. On the other hand, the decomposition products are usually much less efficient individually than the parent molecule in the energization step. These two effects act in opposition so that in practice the products are together about as efficient as the reactant, and the time-development of the reaction is again approximately first-order in the pressure of reactant.

To return to the main argument, it is seen that the Lindemann theory predicts a change in the order of the initial rate of reaction with respect to concentration at low pressures. This is reflected in a decline in the pseudo-first-order rate-constant k_{uni} with concentration, where k_{uni} is defined as in (1.23). The extent of this 'fall-off' is conveniently measured in terms

$$k_{uni} = \frac{1}{[A]}\left(\frac{-d[A]}{dt}\right) = \frac{k_1[M]}{1 + k_2[M]/k_3} \tag{1.23}$$

of the ratio k_{uni}/k_∞, given by (1.24), and a *transition-pressure* $p_{\frac{1}{2}}$ at which

$$k_{uni}/k_\infty = \frac{1}{1 + k_3/k_2[M]} \tag{1.24}$$

$k_{uni}/k_\infty = \frac{1}{2}$ is then found by putting $k_3 = k_2[M]$. Replacing [M] by the total pressure, we find that $p_{\frac{1}{2}}$ is given by (1.25).

$$p_{\frac{1}{2}} = k_3/k_2 = k_\infty/k_1 \tag{1.25}$$

Although the 'fall-off curve' of k_{uni} against pressure is not always obtainable over the whole range of pressure, it is sometimes useful to quote the transition pressure for a particular reaction at a particular temperature (see, for example, Tables 7.34 and 7.36).

It is easily seen from (1.24) that the predicted effect of adding inert gas below the high-pressure limit of a unimolecular reaction (i.e. in the fall-off region) is to increase the rate constant back to its high-pressure value. The effect is comparable to increasing the pressure of the reactant, although added gases are often less efficient than the reactant itself in the energization process. Studies of the efficiency of collisional energy transfer and their significance for unimolecular rate theories are treated in more detail in Chapter 10.

The simple conclusion from Lindemann theory concerning the high-pressure rate-constant k_∞ is that this is a true constant independent of pressure. No effect of added inert gases would therefore be predicted for a unimolecular reaction in its first-order region, since here the equilibrium proportion of energized molecules is already achieved. It is, however, possible that a small change in k_∞ may occur at very high pressures, due to a non-zero volume of activation ΔV^+; the isomerization of ethylcyclobutane[26] provides a recent example of this phenomenon.

1.3.2   Comparison of Lindemann theory with experiment

Early tests of the Lindemann theory were made by comparing the calculated rates of energization with the observed overall rates of unimolecular reactions. Since many of the examples chosen for such tests are now known not to be simple unimolecular reactions, calculations involving these reactions have now lost their original value. It is instructive instead to apply the simple Lindemann equation to some reliable unimolecular reaction data and to compare the experimental and theoretical results. From equations (1.24) and (1.25) the general first-order rate-constant can be written as (1.26) in which [M] has been replaced by the total pressure p.

$$k_{uni} = \frac{k_\infty}{1 + k_\infty/k_1 p} \qquad (1.26)$$

The rate-constant k_{uni} was calculated from equation (1.26) for the isomerization of cis- to trans-but-2-ene at 469 °C using the experimental data of Rabinovitch and Michel.[27] The experimental high-pressure rate-constant at 469 °C is $k_\infty = 1.90 \times 10^{-5}$ s^{-1} and the high-pressure activation energy of $E_\infty = 62.8$ kcal mol^{-1} is simply equated with E_0.† The rate-constant k_1 was calculated from the collision theory expression (1.27) with

$$k_1 = Z_1 \exp(-E_0/kT) \qquad (1.27)$$

$Z = 7.66 \times 10^6$ Torr^{-1} s^{-1}, calculated from the collision theory expression (6.21) with the collision diameter σ_d for cis-but-2-ene taken as 5×10^{-8} cm.

In Figure 1.5, k_{uni} is plotted logarithmically against reactant pressure p. It will be seen that the transition pressure calculated in this way is about 9×10^6 Torr, whereas the experimental data (see Figure 1.7) give a value of 0.04 Torr. Thus although the simple Lindemann theory correctly predicts a fall-off in k_{uni} with pressure, there is a gross discrepancy between the calculated and experimental fall-off curves. This is a general observation for all calculations using the simple Lindemann theory, and the principal reason is the erroneous calculation of k_1 by equation (1.27). This equation gives the frequency of collisions in which the relative kinetic energy along the line of centres is $\geqslant E_0$ and takes no account of the internal energy of the molecules. It will be seen later that a much more realistic model is obtained when this internal energy is taken into account in calculating the rate of energization. A further consequence of the use of (1.27) to calculate k_1 is seen by rearranging (1.25) to give $k_3 = k_\infty(k_2/k_1)$.

---

† In this book we use the non-SI units kcal and Torr, since these units (with Torr $\approx$ mmHg) have been adopted in most of the published work on unimolecular reactions up to the present time. The units have not generally been defined precisely, but where the definition is within our control, kcal = 4.184 kJ and Torr = (101.325/760) kN m^{-2}.

Since $k_2 = Z_2 \approx Z_1$ and $E_\infty \approx E_0$, it follows (1.28) that $k_3 \approx A_\infty$. The high-pressure A-factor is normally found experimentally to be about $10^{13\cdot5}$ s^{-1};

$$(k_3)_{\text{Lindemann}} = [A_\infty \exp(-E_\infty/kT)]Z_2/[Z_1 \exp(-E_0/kT)]$$
$$\approx A_\infty \qquad (1.28)$$

hence the theory in this form does not require a time-lag since k_3 would similarly be commensurate with a vibration frequency.

The 'time-lag' concept therefore requires k_3 to be much less than $10^{13\cdot5}$ s^{-1} and conversely k_1 to be much larger than its value calculated from

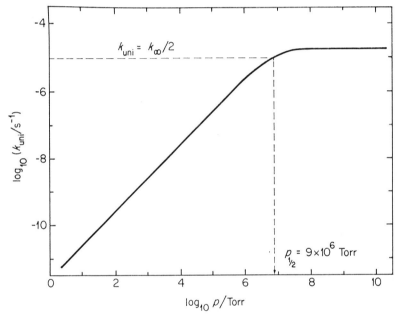

Figure 1.5 Theoretical Lindemann plot for *cis-* $\longrightarrow$ *trans*-but-2-ene at 469 °C

the collision theory expression. A modification of the theory, due to Hinshelwood, calculates k_1 with allowance for energy in the internal degrees of freedom of the molecule, and is described in the next section.

1.3.3 The Hinshelwood modification

The failure of Lindemann theory in its simple form has been illustrated by the calculation of transition pressures $p_{\frac{1}{2}}$ which are much too high to agree with experiment. Alternatively it could be shown that the rate of energization given by equation (1.27) is too low to account for the observed rate of reaction. Both situations are remedied by an increase in the rate-constant k_1. According to simple collision theory, the rate is calculated from the product of the collision number and the chance that two colliding molecules have relative translational energy $\geqslant E_0$ along the line of centres, i.e. have energy $\geqslant E_0$ in two classical degrees of freedom. Hinshelwood[28] developed a suggestion of Lindemann's that a more realistic model would be obtained by assuming that the required energy could be drawn in part from the internal degrees of freedom (mainly vibrational) of the reactant molecule. The chance of a molecule containing energy $\geqslant E_0$ clearly increases with the number of degrees of freedom which contribute, and the rate of energization is thus increased, as is required. Hinshelwood showed that the chance of a molecule possessing total energy $\geqslant E_0$ in s classical degrees of freedom is much higher than $\exp(-E_0/kT)$, and is in fact approximately $\exp(-E_0/kT)(E_0/kT)^{s-1}/(s-1)!$. The rate-constant k_1 in the modified Hinshelwood–Lindemann theory is therefore given by (1.29), which even for moderate values of s leads to much bigger values of k_1 than does (1.27).

$$k_1 = \frac{Z_1}{(s-1)!} \left(\frac{E_0}{kT}\right)^{s-1} \exp(-E_0/kT) \qquad (1.29)$$

To derive the Hinshelwood expression for k_1, some results from statistical mechanics have to be assumed. The derivation starts from the statement that the proportion of molecules possessing energy between E and $E + \delta E$ in one degree of vibrational freedom (two square terms) is given by (1.30).

$$f = (1/kT)\exp(-E/kT)\,\delta E \qquad (1.30)$$

Now suppose there are s degrees of freedom in a molecule and the energy is distributed in such a way that the amount of energy in the first degree of freedom is between E_1 and $E_1 + \delta E_1$, the amount in the second is between E_2 and $E_2 + \delta E_2$ and so on. The probability of this happening is found by the multiplication law of probabilities—the probability of the simultaneous occurrence of two or more events is the product of the probabilities of the individual events. If E is the total energy (per molecule) in all degrees of freedom, i.e. $E = E_1 + E_2 + \ldots + E_s$, then the fraction of molecules having energy E distributed in this way among s degrees of freedom is given by (1.31). To obtain the fraction of molecules containing total energy E in a

$$f = (1/kT)^s \exp(-E_1/kT)\exp(-E_2/kT)\ldots \delta E_1 \ldots \delta E_s$$
$$= (1/kT)^s \exp(-E/kT)\,\delta E_1 \ldots \delta E_s \qquad (1.31)$$

specified range E to $E + \delta E$, where bold type is used to distinguish the limiting values of E, this expression has to be integrated for all possible distributions of the energy among the degrees of freedom, subject only to the condition that $E \leqslant E \leqslant E + \delta E$, and the required result is thus (1.32).

$$f_{E \leqslant E \leqslant E + \delta E} = (1/kT)^s \exp(-E/kT) \int_{E \leqslant E \leqslant E + \delta E} \cdots\cdots\cdots \int dE_1 \ldots dE_s \quad (1.32)$$

The multiple integral in (1.32) may be evaluated by considering first the energy range $0 \leqslant E \leqslant E$. It can be shown that

$$\int_{0 \leqslant E \leqslant E} \cdots\cdots\cdots \int dE_1 \ldots dE_s = E^s/s!$$

[see Appendix 6 for a closely related integral, differing only in division by the constant $\prod(h\nu_i)$]. The corresponding integral in (1.32) is obtained as the difference between the integral from 0 to E and that from 0 to $E + \delta E$, i.e.

$$\int_{E \leqslant E \leqslant E + \delta E} \cdots\cdots\cdots \int dE_1 \ldots dE_s = \frac{(E + \delta E)^s}{s!} - \frac{E^s}{s!} = \frac{E^{s-1} \delta E}{(s-1)!} \quad (1.33)$$

The fraction of molecules with energies in the range E to $E + \delta E$, distributed in any way among s degrees of freedom, is then obtained by substitution of (1.33) into (1.32):

$$f_{E \leqslant E \leqslant E + \delta E} = \left(\frac{E}{kT}\right)^{s-1} \exp(-E/kT) \frac{1}{(s-1)!} \left(\frac{\delta E}{kT}\right) \quad (1.34)$$

In the Hinshelwood–Lindemann mechanism, k_1 is a function of energy, and attention is first concentrated upon the rate-constant $k_{1(E \to E + \delta E)}$ for energization to a small energy range E to $E + \delta E$. The rate-constant k_2 for de-energization is still assumed to be independent of energy. The Hinshelwood–Lindemann mechanism is then as follows:

$$A + M \xrightarrow{k_{1(E \to E + \delta E)}} A^*_{(E \to E + \delta E)} + M$$

$$A^*_{(E \to E + \delta E)} + M \xrightarrow{k_2} A + M$$

$$A^*_{(E)} \xrightarrow{k_3} \text{Products}$$

As in the simple Lindemann treatment, the high-pressure limit corresponds to a situation in which the energized molecules are present at their equilibrium proportion, given now by (1.34) for any specified small energy range. The equilibrium proportion of molecules energized in this range is also given by $\delta k_{1(E \to E + \delta E)}/k_2$ and if this is equated to the right-hand side of (1.34), an equation for $\delta k_1/k_2$ is obtained. By integrating between the limits $E = E_0$ and ∞, where E_0 is the critical energy per molecule required to bring about unimolecular reaction, equation (1.35) for k_1/k_2

$$\frac{k_1}{k_2} = \int_{E=E_0}^{\infty} \frac{dk_1}{k_2} = \int_{E=E_0}^{\infty} \frac{1}{(s-1)!} \left(\frac{E}{kT}\right)^{s-1} \exp\left(-E/kT\right) d(E/kT) \quad (1.35)$$

is then obtained. Changing the variable to $\exp(-E/kT)$ and integrating by parts, one obtains (1.36).

$$\frac{k_1}{k_2} = \left[\frac{1}{(s-1)!} \left(\frac{E_0}{kT}\right)^{s-1} + \frac{1}{(s-2)!} \left(\frac{E_0}{kT}\right)^{s-2} + \ldots\right] \exp\left(-E_0/kT\right) \quad (1.36)$$

Some calculations of the relative magnitudes of the first and second terms in equation (1.36) for some unimolecular reactions are given in Table 1.1,

Table 1.1 Comparison of terms in equation (1.36)

Unimolecular reaction	$N_A E_0/$ kcal mol^{-1}	$RT/$ kcal mol^{-1}	$(N_A E_0)/RT$	s	2nd term in (1.36) / 1st term in (1.36)
$N_2O \to N_2 + O$	60	1·78	34	2^a	0·03
$O_3 \to O_2 + O$	25	0·75	34	3^a	0·06
$H_2O_2 \to 2OH$	48	1·45	33	6^a	0·15
$N_2O_5 \to NO_3 + NO_2$	25	0·60	41	7^a	0·15
Cyclopropane $\to$ propene	65	1·58	41	7^b	0·15
$C_2H_5Cl \to C_2H_4 + HCl$	60	1·55	39	8^b	0·18

[a] Value for which HL theory predicts approximately the correct value of k_{bim}.
[b] Value for which HL theory predicts approximately the correct value of $p_{\frac{1}{2}}$.

and it can be seen that since usually $E_0 \gg (s-1)kT$, the second term in (1.36) can normally be neglected compared with the first. Neglect of the second term may cause an error in k_1 by a factor of 1.2, but a much larger error in k_1 actually occurs due to the uncertainty in the value of s which is required by the theory. If the second term is neglected, then k_1/k_2 is given by (1.37) and, if $k_2 = Z_2 \approx Z_1$, one obtains the Hinshelwood expression

$$\frac{k_1}{k_2} = \frac{1}{(s-1)!} \left(\frac{E_0}{kT}\right)^{s-1} \exp\left(-E_0/kT\right) \quad (1.37)$$

for k_1 given previously (equation 1.29). The high-pressure rate-constant according to the Hinshelwood–Lindemann theory is given by

$$k_{\infty} = \frac{k_1 k_3}{k_2} = \frac{k_3}{(s-1)!} \left(\frac{E_0}{kT}\right)^{s-1} \exp\left(-E_0/kT\right) \quad (1.38)$$

Equations (1.29) for k_1 and (1.38) for k_{∞} are thus the basic Hinshelwood–Lindemann equations for the rate-constants at low and high pressures, and the essential features of the results will be illustrated in section 1.3.4.

From equations (1.29) and (1.38) and utilizing (1.6) we can derive the Arrhenius activation energies E_{bim} and E_∞ predicted by the theory for the low- and high-pressure regions. The activation energy at low pressures, associated with k_1 in concentration units $(Z_1 \propto T^{\frac{1}{2}})$ is thus given by (1.39),

$$E_{\text{bim}} = E_0 - (s - \tfrac{3}{2})kT \qquad (1.39)$$

and that at high pressures is given by (1.40). The activation energies

$$E_\infty = E_0 - (s - 1)kT \qquad (1.40)$$

predicted by (1.39) and (1.40) are very similar, and thus virtually no change in activation energy with pressure is predicted. Unimolecular reactions do in practice show a substantial decline in activation energy with pressure (section 4.8), and this behaviour is predicted by the later theories (sections 2.2.4 and 4.8).

It was pointed out previously that an increase in k_1 at constant k_∞ produces a corresponding decrease in the rate-constant k_3. According to the Hinshelwood–Lindemann theory, the rate-constant k_3 is independent of the energy of the energized molecule but does depend upon the molecular complexity since it is a function of the number of vibrational modes, s.

From equation (1.25), $k_3 = k_\infty k_2/k_1$. Calculations using equations (1.37) and (1.40) with the typical parameters $A_\infty = 10^{13}$ s^{-1}, $E_\infty/kT = 40$ yield the following results:[29]

$$
\begin{array}{cccc}
s & = 1 & 5 & 10 \\
k_3/\text{s}^{-1} = & 10^{13} & 10^{9 \cdot 5} & 10^{6 \cdot 5}
\end{array}
$$

It is therefore clear that the lifetimes of the energized molecules (which are given by $\tau \approx 1/k_3$) become greater when the molecule can store energy among a greater number of degrees of freedom. If only one degree of freedom is involved, e.g. for a diatomic molecule, the energized molecule has a very short lifetime and disappears during the course of a single vibration.

1.3.4 Comparison of Hinshelwood–Lindemann theory with experiment

Calculations of a similar kind to those described in section 1.3.2 can now be carried out for the general rate-constant k_{uni} at various pressures, using equation (1.26) but with k_1 values calculated from equation (1.29).

$$k_{\text{uni}} = \frac{k_\infty}{1 + k_\infty/k_1 p} \qquad (1.26)$$

Figure 1.6 illustrates such calculations for the isomerization of cis- to trans-but-2-ene at 469 °C, using the experimental data of Rabinovitch and Michel as before. It can be seen that the inclusion of moderate values of s in equation (1.29) has the effect of reducing dramatically the calculated transition pressure. For $s = 18$, the calculated and observed transition pressures are in good agreement. It may be noted that the number of degrees of vibrational freedom for this molecule is given by $3N - 6 = 30$, which gives a maximum possible value of $s = 30$.

The example given here is again typical, and it is generally possible to obtain agreement between theory and experiment at the transition pressure

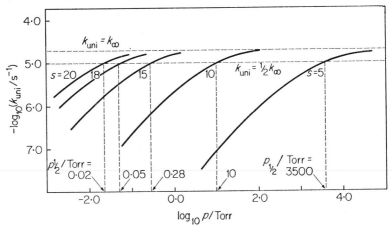

Figure 1.6 Hinshelwood–Lindemann plots for *cis*- ⟶ *trans*-but-2-ene at 469 °C

by choosing a suitable value of s. One unsatisfactory feature which remains, however, is that there is no *a priori* method of calculating s for a particular reaction on this theory. Although the values of s found are feasible in the sense that they are less than (usually about one-half) the number of modes of vibration, no significance can be attached to their actual values. In addition, although the theoretical curve may be made to give approximate agreement with experiment at the transition pressure, it is a poor fit to the experimental curve over the whole pressure range. This is well shown by plotting $1/k_{uni}$ versus $1/p$, and such a plot for *cis*-but-2-ene isomerization is shown in Figure 1.7. Whereas the Hinshelwood–Lindemann equation (1.26) can be rearranged into the linear form (1.41), the experimental plot is strongly curved and is only poorly approximated

$$\frac{1}{k_{uni}} = \frac{1}{k_1} \cdot \frac{1}{p} + \frac{1}{k_\infty} \qquad (1.41)$$

by the Hinshelwood–Lindemann line irrespective of whether the parameters are chosen to fit the results at high or at low pressures. For this reason the experimental value of k_∞ was derived for this reaction using instead an empirical plot of $1/k_{uni}$ versus $1/p^{\frac{1}{2}}$.[27]

The reason for the curvature in this plot lies in the fact that the constant k_3 as well as k_1 is really a function of the energy possessed by the molecule, whereas in the simple treatment given so far, it has been assumed to be a true constant for molecules of all energies above the critical energy. It is

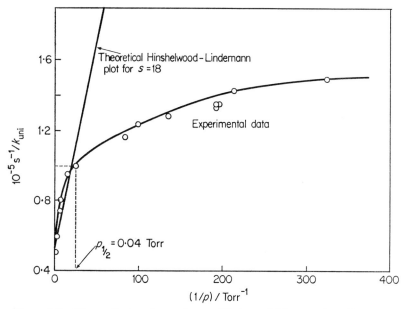

Figure 1.7 Theoretical and experimental[27] plots of $1/k_{uni}$ against $1/$pressure for cis- $\longrightarrow$ $trans$-but-2-ene at 469 °C

intuitively more plausible that k_3 should increase with the energy possessed by the molecule in excess of the critical energy. Now that molecules can be made to undergo unimolecular reaction at different energy levels, for example by chemical activation (see Chapter 8), there is clear experimental evidence for the energy-dependence of k_3. The major subsequent theoretical developments are concerned to a large extent with the calculation of the rate of unimolecular reaction of an energized molecule in terms of its energy content.

1.4 THE FURTHER DEVELOPMENT OF UNIMOLECULAR REACTION RATE THEORIES

For the further development of unimolecular reaction rate theories, both the rate-constants k_1 and k_3 are made energy-dependent. The rate-constant k_1 is considered, as in Hinshelwood–Lindemann theory, by the contribution $\delta k_{1(E \to E + \delta E)}$ for energization to the small energy range E to $E + \delta E$. The rate-constant k_3 is replaced by the energy-dependent rate-constant $k_a(E)$ (or the related L in Slater's theory, see below), and the reaction scheme becomes (1.42)–(1.44). Application of the steady-state

$$A + M \xrightarrow{\delta k_1} A^*_{(E \to E + \delta E)} + M \qquad (1.42)$$

$$A^*_{(E \to E + \delta E)} + M \xrightarrow{k_2} A + M \qquad (1.43)$$

$$A^*_{(E)} \xrightarrow{k_a(E)} \text{Products} \qquad (1.44)$$

treatment to the concentration of $A^*_{(E \to E + \delta E)}$ leads to the expression (1.45) for the contribution to the unimolecular rate-constant k_{uni} from energized molecules in this energy range, and the total rate-constant k_{uni} is then obtained by integration as in (1.46).

$$\delta k_{uni(E \to E + \delta E)} = \frac{k_a(E)(\delta k_1 / k_2)_{(E \to E + \delta E)}}{1 + k_a(E)/k_2[M]} \qquad (1.45)$$

$$k_{uni} = \int_{E=E_0}^{\infty} \frac{k_a(E) \, dk_{1(E \to E + dE)}/k_2}{1 + k_a(E)/k_2[M]} \qquad (1.46)$$

In the Slater theory (Chapter 2) the rate-constant k_a is related to a 'specific dissociation probability' L; this is the frequency with which a chosen critical coordinate in the molecule reaches a critical value, and can be calculated for the case in which the vibrations of the molecule are assumed to be harmonic. The specific dissociation probability L for a system of independent harmonic oscillators is actually a function of the energies in the individual oscillators [denoted $L(E_1, ..., E_n)$, cf (2.39)], and not simply of the total energy E of the molecule. The single integration in (1.46) is therefore replaced in Slater theory by a multiple integral over all the possible energy ranges of the individual oscillators.

In RRK theory (Chapter 3) the assumption is made that k_a is related to the chance that the critical energy E_0 is concentrated in one part of the molecule, e.g. in one oscillator (Kassel theory) or in one squared term (Rice–Ramsperger theory). This probability is clearly a function of the total energy E of the energized molecule; one of the resulting expressions for k_a is (3.4), for example.

In RRKM theory, which is discussed in detail in Chapter 4, the main developments are the calculation of the rate-constant k_1 by quantum-statistical mechanics and the application of ideas related to the Absolute Rate Theory for the calculation of k_a. For the latter it is convenient to replace (1.44) by (1.47) in which distinction is made between an energized

$$A^*_{(E)} \xrightarrow{k_a(E)} A^+ \xrightarrow{k^+} \text{Products} \qquad (1.47)$$

molecule A* and an activated complex A$^+$. The rate-constant $k_a(E)$ is then the rate-constant for conversion of molecules of a specific energy E into activated complexes A^+. The rate-constant k^+ refers to the absolute rate at which activated complexes pass into products and is equivalent to the kT/h of Absolute Rate Theory.

References

1. C. H. Bamford and C. F. H. Tipper, *Comprehensive Chemical Kinetics*, Vol. 1, Elsevier, Amsterdam, 1969; H. W. Melville and B. G. Gowenlock, *Experimental Methods in Gas Reactions*, MacMillan, London, 1964; A. Maccoll, 'Homogeneous Gas-phase Reactions', in *Technique of Organic Chemistry* (Ed. A. Weissberger), Vol. 8, Part 1, Wiley, New York, 1961, p. 427.
2. S. H. Bauer, *Science*, **141**, 867 (1963); E. F. Greene and J. P. Toennies, *Chemical Reactions in Shock Waves*, Arnold, London, 1964; J. N. Bradley, *R.I.C. Lec. Series*, No. 6 (1963); J. N. Bradley, *Shock Waves in Chemistry and Physics*, Methuen, London, 1962.
3. J. D. Lambert, *Quart. Rev.*, **21**, 67 (1967).
4. D. R. Herschbach, *Advan. Chem. Phys.*, **10**, 319 (1966).
5. J. C. McCoubrey and W. D. McGrath, *Quart. Rev.*, **11**, 87 (1957).
6. B. Stevens, *Collisional Activation in Gases*, Pergamon, Oxford, 1967.
7. F. O. Rice and K. F. Herzfeld, *J. Amer. Chem. Soc.*, **56**, 284 (1934).
8. K. A. Holbrook, *Symposium on Kinetics of Pyrolytic Reactions*, Chemical Institute of Canada, Ottawa, 1964, p. W-1.
9. B. W. Wojciechowski and K. J. Laidler, *Trans. Faraday Soc.*, **59**, 369 (1963).
10. S. H. Bauer, *J. Amer. Chem. Soc.*, **91**, 3688 (1969).
11. M. Niclause, R. Martin, A. Combes and M. Dziezynski, *Can. J. Chem.*, **43**, 1120 (1965).
12. J. H. Purnell and C. P. Quinn, *J. Chem. Soc.*, 4128 (1961).
13. K. A. Holbrook and J. J. Rooney, *J. Chem. Soc.*, 247 (1965).
14. K. A. Holbrook, *Proc. Chem. Soc.*, 418 (1964).
15. M. R. Bridge and J. L. Holmes, *J. Chem. Soc. (B)*, 713 (1966); J. L. Holmes and L. S. M. Ruo, *J. Chem. Soc. (A)*, 1231 (1968).
16. R. C. Tolman, *Statistical Mechanics with Applications to Physics and Chemistry*, Chemical Catalog Co., New York, 1927.
17. K. J. Laidler, *Theories of Chemical Reaction Rates*, McGraw–Hill, London, 1969, Chap. 2.
18. Ref. 17, Chap. 3.
19. S. Glasstone, K. J. Laidler, and H. Eyring, *The Theory of Rate Processes*, McGraw–Hill, New York, 1941.
20. J. Perrin, *Ann. Phys.*, **11**, 1 (1919).
21. I. Langmuir, *J. Amer. Chem. Soc.*, **42**, 2190 (1920).
22. J. A. Christiansen and H. A. Kramers, *Zeit. Phys. Chem.*, **104**, 451 (1923).
23. F. A. Lindemann, *Trans. Faraday Soc.*, **17**, 598 (1922).
24. J. A. Christiansen, *Ph.D. Thesis*, Copenhagen, 1921.
25. M. Volpe and H. S. Johnston, *J. Amer. Chem. Soc.*, **78**, 3910 (1956).
26. J. Aspden, N. A. Khawaja, J. Reardon and D. J. Wilson, *J. Amer. Chem. Soc.*, **91**, 7580 (1969).
27. B. S. Rabinovitch and K. W. Michel, *J. Amer. Chem. Soc.*, **81**, 5065 (1959).
28. C. N. Hinshelwood, *Proc. Roy. Soc. (A)*. **113**, 230 (1927).
29 A F. Trotman-Dickenson, *Gas Kinetics*, Butterworths, London, 1955, p. 55.

2 The Slater Theory

The theory of Slater, first put forward in 1939[1] and explained in detail in his book,[2] was the first serious attempt to relate the kinetics of unimolecular reactions to our knowledge of molecular vibrations. The theory accepts the basic Hinshelwood–Lindemann mechanism of collisional energization (outlined in Chapter 1) with, however, a more restricted definition of the energized molecule. The molecule undergoing reaction is pictured as an assembly of harmonic oscillators of particular amplitudes and phases. Reaction is said to occur when a chosen coordinate in the molecule (such as a combination of bond distances and/or bond angles) attains a critical extension. The rate-constant k_a of the scheme given previously is replaced by a 'specific dissociation probability' L which is the frequency with which the 'critical coordinate' attains the critical extension. By calculating L in this way and integrating the necessary equations, Slater was able to derive expressions for the rate-constant k_{uni} at high, low and intermediate pressures.

In order to understand more fully the derivation of these equations and the assumptions involved, it is first necessary to consider the vibrational analysis of polyatomic molecules. This topic is therefore treated in some detail in section 2.1. The reader who is familiar with the methods involved may proceed directly to the outline of Slater's harmonic theory in section 2.2, and the discussions of its assumptions and applications which follow in sections 2.3 and 2.4 respectively.

2.1 THE VIBRATIONAL ANALYSIS OF POLYATOMIC MOLECULES

2.1.1 Development of the secular equation

Polyatomic molecules at normal temperatures undergo complicated vibrational motions. For a given non-linear molecule with N atoms it can be shown that, for small displacements of the atoms from their mean positions, the motion can be resolved into $3N - 6$ constituent *normal modes of vibration*. In a normal mode of vibration the atoms all execute simple harmonic motion about their mean positions with the same phase and with the same frequency.

Vibrational analysis is the treatment of the vibrations of a polyatomic molecule by the methods of classical mechanics. The motions of N atoms of masses $m_1, ..., m_N$ can be described by $3N$ Cartesian displacement coordinates $x_1, ..., x_{3N}$. These might be given the symbols $x_1, y_1, z_1, x_2, y_2, z_2, ..., z_N$, but the present terminology leads to simpler equations. The potential energy V is then given in general form, for small displacements, by (2.1), where the k_{ij} are constants, and the kinetic energy T is given by (2.2).

$$2V = k_{11} x_1^2 + k_{12} x_1 x_2 + ... + k_{3N,3N} x_{3N}^2 \qquad (2.1)$$

$$2T = m_1(\dot{x}_1^2 + \dot{x}_2^2 + \dot{x}_3^2) + m_2(\dot{x}_4^2 + \dot{x}_5^2 + \dot{x}_6^2) + ... + m_N(... \dot{x}_{3N}^2) \qquad (2.2)$$

Equations of motion can then be written for each coordinate x_i. In Lagrangian form these are (2.3), where the Lagrange function $\mathscr{L}$ is related to the kinetic energy T and the potential energy V by (2.4).

$$\mathrm{d}/\mathrm{d}t(\partial\mathscr{L}/\partial\dot{x}_i) - \partial\mathscr{L}/\partial x_i = 0 \qquad (2.3)$$

$$\mathscr{L} = T - V \qquad (2.4)$$

Hence

$$\mathrm{d}/\mathrm{d}t(\partial T/\partial\dot{x}_i) - (\partial T/\partial x_i) + (\partial V/\partial x_i) = 0$$

Since, from (2.1) and (2.2),

$$\partial V/\partial x_1 = k_{11} x_1 + k_{12} x_2 + k_{13} x_3 + ... + k_{1,3N} x_{3N}$$

$$\partial T/\partial\dot{x}_1 = m_1 \dot{x}_1 \quad \text{and} \quad \partial T/\partial x_1 = 0$$

the equation of motion involving x_1 is then (2.5) which is simply an extension of the Newtonian equation, *mass × acceleration = force*.

$$m_1 \ddot{x}_1 + k_{11} x_1 + k_{12} x_2 + ... + k_{1,3N} x_{3N} = 0 \qquad (2.5)$$

There will be $3N$ equations of this kind, applicable to the $3N$ coordinates $x_1, ..., x_{3N}$. In a normal mode of vibration (see above), each atom executes simple harmonic motion about its mean position with the same frequency ν, and so it is necessary to apply the conditions defining simple harmonic motion in each coordinate, viz.

$$\ddot{x}_i = -\lambda x_i \quad (\text{where } \lambda = 4\pi^2 \nu^2)$$

This leads to a new set of simultaneous equations (2.6) which are linear in the x_i.

$$\left. \begin{array}{l} (k_{11} - m_1 \lambda) x_1 + k_{12} x_2 + k_{13} x_3 + ... + k_{1,3N} x_{3N} = 0 \\ \quad \cdot \quad \cdot \quad \cdot \quad \cdot \quad \cdot \quad \cdot \quad \cdot \quad \cdot \quad \cdot \\ k_{3N,1} x_1 + k_{3N,2} x_2 + ... + (k_{3N,3N} - m_N \lambda) x_{3N} = 0 \end{array} \right\} \qquad (2.6)$$

The condition for the solution of these equations is that the *secular determinant* is equal to zero:

$$\begin{vmatrix} k_{11}-m_1\lambda & k_{12} & k_{13} & \cdots & k_{1,3N} \\ \vdots & \vdots & \vdots & \vdots & \vdots \\ k_{3N,1} & k_{3N,2} & k_{3N,3} & \cdots & k_{3N,3N}-m_N\lambda \end{vmatrix} = 0$$

This secular equation is a polynomial equation in λ^{3N}, and thus leads to $3N$ values of the roots λ. It can be shown, however, that for a non-linear polyatomic molecule, six of the roots are zero, corresponding to the three translations and three overall rotations of the molecule. The equation thus reduces to one of order $3N-6$ and there are $3N-6$ roots λ_k corresponding to $3N-6$ normal mode frequencies ν_k. For a linear polyatomic molecule there are $3N-5$ non-zero roots and hence $3N-5$ normal mode frequencies.

Since the equations (2.6) are homogeneous, it is not possible to obtain absolute values of the coordinates x_i but only their ratios. Substitution of a particular value of a non-zero root λ_k into the secular equation enables the ratios $x_1 : x_2 : \ldots : x_{3N}$ to be determined. These ratios determine the form of the normal mode vibration of frequency ν_k and indicate the relative displacements and directions of displacement of the individual atoms when the molecule is vibrating in this particular normal mode.

As a simple example, the water molecule is a non-linear triatomic molecule with $3\times3-6=3$ modes of vibration. Vibrational analysis enables the ratios of the displacements in each normal mode to be calculated. These determine the form of the normal mode vibrations which are illustrated in Figure 2.1.

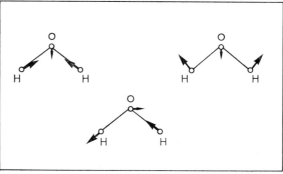

Figure 2.1 Normal modes of vibration of the water molecule. The directions of the arrows represent the directions of the relative motions of the nuclei. The magnitudes of the displacements are not to scale however; those of the oxygen atom have been exaggerated for clarity

2.1.2 Normal coordinates

The solutions of the secular equation which yield $3N-6$ non-zero values of λ_k can be described in terms of $3N-6$ normal coordinates Q_k. If normal coordinates Q_k can be found which result in potential and kinetic energy expressions of the form (2.7) and (2.8), where $\lambda_k = 4\pi^2 \nu_k^2$, then the

$$V = \sum_{k=1}^{3N-6} Q_k^2 \tag{2.7}$$

$$T = \sum_{k=1}^{3N-6} \dot{Q}_k^2 / \lambda_k \tag{2.8}$$

equations of motion become $3N-6$ separable equations of the form

$$\ddot{Q}_k + \lambda_k Q_k = 0 \tag{2.9}$$

for which there are $3N-6$ independent solutions of the form

$$Q_k = A_k \cos(2\pi \nu_k t + \psi_k)$$

Each solution corresponds to a normal mode of vibration with the corresponding normal mode frequency ν_k, amplitude A_k and phase ψ_k. For use in Slater theory the Q_k are normalized so that the coefficients of Q_k^2 in the potential energy expression (2.7) are unity. This has the useful corollary that, as may readily be verified, $A_k = \sqrt{E_k}$ where E_k is the energy in the kth mode.

The ideas of normal coordinates and normal mode analysis will now be illustrated by a few simple applications. The main argument continues in the outline of Slater's harmonic theory in section 2.2.

2.1.3 Application to two independent oscillators

Suppose we have two masses m_1 and m_2 at the end of two separate springs independently fixed to a firm support. If they are given displacements x_1 and x_2 from their mean positions, the total Lagrangian function (2.4) for the system can be written as (2.10).

$$\mathscr{L} = \tfrac{1}{2}m_1 \dot{x}_1^2 + \tfrac{1}{2}m_2 \dot{x}_2^2 - \tfrac{1}{2}k_1 x_1^2 - \tfrac{1}{2}k_2 x_2^2 \tag{2.10}$$

Application of (2.3) then gives two separate equations of motion, (2.11) and (2.12).

$$m_1 \ddot{x}_1 + k_1 x_1 = 0 \tag{2.11}$$

$$m_2 \ddot{x}_2 + k_2 x_2 = 0 \tag{2.12}$$

Since the oscillators are independent, no cross-terms occur in (2.10), and hence (2.11) and (2.12) are completely separable. The solutions to (2.11) and (2.12) are each of the form $x_i = A_i \cos(2\pi\nu_i t)$; thus the coordinates x_1 and x_2 in this simple case are themselves normal coordinates, and each oscillator executes simple harmonic motion independently of the other oscillator. For consistency with the forms (2.7) and (2.8) the renormalized coordinates $Q_1 = x_1 \sqrt{(k_1/2)}$ and $Q_2 = x_2 \sqrt{(k_2/2)}$ would be used.

2.1.4   Application to a diatomic molecule

Suppose we have a diatomic molecule with atomic masses m_1 and m_2, and that the atoms are subjected to displacements x_1 and x_2 from their mean positions, as shown in Figure 2.2. The kinetic energy is given by

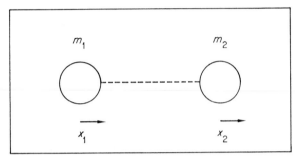

Figure 2.2   Coordinates for the diatomic molecule

(2.13) and the potential energy can be expressed in terms of the net extension of the internuclear distance $(x_2 - x_1)$ by (2.14).

$$T = \tfrac{1}{2}m_1 \dot{x}_1^2 + \tfrac{1}{2}m_2 \dot{x}_2^2 \tag{2.13}$$

$$V = \tfrac{1}{2}k(x_2 - x_1)^2 = \tfrac{1}{2}kx_1^2 + \tfrac{1}{2}kx_2^2 - kx_1 x_2 \tag{2.14}$$

From (2.13) and (2.14),

$$\mathscr{L} = T - V = \tfrac{1}{2}m_1 \dot{x}_1^2 + \tfrac{1}{2}m_2 \dot{x}_2^2 - \tfrac{1}{2}k(x_1^2 + x_2^2 - 2x_1 x_2)$$

Hence from (2.3) the equations of motion are (2.15) and (2.16)

$$m_1 \ddot{x}_1 + kx_1 - kx_2 = 0 \tag{2.15}$$

$$m_2 \ddot{x}_2 + kx_2 - kx_1 = 0 \tag{2.16}$$

Unlike (2.11) and (2.12) these equations are not completely separable since (2.15) involves x_2 and (2.16) involves x_1. It is possible, however, by suitable choice of coordinates, to write equations from which cross-terms are eliminated and which then become separable. Using new coordinates defined by

$$r = x_2 - x_1 \quad \text{and} \quad R = (m_1 x_1 + m_2 x_2)/M$$

where
$$M = m_1 + m_2$$
then the kinetic and potential energies become
$$T = \tfrac{1}{2}M\dot{R}^2 + \tfrac{1}{2}\mu\dot{r}^2 \quad \text{and} \quad V = \tfrac{1}{2}kr^2$$
where
$$\mu = m_1 m_2 / M$$
The Lagrangian function (2.4) is then
$$\mathscr{L} = T - V = \tfrac{1}{2}M\dot{R}^2 + \tfrac{1}{2}\mu\dot{r}^2 - \tfrac{1}{2}kr^2$$
which leads to two equations of motion. The first involves only R;
$$\mathrm{d}(M\dot{R})/\mathrm{d}t = 0 \quad \text{or} \quad m_1\dot{x}_1 + m_2\dot{x}_2 = \text{constant} \qquad (2.17)$$
This is simply a statement of the law of conservation of momentum, and if the centre of mass of the molecule is taken to be at rest the constant in (2.17) is in fact zero. The second equation involves only r;
$$\mathrm{d}/\mathrm{d}t(\partial\mathscr{L}/\partial\dot{r}) - \partial\mathscr{L}/\partial r = 0$$
whence
$$\mu\ddot{r} + kr = 0$$
for which the solution is
$$r = A\cos(\sqrt{k/\mu}\,t + \psi)$$
which corresponds to the normal mode of vibration of a diatomic molecule. Thus the coordinates R and r in this case are both normal coordinates although only one (r) represents a genuine vibration. Again, normalization to fit (2.7) and (2.8) requires definition of a coordinate $Q = r\sqrt{(k/2)}$, in which case
$$Q = A'\cos(\sqrt{k/\mu}\,t + \psi)$$
$$= \sqrt{E_{\mathrm{vib}}}\cos(\sqrt{k/\mu}\,t + \psi) \qquad (2.18)$$

2.1.5 Application to a linear triatomic molecule

We consider in this section an extension of the above method to the stretching vibrations of a linear symmetric triatomic molecule ABA, i.e. the modes which involve displacements along the internuclear axis only. In restricting motion to the internuclear axis, we are ignoring bending vibrations and also rotations. If the masses and displacements of the atoms are defined as in Figure 2.3, then
$$T = \tfrac{1}{2}m_1\dot{x}_1^2 + \tfrac{1}{2}m_2\dot{x}_2^2 + \tfrac{1}{2}m_1\dot{x}_3^2$$

Defining new coordinates by

$$r_1 = x_2 - x_1, \quad r_2 = x_3 - x_2$$

$$R = (m_1 x_1 + m_2 x_2 + m_1 x_3)/M$$

where

$$M = 2m_1 + m_2$$

then

$$V = \tfrac{1}{2}kr_1^2 + \tfrac{1}{2}kr_2^2 + ar_1 r_2$$

where a is a constant, and

$$T = \tfrac{1}{2}M\dot{R}^2 + \frac{\tfrac{1}{2}m_1(m_1 + m_2)}{M}(\dot{r}_1^2 + \dot{r}_2^2) + \frac{m_1^2}{M}\dot{r}_1 \dot{r}_2$$

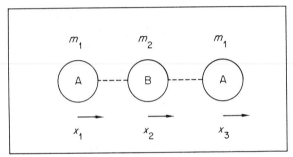

Figure 2.3 Coordinates for the linear triatomic molecule ABA

Ignoring the coordinate R, which as before leads to an equation referring to translation of the molecule as a whole, the Lagrangian in terms of coordinates r_1 and r_2 is given by (2.19).

$$\mathcal{L} = \frac{\tfrac{1}{2}m_1(m_1 + m_2)}{M}(\dot{r}_1^2 + \dot{r}_2^2) + \frac{m_1^2}{M}\dot{r}_1 \dot{r}_2 - \tfrac{1}{2}kr_1^2 - \tfrac{1}{2}kr_2^2 - ar_1 r_2 \quad (2.19)$$

This equation is not directly separable in terms of the coordinates r_1 and r_2 since cross-products appear. It is, however, possible to introduce new coordinates to eliminate the cross-products; suitable expressions are the 'symmetry coordinates' (2.20) and (2.21)

$$Y_1 = (r_1 + r_2)/\sqrt{2} \quad (2.20)$$

$$Y_2 = (r_1 - r_2)/\sqrt{2} \quad (2.21)$$

Rotation of the molecule by $180°$ about the y or z axes (see Figure 2.4) changes r_1 into r_2 and r_2 into r_1. Under these conditions Y_1 remains unchanged but Y_2 changes sign. Since the potential and kinetic energies must remain invariant towards such 'symmetry operations', terms involving the product $Y_1 Y_2$ cannot occur in the expressions for T and V in terms of Y_1 and Y_2.

Further, since

$$r_1{}^2 + r_2{}^2 = Y_1{}^2 + Y_2{}^2$$

and

$$Y_1{}^2 - Y_2{}^2 = 2r_1 r_2$$

the Lagrangian (2.19) becomes

$$\mathscr{L} = \tfrac{1}{2}m_1 \dot{Y}_1{}^2 + \tfrac{1}{2}\mu \dot{Y}_2{}^2 - \tfrac{1}{2}(k+a) Y_1{}^2 - \tfrac{1}{2}(k-a) Y_2{}^2$$

where

$$\mu = m_1 m_2 / M$$

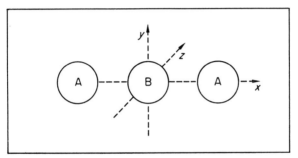

Figure 2.4   Rotation axes for the linear triatomic molecule

This is now separable and application of (2.3) leads to

$$m_1 \ddot{Y}_1 - (k+a) Y_1 = 0 \tag{2.22}$$

$$\mu \ddot{Y}_2 - (k-a) Y_2 = 0 \tag{2.23}$$

Hence Y_1 and Y_2 are normal coordinates and describe two different stretching modes of vibration. For use in Slater theory we require coordinates normalized to fit the simple forms (2.7) and (2.8), and thus we write:

$$Q_1 = [(k+a)/2]^{\frac{1}{2}} Y_1 = [(k+a)^{\frac{1}{2}}/2](r_1 + r_2) \tag{2.24}$$

$$Q_2 = [(k-a)/2]^{\frac{1}{2}} Y_2 = [(k-a)^{\frac{1}{2}}/2](r_1 - r_2) \tag{2.25}$$

Then (2.22) and (2.23) become

$$\ddot{Q}_1 - [(k+a)/m_1] Q_1 = 0$$

$$\ddot{Q}_2 - [(k-a)/m_2] Q_2 = 0$$

whence

$$\lambda_1 = 4\pi^2 \nu_1{}^2 = (k+a)/m_1$$

$$\lambda_2 = 4\pi^2 \nu_2{}^2 = (k-a)/\mu$$

The variation of Q_1 with time in the normal vibration of frequency ν_1 is given by (2.26). Since $Q_1 = \text{constant} \times (r_1 + r_2)$ and r_1 and r_2 both represent increases in the interatomic distances, it can be seen from (2.26) that Q_1 is the normal coordinate describing the symmetrical mode of vibration. Similarly Q_2 represents the asymmetric mode in which $(r_1 - r_2)$ increases and decreases with time according to (2.27). From (2.24) and (2.25) we can

$$Q_1 = \sqrt{E_1} \cos(2\pi\nu_1 t + \psi_1) \tag{2.26}$$

$$Q_2 = \sqrt{E_2} \cos(2\pi\nu_2 t + \psi_2) \tag{2.27}$$

see how the internal coordinates r_1 and r_2 vary with time when the modes represented by Q_1 and Q_2 are both present; for r_1, for example,

$$r_1 = (k+a)^{-\frac{1}{2}} Q_1 + (k-a)^{-\frac{1}{2}} Q_2 \tag{2.28}$$

$$= [E_1{}^{\frac{1}{2}}/(k+a)^{\frac{1}{2}}] \cos(2\pi\nu_1 t + \psi_1) + [E_2{}^{\frac{1}{2}}/(k-a)^{\frac{1}{2}}] \cos(2\pi\nu_2 t + \psi_2) \tag{2.29}$$

Equation (2.29) represents a complicated behaviour of the coordinate r_1 with time. Such behaviour for a typical internal coordinate q_1 of any polyatomic molecule is schematically represented by Figure 2.5. It can be seen that the coefficients multiplying the amplitude factors [$\sqrt{E_1}$ and $\sqrt{E_2}$ in (2.29)] are the coefficients of the transformation from internal to normal coordinates in (2.28).

2.1.6   More complicated molecules

In the preceding sections it has been shown how it is possible to derive the form of normal mode vibrations described by the normal coordinates Q_k from solutions of the equations of motion. The motions of the atoms were first described in terms of Cartesian displacement coordinates (x), then internal coordinates (r) relating to displacements of the bonds within the molecule, and finally it was shown how, in some simple cases, internal coordinates could be combined to give normal coordinates.

For most molecules it is not possible to guess the relationship between the internal coordinates and normal coordinates which permits direct solution of the then independent equations of motion. Instead, use is made of the known symmetry of the molecule to construct symmetry coordinates from internal coordinates and these reduce the order of the secular equation which has to be solved. The method is described in detail in the book by Wilson, Decius and Cross.[3] There are as many symmetry coordinates of a given symmetry type as there are genuine vibrations of that type, and the symmetry coordinates behave towards the symmetry operations of the relevant point group in exactly the same way as do vibrations of the same symmetry type. This eliminates all cross-products between coordinates belonging to different symmetry species, since cross-products would result in a variation of the potential and kinetic energies with certain symmetry operations, and this is physically impossible. As a result the secular determinant can be factorized into blocks, each block corresponding to a different symmetry species.

2.1.7 The results of vibrational analysis

Vibrational analysis is most often used in order to determine the normal mode frequencies v_k for a molecule. In order to do this it is necessary to assume a potential energy expression and to assign values to unknown force constants in such an expression. Agreement between observed and calculated vibration frequencies then constitutes some justification for the assumed potential function for a molecule.

For the application of Slater theory, we are more interested in the form of the normal mode vibrations and the way in which they influence the behaviour of a particular internal coordinate with time. We need therefore to know both the normal mode frequencies v_k and the force constants which occur in the secular equation. Substitution of any particular value of $\lambda_k = 4\pi^2 v_k^2$ in the secular equation allows the ratios of the displacement coordinates to each other in the corresponding normal mode to be found.

If the secular equation has been written in terms of $3N-6$ coordinates q_i (which are chosen so that the transformation to normal coordinates is orthogonal) then for mode 1, for example, the ratios of the q_i will be obtained from the equation:

$$q_1 : q_2 : q_3 : \ldots : q_{3N-6} = a_{11} : a_{12} : a_{13} : \ldots : a_{1,3N-6}$$

where the quantities a_{1i} are related to the normal coordinate Q_1 by (2.30).

$$Q_1 = a_{11}q_1 + a_{12}q_2 + a_{13}q_3 + \ldots + a_{1,3N-6}q_{3N-6} \qquad (2.30)$$

In terms of symmetry coordinates S_i, the result of vibrational analysis similarly yields ratios of symmetry coordinates to each other in a given normal mode, so that Q_1 for example is given by (2.31).

$$Q_1 = L_{11}S_1 + L_{12}S_2 + \ldots + L_{1,3N-6}S_{3N-6} \qquad (2.31)$$

In general there will be less than $3N-6$ symmetry coordinates involved in a particular normal coordinate Q_1, since only those of the same symmetry as Q_1 have non-zero L_{1j} values, and this leads to considerable simplification.

For the purposes of Slater theory, we need the contribution of each normal coordinate to a particular internal coordinate or to a particular symmetry coordinate. These contributions are found from the inverse relations corresponding to (2.30) and (2.31). Thus if (2.30) is written in matrix notation as $Q = aq$ the inverse relation is given by (2.32).

$$q = a^{-1}Q = \alpha Q \qquad (2.32)$$

That is,

$$\left.\begin{array}{l} q_1 = \alpha_{11}Q_1 + \alpha_{12}Q_2 + \ldots + \alpha_{1,3N-6}Q_{3N-6} \\[4pt] \quad\cdot\quad\cdot\quad\cdot\quad\cdot\quad\cdot\quad\cdot\quad\cdot \\[4pt] q_{3N-6} = \alpha_{3N-6,1}Q_1 + \ldots + \alpha_{3N-6,3N-6}Q_{3N-6} \end{array}\right\} \qquad (2.33)$$

Similarly, if there are m symmetry coordinates, which are related to m normal coordinates of a given symmetry type by equations of the type (2.31), or in matrix notation $Q = LS$, then

$$S = L^{-1}Q = L'Q$$

That is,

$$S_1 = L'_{11}Q_1 + L'_{12}Q_2 + \ldots + L'_{1m}Q_m$$

$$\quad\cdot\quad\cdot\quad\cdot\quad\cdot\quad\cdot\quad\cdot\quad\cdot$$

$$S_m = L'_{m1}Q_1 + L'_{m2}Q_2 + \ldots + L'_{mm}Q_m$$

When direct solution of the secular equation yields the a matrix or the L matrix, then the inverse matrices α or L' must be found by the standard techniques of matrix inversion.

2.2 SLATER'S HARMONIC THEORY

For the example of a linear triatomic molecule discussed in the preceding section, we derived an equation (2.29) for the time-dependence of internal coordinate r_1 when both normal coordinates Q_1 and Q_2 were non-zero. This equation was of the form

$$r_1 = \alpha_{11}A_1 \cos(2\pi\nu_1 t + \psi_1) + \alpha_{12}A_2 \cos(2\pi\nu_2 t + \psi_2)$$

where the coefficients α relate r_1 to Q_1 and Q_2 by the equation

$$r_1 = \alpha_{11}Q_1 + \alpha_{12}Q_2$$

For any general internal coordinate q_i, related to $n = 3N - 6$ normal coordinates by the equation:

$$q_i = \sum_{k=1}^{n} \alpha_{ik} Q_k$$

it can be similarly shown that the time-dependence of q_i is given by (2.34).

$$q_i = \sum_{k=1}^{n} \alpha_{ik} A_k \cos\left(2\pi\nu_k\, t + \psi_k\right) \qquad (2.34)$$

Slater's theory regards the behaviour of coordinate q_i as a superposition of n independent contributions from the normal modes of vibration. The function q_i varies in a complicated way with time, only occasionally attaining high values (Figure 2.5). An internal coordinate q_1 (or possibly a

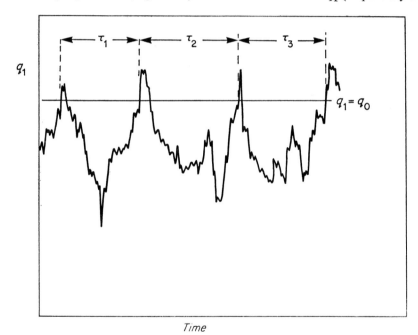

Figure 2.5 Variation of a typical internal coordinate q_1 with time, showing gaps between successive attainments of a critical value q_0

combination of internal coordinates) is chosen as the critical coordinate. Unimolecular reaction is then said to occur if the chosen coordinate exceeds a critical value q_0, harmonic motion being assumed up to the point of rupture. From (2.34) the maximum value of a particular coordinate q_1

is given by (2.35), where E_k is the total energy in normal mode k and $A_k = \sqrt{E_k}$ by virtue of the special normalization of (2.7)–(2.8).

$$(q_1)_{\max} = \sum_{k=1}^{n} |\alpha_{1k}| A_k = \sum_{k=1}^{n} |\alpha_{1k}| \sqrt{E_k} \qquad (2.35)$$

A molecule is said to be energized if the last collision has given it sufficient energy to satisfy the inequality (2.36)

$$\sum_{k=1}^{n} |\alpha_{1k}| \sqrt{E_k} \geqslant q_0 \qquad (2.36)$$

It can then be shown that the minimum total energy E_0 for which this is possible is given by (2.37).

$$E_0 = q_0^2/\alpha^2 \qquad (2.37)$$

where

$$\alpha = \left(\sum_{k=1}^{n} \alpha_{1k}^2 \right)^{\frac{1}{2}}$$

This energy is the critical energy for reaction and will be seen later to be equal to the high-pressure activation energy. Equation (2.37) represents the minimum energy which the molecule must possess in order to undergo unimolecular reaction, although it will not do so unless the energy is distributed among the normal modes in such a way that (2.36) is also satisfied. Whether the energized molecule reacts or not is then dependent upon the relative chance of the critical coordinate exceeding the value q_0 and the chance of deactivation by collision. The frequency with which the former occurs is called by Slater the specific dissociation rate or probability L.

The Slater calculation of L is based upon the behaviour of a sum of harmonic vibrations. If the average behaviour of a coordinate over a long time interval is required, the initial phase is irrelevant and (2.34) becomes, for a particular coordinate q_1,

$$q_1 = \sum_{k=1}^{n} \alpha_{1k} \sqrt{E_k} \cos(2\pi \nu_k t) \qquad (2.38)$$

Slater makes use of a formula due to Kac[4] for the frequency of 'up-zeros' of the function

$$\sum_{k=1}^{n} \alpha_{1k} A_k \cos(2\pi \nu_k t) - q_0$$

where there are n independent frequencies ν_k.

Unfortunately, the precise equation for L is too cumbersome to deal with in Slater theory, except at the high-pressure limit, and an approximate expression is used instead. Slater has shown that when the sum of the amplitudes $\sum \alpha_{1k} A_k$ is not much greater than q_0, then the equation for L becomes (2.39). It will be noted that (2.39) gives L for a specified distribu-

$$L(E_1, E_2, ..., E_n) = \frac{1}{\Gamma(\frac{1}{2}n + \frac{1}{2})} \left(\frac{\sum\limits_{k=1}^{n} a_k \nu_k^2}{a_1 a_2 ... a_n} \right)^{\frac{1}{2}} \left(\frac{a - q_0}{2\pi} \right)^{\frac{1}{2}(n-1)} \quad (2.39)$$

where

$$a_k = \alpha_{1k} \sqrt{E_k} \quad \text{and} \quad a = \sum_{k=1}^{n} a_k$$

tion of energy among the normal modes and not simply for a given total energy as in most other theories. The corresponding rate-constant $k_a(E)$ for a given total energy E, which is the term used in most other theories, may be calculated by averaging L over all the possible distributions of the energy (see ref. 2, p. 92).

2.2.1 The general-pressure rate-constant

The Slater theory makes use of earlier ideas concerning the overall mechanism of unimolecular reactions in order to express the rate-constant at any pressure. These may be conveniently discussed in terms of the basic mechanism given in Chapter 1 by equations (1.42)–(1.44):

$$A + M \xrightarrow{\delta k_1} A^*_{(E \to E + \delta E)} + M \quad (1.42)$$

$$M + A^*_{(E \to E + \delta E)} \xrightarrow{k_2} A + M \quad (1.43)$$

$$A^*_{(E)} \xrightarrow{k_a(E)} \text{Products} \quad (1.44)$$

and this leads to the expression (1.46) for the first-order rate-constant k_{uni}. In the Hinshelwood–Lindemann treatment $\delta k_1/k_2$ is equated to the

$$k_{\text{uni}} = \int_{E = E_0}^{\infty} \frac{k_a(E) \, dk_{1(E \to E + dE)}/k_2}{1 + k_a(E)/k_2[M]} \quad (1.46)$$

equilibrium fraction of molecules having total energy in the range E to $E + \delta E$, and this is obtained by integrating the fraction of molecules having energy E distributed in a specified way among the s degrees of freedom (1.31), subject to the restriction $E \leqslant E \leqslant E + \delta E$ on the total energy (see section 1.3.3). Because of the different meaning of the term 'energization' in Slater theory a different approach is required here; it is necessary to integrate over the relevant ranges of energy for each degree of freedom rather than the corresponding range of total energy. Thus the expression for k_{uni} is evaluated by replacing $k_a(E)$ in (1.46) by $L(E_1, ..., E_n)$, and replacing dk_1/k_2 by the expression $(1/kT)^n \exp(-E/kT) dE_1 ... dE_n$ (1.31). This expression gives the equilibrium fraction of molecules having E_i to $E_i + dE_i$ in the ith mode of vibration, where $E = E_1 + E_2 + ... + E_n$ is the total energy in the n contributing modes of vibration. Integration over all the relevant energy distributions then gives the expression (2.40) for k_{uni}.

$$k_{\text{uni}} = \int \int ... \int \frac{L(E_1, ..., E_n) \exp(-E/kT)}{1 + L(E_1, ..., E_n)/k_2[M]} \prod_{k=1}^{n} \left(\frac{dE_k}{kT} \right) \qquad (2.40)$$

The integrations with respect to $dE_1 ... dE_n$ in (2.40) are now over the restrictive energy ranges governed by (2.36) and (2.37), which reduce to the condition

$$\sum_{k=1}^{n} \mu_k \sqrt{E_k} \geqslant \sqrt{E_0} \qquad (E_k \geqslant 0) \qquad (2.41)$$

where

$$\mu_k = |\alpha_{1k}|/\alpha = |\alpha_{1k}| \Big/ \left(\sum_{k=1}^{n} \alpha_{1k}^2 \right)^{\frac{1}{2}} \qquad (2.42)$$

The integral (2.40) reduces to a manageable form if the approximate expression (2.39) is inserted for L, and the result of the integration is (2.43).

$$k_{\text{uni}} = \nu \exp(-b) I_n(\theta) \qquad (2.43)$$

where

$$I_n(\theta) = \frac{1}{\Gamma(\frac{1}{2}n + \frac{1}{2})} \int_{x=0}^{\infty} \frac{x^{\frac{1}{2}(n-1)} \exp(-x) dx}{1 + x^{\frac{1}{2}(n-1)} \theta^{-1}} \qquad (2.44)$$

$$\nu^2 = \sum_{k=1}^{n} \alpha_k^2 \nu_k^2 \Big/ \alpha^2 = \sum_{k=1}^{n} \mu_k^2 \nu_k^2 \qquad (2.45)$$

$$b = E_0/kT \qquad (2.46)$$

$$\theta = (k_2 p/\nu) b^{\frac{1}{2}(n-1)} f_n \qquad (2.47)$$

$$f_n = (4\pi)^{\frac{1}{2}(n-1)} \Gamma(\frac{1}{2}n + \frac{1}{2}) \prod_{k=1}^{n} \mu_k \qquad (2.48)$$

In (2.47), p is the total pressure and k_2, the rate-constant for collisional de-energization, is given by the collision-theory expression (2.49).

$$k_2 = 4\sigma_d^2 (\pi RT/M)^{\frac{1}{2}} \qquad (2.49)$$

2.2.2 The limiting forms at high and low pressures

When the pressure p tends to infinity, so does θ, and $I_n(\theta)$ tends to unity, so that k_∞ is given by (2.50). Slater's theory thus predicts that the high-pressure activation energy is equal to the critical energy E_0 and that the

$$k_\infty = \nu \exp(-b) = \left(\sum_{k=1}^{n} \mu_k^2 \nu_k^2\right)^{\frac{1}{2}} \exp(-E_0/kT) \qquad (2.50)$$

high-pressure A-factor is a weighted mean of the vibration frequencies in the molecule.

At low pressures, θ tends to zero and k_{uni} becomes proportional to the first power of the pressure, i.e. the rate is second-order. The rate-constant k_{uni} then becomes

$$k_{\text{uni}} = k_{\text{bim}}\, p = k_2 p(4\pi E_0/kT)^{\frac{1}{2}(n-1)} \exp(-E_0/kT) \prod_{k=1}^{n} \mu_k$$

and the second-order rate-constant k_{bim} is given by (2.51).

$$k_{\text{bim}} = k_2(4\pi E_0/kT)^{\frac{1}{2}(n-1)} \exp(-E_0/kT) \prod_{k=1}^{n} \mu_k \qquad (2.51)$$

2.2.3 The theoretical fall-off curve

From (2.43) and (2.50), the fall-off curve is defined by

$$k_{\text{uni}}/k_\infty = I_n(\theta)$$

The function $I_n(\theta)$, the 'Slater integral' defined in (2.44), depends solely upon n (the number of contributing modes of vibration) and upon θ, which is proportional to p at a constant temperature T. At a given temperature there is thus a single curve of $\log(k_{\text{uni}}/k_\infty)$ against $\log p$ for each value of n. The effect on this curve of increasing the temperature from T_1 to T_2, for a given value of n, is to translate the curve to higher pressures at the higher temperature by an amount

$$\Delta \log p = \tfrac{1}{2} n \log(T_2/T_1) \qquad (2.52)$$

From (2.47) and (2.48), it can be seen that the parameter θ which determines k_{uni}/k_∞ depends upon the value of n. For a given value of E_0/kT, θ is lower and k_{uni}/k_∞ is therefore lower the smaller the value of n. For two molecules having similar values of n, k_{uni}/k_∞ will be lower for the one having the lower value of E_0/kT. These considerations are helpful in deciding which of several possible unimolecular rate-constants is likely to decline first from its high pressure limit in a decomposition occurring by a complex mechanism. For example, in the generally accepted mechanism for the pyrolysis of ethane, there are two unimolecular steps (2.53) and (2.54). From the above considerations, one would expect reaction (2.54),

$$C_2H_6 \longrightarrow 2CH_3 \tag{2.53}$$

$$C_2H_5 \longrightarrow C_2H_4 + H \tag{2.54}$$

which has a lower activation energy than (2.53), to depart from first-order behaviour before (2.53) on lowering the total pressure. This is confirmed by experimental evidence.[5]

2.2.4 The change in activation energy with pressure

Since v is a temperature-independent constant for a given reaction [see (2.45)], the activation energy at high pressure, E_∞, is predicted from (2.50) to be equal to the critical energy E_0. At low pressure, Slater theory predicts a decline in activation energy. The experimental activation energy is defined by

$$E_{Arr} = kT^2(\mathrm{d}\ln k/\mathrm{d}T) \tag{1.6}$$

so that at low pressure, from (2.51),

$$E_{bim} = kT^2(\mathrm{d}\ln k_{bim}/\mathrm{d}T)$$
$$= E_0 - \tfrac{1}{2}(n-1)kT + kT^2(\mathrm{d}\ln k_2/\mathrm{d}T)$$

The evaluation of E_{bim} depends on whether the rate-constants k_2 and k_{bim} are expressed in concentration units or in pressure units. Equation (2.49) defines k_2 in concentration units (e.g. $cm^3\,mol^{-1}\,s^{-1}$), from which it is seen that in these units, $k_2 \propto T^{\frac{1}{2}}$. Similarly, in pressure units, $k_2 \propto T^{-\frac{1}{2}}$. Thus if E_{bim} is determined from experimental values of k_{bim} measured in concentration units, it is related to E_∞ by (2.55), but if E_{bim} is determined from k_{bim} values in terms of pressure units, the relationship

$$E_{bim} = E_\infty - (\tfrac{1}{2}n - 1)kT \quad \text{(concentration units)} \tag{2.55}$$

$$E_{bim} = E_\infty - \tfrac{1}{2}nkT \quad \text{(pressure units)} \tag{2.56}$$

is (2.56). In both cases it is predicted that $E_{bim} < E_\infty$ by an amount of several kilocalories per mole for moderately complex molecules ($n \approx 10\text{--}15$) at temperatures around 400–500 °C.

2.3 THE ASSUMPTIONS OF SLATER THEORY

The description of Slater theory given here follows the early classical harmonic form of the theory. It has long been realized, however, that molecular vibrations are neither classical nor purely harmonic under the conditions of unimolecular reactions. Considerable work has been done to produce modifications to the theory avoiding these assumptions, but, as Slater has noted,[6] it is difficult to abandon all the assumptions simultaneously.

2.3.1   The harmonic assumption

The assumption of harmonicity has the useful corollary of allowing the molecular vibrations to be treated in terms of normal modes of vibration. Qualitatively, the inclusion of anharmonicity provides a means whereby energy can flow between normal modes. Thus at sufficiently low pressures, the restrictive Slater condition (2.36) on the normal mode energies no longer applies, and every molecule with energy $\geqslant E_0$ is capable of reacting. The low-pressure second-order rate-constant is then found by integrating (2.40) subject only to the condition

$$\sum_{k=1}^{n} E_k = E \geqslant E_0$$

The resulting form of the second-order rate constant is then

$$k_{\mathrm{bim}} = k_2 (E_0/kT)^{(n-1)} \exp\left(-E_0/kT\right)/(n-1)!$$

instead of (2.51), and this is closely similar to Hinshelwood's expression (1.29) for the rate of energization.

It is difficult to evaluate quantitatively the effect on the general rate-constant of using specific expressions for the potential energy of the molecule involving anharmonic cross-terms. The difficulty arises partly from lack of knowledge of the precise potential functions applicable and partly from the difficulty of handling the resultant equations of motion when these are not amenable to the normal mode approximation. Some progress has been made by machine integration of the equations for model systems such as linear triatomic[7] and tetratomic[8] molecules with Morse potentials.

A quantum-mechanical version of the harmonic theory has been worked out by Slater,[9] and results in the same form (2.43)–(2.45), (2.47) and (2.48) of the results but with b and the μ_k redefined. The main difference is that the new b has a higher value than that from (2.46). The fall-off curve is given by the same function and therefore has the same shape for both the quantum and classical theories, but a given extent of fall-off occurs at lower pressures according to the quantum theory. A major difficulty of the theory is that it predicts an A-factor which is lower than the mean of the vibration frequencies in the molecule, and it has not been tested in any detail.

2.3.2   The random-gap assumption

A further assumption of Slater theory is the 'random-gap' assumption. The 'gap length' is the time between successive crossings by the critical coordinate q_1 of the critical boundary $q_1 = q_0$ (see Figure 2.5). It is sometimes convenient to formulate unimolecular reaction as the movement of a representative point on a hypersurface defined by the coordinates and momenta of the molecule (see Appendix 2, section A2.7). Under these conditions, the gap length becomes the time (τ) between successive crossings of a critical surface. If the incidence of this event is random, the probability of a particular value of τ is given by

$$P(\tau) = L \exp(-L\tau)$$

The validity of this assumption has been discussed by Bunker[10] and Thiele,[11] and shown to be inherent in many theories of unimolecular reactions; it will be discussed in more detail in connection with RRKM theory in section 4.12.4. Slater's 'new approach' to rate theory examines the consequences of adopting different gap distributions and concludes that despite differences at intermediate pressures, the limiting high- and low-pressure rate-constants are the same for all gap distributions.

2.3.3   The strong-collision assumption

In the Slater, RRK and basic RRKM theories of unimolecular reactions the implicit assumption is made that de-energization occurs upon every collision between an energized and a non-energized molecule. For thermal reactions this assumption appears to be well founded except for very small molecules or for reacting molecules dispersed in a heat bath of very small molecules. The evidence will be discussed in detail in connection with RRKM theory in section 4.12.2 and in Chapter 10.

2.4   THE APPLICATION OF SLATER THEORY

The Slater theory is normally tested by comparing theoretical and experimental data in the form of fall-off curves, i.e. plots of $\log(k_{uni}/k_\infty)$ versus $\log p$. In order to construct the theoretical curve for a particular reaction, one needs to have the α_{1k} factors relating the critical coordinate q_1 to the normal coordinates [cf (2.33)]. These α_{1k} can be used to calculate the μ_k from (2.42) and hence to determine θ from (2.47) and (2.48). From a knowledge of θ and n (the number of contributing modes of vibration), $I_n(\theta)$ is determined by numerical integration or from tables[12] and this gives directly the ratio k_{uni}/k_∞.

The complete Slater treatment can therefore only be carried out if vibrational analysis has previously yielded the necessary α_{1k} factors. A value is also needed for the critical energy E_0, and the experimental high-pressure activation energy will be used if this is available. A decision must also be made concerning the appropriate critical coordinate; this choice is effectively an estimate of the configuration of the activated complex and has a profound effect on the results of the Slater calculations. The factor n in Slater theory corresponds to the number of normal modes which produce changes in the chosen critical coordinate. It is not therefore possible to include in the harmonic form of Slater theory any normal modes of vibration which belong to symmetry species other than that of the reaction coordinate. This rather restrictive condition is responsible for the poor agreement found in general between Slater theory and the experimental results for small molecules.

Among the n modes of vibration of the same symmetry as the critical coordinate, certain modes may have relatively little effect upon that coordinate, i.e. for these modes μ_k is small. Slater has suggested that those normal modes need not be included, and the corresponding μ_k factors should be dropped from the relevant equations when

$$\mu_k < (4\pi E_0/kT)^{-\frac{1}{2}}$$

In view of more recent developments it is not generally regarded as profitable to make lengthy Slater calculations in order to apply the purely harmonic theory to experimental results. It is useful, however, to know whether such a treatment, if carried out, could under any circumstances produce agreement. This can be done fairly readily by making an approximate estimate of the product $\prod \mu_k$. It can be shown that since the μ_k are normalized, i.e.

$$\sum \mu_k^2 = 1$$

the maximum possible value of their product is given by

$$\left(\prod_{k=1}^{n} \mu_k\right)_{\max} = (n)^{-\frac{1}{2}n}$$

Substitution of this value into (2.51) enables one to calculate the maximum possible value of the second-order rate-constant k_{bim} and hence the minimum possible transition pressure (k_∞/k_{bim}, see section 1.3.1). It has been suggested by Slater[13] that a more realistic estimate of $\prod \mu_k$ may be obtained by assuming the μ_k to be in geometric progression with $\mu_{\max} \approx 5\mu_{\min}$; the product obtained is about one-tenth of the maximum value $n^{-\frac{1}{2}n}$ when $n = 10$. For comparison with experiment one can make use of the fact that the experimental curve of $\log(k_{\text{uni}}/k_\infty)$ versus log pressure

3

should be the same shape as the theoretical curve of $\log I_n(\theta)$ versus $\log \theta$. From the lateral displacement required to make the curves coincide, one can determine the value of the product $\prod \mu_k$ which gives agreement. If this value is reasonable on the above basis it can be concluded that agreement with Slater theory may be possible.

Very few detailed Slater treatments have been carried out. The best known is that of Slater himself[14] for the isomerization of cyclopropane to propene (2.57). This reaction was studied by Chambers and Kistiakowsky[15]

$$\begin{array}{c} CH_2 \\ / \quad \backslash \\ CH_2-CH_2 \end{array} \longrightarrow CH_3-CH=CH_2 \qquad (2.57)$$

who found the reaction to be unimolecular and suggested two possible mechanisms: (a) rupture of a C—C bond to give the trimethylene biradical, or (b) simultaneous H migration and C—C rupture without the intermediate formation of the trimethylene biradical. Slater's calculations were based upon critical coordinates involving (a) a C—C bond length, and (b) the distance between a hydrogen atom and a neighbouring (non-bonded) carbon atom. There are twelve equivalent H-(non-bonded C) distances in the molecule and this multiplicity of reaction coordinates is taken into account by multiplying the theoretical pre-exponential factor by the statistical factor of twelve (see section 4.9). For the C—C coordinate the corresponding factor is three. The critical energy was taken as $65 \cdot 0 \text{ kcal mol}^{-1}$, the experimental E_∞ found by Pritchard, Sowden and Trotman-Dickenson.[17] A complication in the application of Slater theory is the occurrence of degenerate modes of vibration (i.e. pairs of normal modes with the same frequency). Slater theory requires each pair of degenerate modes to be treated as a single mode[16] and hence the maximum value of n for the C—H critical coordinate, for example, is 14 and not the total number 21. In addition, the vibrational analysis showed that one μ_k factor was negligible, and hence the value of n used in the calculation was 13.

The calculated A-factors are compared with the experimental value in Table 2.1, and the fall-off curve calculated by Slater is compared with the

Table 2.1 Calculated and experimental A-factors for cyclopropane isomerization

Source of A-factor	A_∞ / s^{-1}
Calculated (C—H coordinate)[14]	$4 \cdot 0 \times 10^{14}$
Calculated (C—C coordinate)[14]	$5 \cdot 7 \times 10^{13}$
Experimental value[17]	$1 \cdot 5 \times 10^{15}$

experimental curve and with the simple Lindemann–Hinshelwood curve in Figure 2.6. Exact coincidence for the Slater and experimental curves required a shift of the experimental points by 0·3 log units to higher pressures. The conclusions reached by Slater were that the C—H coordinate

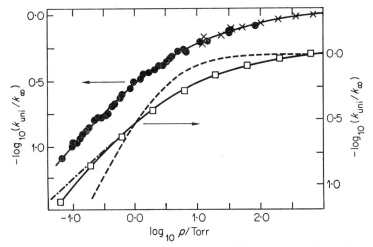

Figure 2.6 Experimental and calculated fall-off curves for cyclo-propane isomerization at 490 °C. Upper curve: experimental points of Pritchard and co-workers[17] (●) and of Chambers and Kistiakowsky[15] (×). Lower curves: —— experimental results displaced by 0·3 log units; – – – – – Hinshelwood–Lindemann curve;[17] —·—·—· classical Kassel curve;[17] □ points calculated by Slater[14]

was the better coordinate and that the calculated shape of the fall-off curve was in good agreement with experiment.

Subsequent work on the pyrolysis of cyclopropane has raised doubts concerning the exact configuration of the transition state and, in particular, has reopened the question of participation of the trimethylene radical (see section 7.11). Experiments on the pyrolysis of *sym*-cyclopropane-d_2 by Rabinovitch and co-workers[18] have been important in this connection. This molecule is less symmetric than 'light' cyclopropane, and as a result does not possess the seven doubly degenerate modes of that molecule. Consequently one would expect for cyclopropane-d_2 a Slater n close to 21 instead of 14. Despite this predicted difference, the fall-off behaviour of the two molecules is very similar, and this provides strong evidence against the harmonic Slater theory. A similar observation has been made for methyl isocyanide isomerization, where the fall-off curve for CH_2DNC lies between those for CH_3NC and CD_3NC despite the different number of non-degenerate modes (see section 9.6). A more recent application of Slater theory to cyclopropane, using new vibrational analysis data, has shown that n is not necessarily increased to 21 by the deuterium substitution. Golike and Schlag[19] have considered a number of possible reaction coordinates and have shown that on the basis of their data some of the μ_k values are small enough to be ignored, with the result that n may only be raised to 17 or 18 instead of 21. This smaller increase is still quite inconsistent with the experimental data however.

The fall-off for methylcyclopropane isomerization (producing the four isomeric butenes) occurs at a pressure which is thirty times lower than the corresponding pressure for cyclopropane. This was interpreted by Chesick[20] as a demonstration of the effect of intramolecular energy transfer on the lifetime of the energized molecule, which is contrary to the assumptions of Slater theory. Complete Slater calculations were not, however, carried out for this reaction. The experimental n for cis-trans isomerization of trans-ethylene-d_2 and cis-but-2-ene are both considerably higher than the number of modes which in Slater theory might reasonably be expected to contribute on the basis of symmetry arguments.[28]

Among the other detailed applications of Slater theory which have been made are those concerning the pyrolyses of ozone,[21] nitrous oxide,[22] hydrogen peroxide,[23] nitryl chloride[24] and methyl chloride.[25] Many of these were reviewed by Gill and Laidler,[23] who concluded that the agreement between Slater theory and experiment is generally poor for small molecules having necessarily few normal modes of vibration. The assumptions of Slater theory were considered to be better founded for molecules of moderate complexity, but more recent work (see section 4.12.1) has invalidated this conclusion.

Approximate Slater calculations have been made by Thiele and Wilson[26] for the pyrolyses of cyclobutane, cyclobutene and nitrogen pentoxide, and by Holbrook and Marsh[27] for ethyl chloride. In the case of ethyl chloride, the approximate Slater curves are a reasonable fit to the experimental data at four temperatures if n is taken to be 13.

References

1. N. B. Slater, *Proc. Camb. Phil. Soc.*, **35**, 56 (1939).
2. N. B. Slater, *Theory of Unimolecular Reactions*, Methuen, London, 1959.
3. E. B. Wilson, J. C. Decius and P. C. Cross, *Molecular Vibrations*, McGraw–Hill, New York, 1955, p. 21.
4. M. Kac, *Amer. J. Math.*, **65**, 609 (1943).
5. C. P. Quinn, *Proc. Roy. Soc. (A)*, **275**, 190 (1963).
6. N. B. Slater, 'Unimolecular Reactions', in *The Transition State, Chem. Soc. Spec. Pub. No. 16*, p. 29 (1962).
7. E. Thiele and D. J. Wilson, *J. Chem. Phys.*, **35**, 1256 (1961).

8. R. J. Harter, E. B. Altermann and D. J. Wilson, *J. Chem. Phys.*, **40**, 2137 (1964).
9. Ref. 2, Chap. 10.
10. D. L. Bunker, *Theory of Elementary Gas Reaction Rates*, Pergamon, Oxford, 1966.
11. E. Thiele, *J. Chem. Phys.*, **36**, 1466 (1962); **38**, 1959 (1963).
12. Ref. 2, Chap. 9.
13. Ref. 2, p. 143.
14. N. B. Slater, *Proc. Roy. Soc. (A)*, **218**, 224 (1953).
15. T. S. Chambers and G. B. Kistiakowsky, *J. Amer. Chem. Soc.*, **56**, 399 (1934).
16. Ref. 2, p. 152.
17. H. O. Pritchard, R. G. Sowden and A. F. Trotman-Dickenson, *Proc. Roy. Soc. (A)*, **217**, 563 (1953).
18. B. S. Rabinovitch, E. W. Schlag and K. B. Wiberg, *J. Chem. Phys.*, **28**, 504 (1958); B. S. Rabinovitch and E. W. Schlag, *J. Amer. Chem. Soc.*, **82**, 5996 (1960).
19. R. C. Golike and E. W. Schlag, *J. Chem. Phys.*, **38**, 1886 (1963).
20. J. P. Chesick, *J. Amer. Chem. Soc.*, **82**, 3277 (1960).
21. E. K. Gill and K. J. Laidler, *Trans. Faraday Soc.*, **55**, 753 (1959).
22. E. K. Gill and K. J. Laidler, *Can. J. Chem.*, **36**, 1570 (1958).
23. E. K. Gill and K. J. Laidler, *Proc. Roy. Soc. (A)*, **250**, 121 (1959).
24. Ref. 2, p. 175.
25. K. A. Holbrook, *Trans. Faraday Soc.*, **57**, 2151 (1961).
26. E. Thiele and D. J. Wilson, *Can. J. Chem.*, **37**, 1035 (1959).
27. K. A. Holbrook and A. R. W. Marsh, *Trans. Faraday Soc.*, **63**, 643 (1967).
28. B. S. Rabinovitch and K. W. Michel, *J. Amer. Chem. Soc.*, **81**, 5065 (1959).

3 The Rice-Ramsperger-Kassel Theories

It was shown in Chapter 1 that in order to make accurate quantitative predictions of the fall-off behaviour of a unimolecular reaction it was essential to take into account the energy dependence of the rate-constant k_a for the conversion of energized molecules into activated complexes and hence products. Chapter 2 dealt with Slater's elegant approach to this calculation; unfortunately, as was seen there, the model on which his theory is based is somewhat artificial, and the theory is therefore at best of limited applicability. We now return to the main theme of developing the theory currently considered to be the most realistic and accurate. This, the RRKM theory, is a statistical theory, and the present chapter will deal with the earlier statistical theories due to O. K. Rice, H. C. Ramsperger and L. S. Kassel. These are known as the RRK or HRRK theories, the H being added in acknowledgement of the fact that Hinshelwood's expression for k_1 was used in most of this work.

The RRK theories, as indicated in Chapter 1, use the basic Hinshelwood–Lindemann mechanism of collisional energization and de-energization, but consider more realistically that the rate of conversion of an energized molecule to products is a function of its energy content. The mechanism is thus written as in (3.1) and (3.2), and the contributions of Rice, Ramsperger and Kassel were directed towards the development of expressions for the energy-dependence of k_a.

$$A + M \xrightarrow[\;\;k_2\;\;]{\;\delta k_{1(E \to E + \delta E)}\;} A^*_{(E \to E + \delta E)} + M \tag{3.1}$$

$$A^*_{(E)} \xrightarrow{\;k_a(E)\;} \text{Products} \tag{3.2}$$

The theories due to Rice and Ramsperger[1,2] and Kassel[3,4,5] were developed virtually simultaneously and are very similar in their approach. Both consider that for reaction to occur a critical amount of energy E_0 must become concentrated in one particular part of the molecule. The total energy E of a molecule under consideration is assumed to be rapidly redistributed round the molecule, so that for any molecule with E greater than E_0 (i.e. for any energized molecule) there is a finite statistical probability that E_0 will be found in the relevant part of the molecule. The RRK theories assume that the rate-constant for conversion of energized molecules to products is proportional to this probability (which obviously increases with E), and hence derive expressions for the energy dependence of k_a. The differences between the Rice–Ramsperger and the Kassel treatments are twofold. Firstly, Rice and Ramsperger used classical statistical mechanics throughout, whereas Kassel also developed a quantum treatment; the latter is, in fact, very much more realistic and accurate. Secondly, different assumptions were made about the part of the molecule into which the critical energy E_0 has to be concentrated. In the calculations of Rice and Ramsperger this was taken to be one squared term in the energy expression, although other possibilities were also mentioned. Kassel assumed that the energy had to be concentrated into one oscillator (i.e. two squared terms), and Kassel's model again seems slightly more realistic. The two classical theories give very similar results in practice. The Kassel version has been used the more widely, and in view of its important quantum version it is the Kassel theory which will be developed in detail here.

3.1   THE KASSEL THEORIES

Kassel's widely used *classical theory* is based on a calculation of the probability that a system of s classical oscillators with total energy E should have energy $\geqslant E_0$ in one chosen oscillator; the derivation is clearly documented elsewhere[3] and the result is (3.3). It is then assumed that the

$$probability \text{ (energy} \geqslant E_0 \text{ in one chosen oscillator)} = \left(\frac{E-E_0}{E}\right)^{s-1} \quad (3.3)$$

$$k_a(E) = A\left(\frac{E-E_0}{E}\right)^{s-1} = A(1-E_0/E)^{s-1} \quad (3.4)$$

rate-constant $k_a(E)$ for conversion of energized molecules to products is proportional to this probability, as in (3.4). At this stage A is no more than a proportionality constant, but it acquires significance when (3.4) is inserted into (1.46), using Hinshelwood's expression for δk_1. The high-pressure limit is then given by (3.5), which on substitution of $x = (E-E_0)/kT$ and use of the defining integral for $\Gamma(s) = (s-1)!$

(Appendix 7) gives (3.6). Thus the theory predicts strict adherence to

$$k_\infty = \int_{E=E_0}^\infty [A(1-E_0/E)^{s-1}]\left[\frac{(E/kT)^{s-1}\exp(-E/kT)}{(s-1)!}\right]\mathrm{d}\left(\frac{E}{kT}\right) \quad (3.5)$$

$$= A\exp(-E_0/kT) \quad (3.6)$$

the Arrhenius equation,† and the constant A is now identified with the high-pressure A-factor for the reaction. The corresponding equation for k_{uni} in the fall-off region is easily written down, and division by (3.6) results in (3.7). This last equation is probably the most widely used result

$$\frac{k_{\mathrm{uni}}}{k_\infty} = \frac{1}{(s-1)!}\int_{x=0}^\infty \frac{x^{s-1}\exp(-x)\,\mathrm{d}x}{1+(A/k_2[M])\,[x/(x+E_0/kT)]^{s-1}} \quad (3.7)$$

of Kassel's work, although this is unfortunate since the quantum version, to be described below, is substantially more realistic. In applying (3.7), $k_2[M]$ is generally replaced by Zp (the total collision frequency at unit pressure of A^*), and the experimental high-pressure A-factor and activation energy are used for A and E_0. The integral is evaluated numerically for various values of s, using methods such as the Gauss or Gauss–Laguerre procedures,[8] or possibly more elaborate techniques if manual work is required.[9]

The *quantum version* of Kassel's theory is in essence very similar to the classical theory outlined above. It assumes, in its simplest form, that there are s identical oscillators in the molecule, all having frequency ν. The energies are expressed in quanta, the critical number of quanta being $m = E_0/h\nu$. The expression corresponding to (3.3) is now the probability that if s oscillators contain a total of n quanta (where $n = E/h\nu$), one chosen oscillator will contain at least m quanta; the result is (3.8) and the corresponding expression for k_a is thus (3.9). The expression used for

$$\begin{array}{cc}\textit{probability} \text{ (energy} \geqslant m \text{ quanta in} \\ \text{chosen oscillator)}\end{array} = \frac{n!\,(n-m+s-1)!}{(n-m)!\,(n+s-1)!} \quad (3.8)$$

$$k_a(n\,h\nu) = A\,\frac{n!\,(n-m+s-1)!}{(n-m)!\,(n+s-1)!} \quad (3.9)$$

$k_1(E)$ is a similar development of Hinshelwood's expression, namely (3.10).[4,5] This is no longer a differential quantity, since it refers to

$$k_1(n\,h\nu) = k_2\,\alpha^n(1-\alpha)^s\,\frac{(n+s-1)!}{n!\,(s-1)!} \quad \text{where } \alpha = \exp(-h\nu/kT) \quad (3.10)$$

† Slater has shown[6,7] that if Hinshelwood's δk_1 is used then (3.4) is the *only* form of $k_a(E)$ which will give a temperature dependence exactly according to the Arrhenius equation, with a temperature-independent A-factor and activation energy. This result is not as useful as it may seem, however, since the Arrhenius equation in this strict form need not necessarily be obeyed and is probably not obeyed by the great majority of reactions.

energization into a specific quantum state rather than into an energy range E to $E+\delta E$. The overall rate-constant is obtained by summing over the discrete energy levels; the high-pressure rate-constant is given by (3.11) and this can be shown to result once more in the Arrhenius form (3.6). The

$$k_{\infty} = \sum_{n=m}^{\infty} \frac{k_1(n\,h\nu)\,k_a(n\,h\nu)}{k_2} \tag{3.11}$$

$$= A\exp\left(-E_0/kT\right) \tag{3.6}$$

interpretation of the constant A is thus the same as in the classical case. The rather cumbersome expression for $k_{\mathrm{uni}}/k_{\infty}$ is similarly derived as (3.12), in which $n-m$ has been replaced by p. Kassel further developed this

$$\frac{k_{\mathrm{uni}}}{k_{\infty}} = (1-\alpha)^s \sum_{p=0}^{\infty} \frac{\alpha^p(p+s-1)!/[p!\,(s-1)!]}{1+(A/k_2[\mathrm{M}])\,(p+m)!\,(p+s-1)!/[(p+m+s-1)!\,p!]} \tag{3.12}$$

type of model to deal with more realistic cases where the s oscillators were not all of the same frequency. Detailed equations were given for the case where there are two frequencies, one an exact multiple of the other, and the extensions to other cases were hinted at.

It will be noted that the limiting case of classical behaviour may be obtained from the quantum theory by letting the quantum interval $h\nu$ become very small, so that n and m are both much greater than s. If (3.9) and (3.10) are written as in (3.13) and (3.14) the approximations involved in replacing them by the corresponding classical results will be more readily appreciated. In (3.14) $k_1(n\,h\nu)$ is divided by the quantum energy $h\nu$

$$k_a(n\,h\nu) = A\frac{\overline{(n-m+1)}\,\overline{(n-m+2)}\ldots\overline{(n-m+s-1)}}{(n+1)\,(n+2)\ldots\overline{(n+s-1)}} \approx A\frac{(n-m)^{s-1}}{n^{s-1}}$$

$$[\equiv A(1-E_0/E)^{s-1}] \quad \text{for } n-m \gg s \tag{3.13}$$

$$\frac{1}{h\nu}\frac{k_1(n\,h\nu)}{k_2} = \frac{\exp\left(-E/kT\right)}{(s-1)!}\frac{(E/kT)^{s-1}}{kT}$$

$$\times \left\{ \frac{(n+1)\,(n+2)\ldots(n+s-1)}{n^{s-1}} \left[\frac{1-\exp\left(-h\nu/kT\right)}{h\nu/kT} \right]^s \right\}$$

$$\approx \frac{\exp\left(-E/kT\right)}{(s-1)!}\frac{(E/kT)^{s-1}}{kT} \quad \text{for } n \gg s \text{ and } h\nu/kT \ll 1 \tag{3.14}$$

for comparison with $dk_1(E)/dE$, the classical rate of energization into a unit energy range. It will be seen in the next section and in Chapter 5 that the classical equations are in fact rather poor approximations, and their use no longer seems a worth-while exercise.

It will be noted that in both forms of the Kassel theory the same value of s has been used in k_1 and in k_a, and that in the quantum theory the same oscillator frequency ν has been assumed for both the energized molecule and the activated complex. The first of these assumptions seems to be perfectly valid in the light of subsequent work with the more sophisticated theories, but the second reflects the extreme simplicity of the model used and is abandoned in later theories. In particular, a Kassel-type formulation by Eyring and co-workers[10] treats the energized molecule as a set of s oscillators with frequencies ν_j ($j = 1$ to s) and the activated complex as a set of $s-1$ oscillators with potentially different frequencies $\nu_i^{\neq}$ ($i = 1$ to $s-1$). Using classical statistical mechanics, $k_a(E)$ for this model is given by (3.15),

$$k_a(E) = \left(\prod_{j=1}^{s} \nu_j \Big/ \prod_{i=1}^{s-1} \nu_i^{\neq} \right) \left(\frac{E-E_0}{E} \right)^{s-1} \tag{3.15}$$

which is identical with Kassel's $k_a(E)$ with the constant A (and therefore the theoretical high-pressure A-factor) now being equated to $\prod^s \nu_j / \prod^{s-1} \nu_i^{\neq}$. Thus A should be of the order of magnitude of a vibration frequency (ca 10^{13} s^{-1}), as is found experimentally to be the case for many unimolecular reactions (see section 7.1). Classical internal rotations could also be included in this treatment.[10] This type of model is clearly more realistic than the degenerate-oscillator type, although the classical treatment is still inadequate and it requires the application of quantum statistical mechanics to produce the currently accepted theory presented in the next chapter.

3.2 APPLICATION OF THE RRK THEORIES

Many workers have applied the RRK theories to their results (see Tables 7.34 and 7.36), usually in the form of the classical Kassel theory [equation (3.7)], and it is not proposed to attempt any comprehensive survey of these applications. Instead a few arbitrarily chosen examples will be discussed with the principal objects of showing how (a) the RRK theories are an improvement over the HL treatment, and (b) the little-used quantum version is a large improvement over the classical version. For these purposes only a rather broad comparison of theory with experiment will be required, and more detailed considerations will only be required in connection with the more sophisticated theories to be described later. One of Kassel's own main examples was the decomposition of azomethane, and this is now known not to be a simple unimolecular reaction. It has been concluded, however,[11] that the fall-off data for this reaction probably reflect in large part the declining rate of the unimolecular initiation process, and the calculations therefore form a useful example for the present purpose, especially since they include one of the few extensive applications of Kassel's quantum theories.

Figure 3.1 therefore shows some of Kassel's curves[3] for azomethane decomposition compared with Ramsperger's experimental data.[12] Also

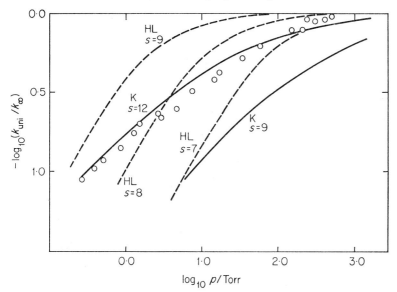

Figure 3.1 Fall-off curves for azomethane at 603 K. Circles are Ramsperger's experimental results.[12] Broken lines are HL curves with $s = 7$, 8 or 9; solid lines are classical Kassel curves[3] with $s = 9$ or 12. $\sigma_d^2 = 150$ Å^2 throughout

shown are the Hinshelwood–Lindemann plots for several values of s [equations (1.26) and (1.29)], and it will be seen immediately that the Kassel curves reproduce much more closely the relatively shallow curvature of the experimental fall-off curve. This conclusion is reinforced by the plots of k_∞/k_{uni} vs $1/p$ shown in Figure 3.2. As was noted in section 1.3.4, the HL treatment gives straight-line plots whereas the experimental data invariably give a curve when plotted in this way. The plot shown for azomethane is typical of the behaviour of unimolecular reactions, e.g. those of CH_3NC, C_2H_5Cl, c-C_3H_6 and numerous others. The HL treatment thus predicts too rapid a variation of k_{uni} with pressure, whereas Figures 3.1 and 3.2 show that the (classical) Kassel theory gives exactly the right sort of curvature. Thus the RRK theories gave for the first time a treatment which, with suitable choice of parameters, could reproduce the experimental results with reasonable accuracy. More detailed comparison of theory and experiment is not very profitable at this stage because the

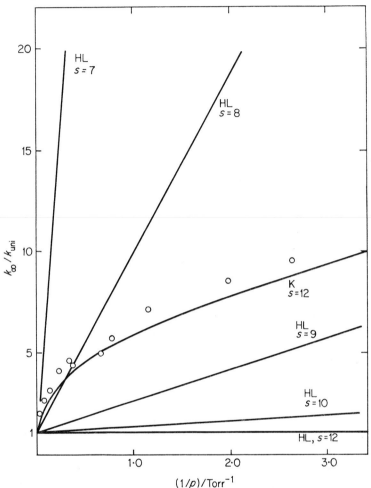

Figure 3.2 Reciprocal plots (k_∞/k_{uni} vs $1/p$) for azomethane decomposition at 603 K. Circles are Ramsperger's experimental points.[12] Lines are HL plots for $s = 7, 8, 9, 10, 12$, and classical Kassel curve[3] for $s = 12$. $\sigma_d^2 = 150$ Å² throughout

choice of parameters is somewhat arbitrary and empirical. In particular, Kassel's value of $\sigma_d^2 = 150$ Å² for the curves in Figures 3.1 and 3.2 is obviously too high, but as the curves for different s are very similar in shape the classical curve for $s = 14$, $\sigma_d^2 = 25$ Å² would also lie close to the experimental results. Variation of the parameters thus makes the theoretical prediction very flexible, and for this reason it does not matter for present purposes that the experimental results shown may not represent in exact detail the behaviour of a unimolecular reaction.

It will be noted that approximate agreement of the HL treatment is obtained with only 8 oscillators, compared with 12 for the classical Kassel theory with the same collision diameter. This is because $k_a(E)$ in the Kassel theory is lower (except at $E = \infty$) than k_a in the HL treatment; see equation (3.4) and Figure 3.3. Thus for a given value of s the Kassel

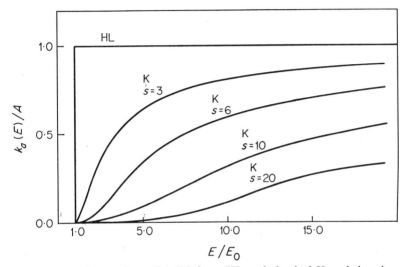

Figure 3.3   Comparison of $k_a(E)$ from HL and classical Kassel theories

theory gives a lower value for k_{uni} than the HL treatment, and the plot of $\log k_{uni}/k_\infty$ vs $\log p$ is displaced to higher pressures. An increase in s for the Kassel curve then serves to bring it into the same pressure region as the HL curve.

One unsatisfactory feature of the classical Kassel theory is that in order to obtain agreement with experiment using a reasonable value of σ_d, the required value of s is usually about half the number of oscillators in the molecule. Admittedly this fraction *can* vary over a wide range,[13] but about one-half is the norm for a large number of reactions. Although it is *a priori* possible that not all the vibrational modes of an energized molecule can contribute their energy to the reaction, it would be a curious coincidence if for most molecules about half did so, and later theoretical and experimental work has in fact indicated that probably *all* modes can contribute in most cases. The above anomaly arises from the serious inadequacy of the classical statistical mechanics used, and the quantum Kassel theory gives satisfactory results assuming all the modes to be 'active'. Figure 3.4 shows two curves calculated[4] from the quantum version, which fit the experimental data just as well as the classical $s = 12$ curve in Figure 3.1. The curve for $s = 24$, $m = 15$, $\sigma_d{}^2 = 6$ Å^2 assumes all

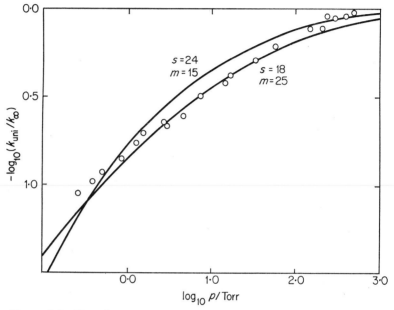

Figure 3.4 Kassel quantum curves[4] for azomethane decomposition at 603 K ($\sigma_d{}^2 = 6$ Å^2). Circles are Ramsperger's experimental results[12]

vibrations to be active and is almost identical to the classical curves for $s = 12$, $\sigma_d{}^2 = 150$ Å^2 or $s = 14$–15, $\sigma_d{}^2 = 6$ Å^2. This value of m corresponds to an oscillator frequency of 1220 cm^{-1}, which must be close to the geometric mean frequency for the molecule, so the model is reasonably realistic. Kassel[4] noted correspondingly that the calculated specific heat was 'not unreasonably large' for this model, although no experimental value was available. Equally satisfactory results were obtained with a two-frequency model, of which the lower curve in Figure 3.4 represents the extreme case where 6 oscillators are of very high frequency ('C—H stretches') and effectively do not contribute; the remaining 18 have a frequency of 730 cm^{-1} which is again reasonable. The quantum Kassel theory can thus be adjusted to fit the results with reasonable parameter values on the assumption that all or most of the vibrational modes contribute their energy to the reaction.

A particularly bad approximation in the classical theory involves the numerator of (3.13) (since $n - m$ is relatively small for thermal energization), and a greater insight into the failure of the classical theory may therefore be obtained by consideration of its predictions for $k_a(E)$ compared with the quantum version and with other theories. One such comparison is made later in section 8.2, where chemical activation studies are shown to give $k_a(E)$ values incompatible with the classical Kassel theory unless an artificially reduced number of oscillators is assumed. Another comparison is illustrated here in Figure 3.5, which refers to the dissociation of nitryl chloride (NO_2Cl).[14, 15] The geometric mean of the vibration frequencies[16] is 745 cm^{-1}, corresponding to $h\nu = 2.13$ kcal mol^{-1}, and E_0 is taken to be 29·8 kcal mol^{-1} so that $m = 14$. The stepwise quantum curve is for $s = 6$, $m = 14$ (i.e. all oscillators active), and classical curves are shown for $s = 3, 4, 5$ and 6. It will be noted that at high energies the classical curve with $s = 6$ is closest to the quantum curve, and this is reasonable since the effects of quantization are least important under these conditions. The classical $s = 6$ curve is very poor at low energies however, k_a being too small by more than an order of magnitude at the energies of interest for a thermal reaction, which are only a few multiples of kT above the critical energy (see for example section 4.8). Under these conditions the quantum curve is probably best approximated by the classical curve for $s = 4$, and this will give the best classical approximation to the fall-off curve. Johnston[14] has also given a curve based on a more sophisticated quantum theory, and the Kassel quantum curve is in general terms a good reproduction of this curve, differing mainly in its coarse stepwise variation compared with the smoother variation of the more realistic formulation.

Similar comparisons have been made by a number of other workers,[13, 17] although most have compared the classical Kassel theory with a non-degenerate oscillator quantum theory rather than Kassel's own quantum formulation. As regards the broad features, however, the comparison with Kassel's quantum theory would yield similar results, and all the comparisons reinforce the above conclusion that the classical Kassel formulation fails on account of the classical approximation (see also section 5.4.1) and not because of any gross inadequacy in the basic model.

In conclusion, the statistical approach of the RRK theories to the calculation of $k_a(E)$ is correct, and this approach later formed the basis of the RRKM theory to be discussed in the next chapter. The quantum statistical mechanical treatment gives results in broad agreement with experiment (and with the RRKM theory) on the basis that all oscillators in the molecule contribute their energy to the reaction. The classical statistical mechanical treatment is able to give agreement with a severely reduced number of oscillators, but this is no more than a curiosity of the

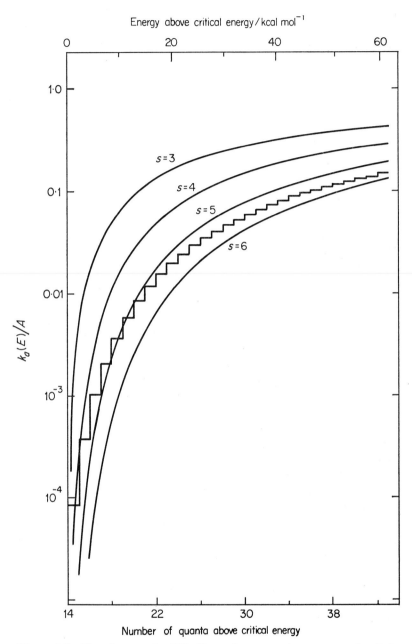

Figure 3.5 Kassel curves for NO_2Cl reaction; smooth curves are classical theory with $s = 3$, 4, 5 and 6; stepped curve is quantum theory.

equations involved. The classical approximation is simply not valid, and no significance should nowadays be attached to any results obtained by its application.

References

1. O. K. Rice and H. C. Ramsperger, *J. Amer. Chem. Soc.*, **49**, 1617 (1927).
2. O. K. Rice and H. C. Ramsperger, *J. Amer. Chem. Soc.*, **50**, 617 (1928).
3. L. S. Kassel, *J. Phys. Chem.*, **32**, 225 (1928).
4. L. S. Kassel, *J. Phys. Chem.*, **32**, 1065 (1928).
5. L. S. Kassel, *Kinetics of Homogeneous Gas Reactions*, Chemical Catalog Co., New York, 1932.
6. N. B. Slater, *Proc. Leeds Phil. Soc.*, **6**, 259 (1955).
7. N. B. Slater, *Theory of Unimolecular Reactions*, Methuen, London, 1959, Chap. 2.
8. H. E. Salzer and R. Zucker, *Bull. Amer. Math. Soc.*, **55**, 1004 (1949).
9. Ref. 7, p. 168.
10. H. M. Rosenstock, M. B. Wallenstein, A. L. Wahrhaftig and H. Eyring, *Proc. Nat. Acad. Sci. U.S.A.*, **38**, 667 (1952); J. C. Giddings and H. Eyring, *J. Chem. Phys.*, **22**, 538 (1954).
11. A. F. Trotman-Dickenson, *Gas Kinetics*, Butterworths, London, 1955, p. 70.
12. H. C. Ramsperger, *J. Amer. Chem. Soc.*, **49**, 1495 (1927).
13. D. W. Placzek, B. S. Rabinovitch and G. Z. Whitten, *J. Chem. Phys.*, **43**, 4071 (1965).
14. H. S. Johnston, *Gas Phase Reaction Rate Theory*, Ronald, New York, 1966, p. 284.
15. H. F. Cordes and H. S. Johnston, *J. Amer. Chem. Soc.*, **76**, 4264 (1954).
16. R. Ryason and M. K. Wilson, *J. Chem. Phys.*, **22**, 2000 (1954).
17. H. M. Rosenstock, *J. Chem. Phys.*, **34**, 2182 (1961); M. Vestal, A. L. Wahrhaftig and W. H. Johnston, *J. Chem. Phys.*, **37**, 1276 (1962); E. Tschuikow-Roux, *J. Phys. Chem.*, **73**, 3891 (1969).

4 RRKM (Marcus-Rice) Theory

The theory presented in this chapter was developed essentially by R. A. Marcus[1] from an earlier paper by Marcus and O. K. Rice,[2] and is known by the names of these authors or very often by the initials RRKM since its basic model is the RRK model discussed in Chapter 3. The theory will be described in considerable detail, for two reasons. Firstly, it is the most realistic and successful current theory, and clearly deserves to be well known and understood, and, secondly, there exists no clear and detailed account of it in any other place. Some useful auxiliary reading will be found in the books by Bunker,[3] Laidler[4] and others,[5] in Wieder's Ph.D. Thesis,[6] and in a review article by Rabinovitch and Setser.[7]

The present chapter is reserved for the basic theory, and the mechanism of its application and the results obtained will be dealt with later. The treatment depends heavily on statistical-mechanical theory, and Appendix 2 reviews the necessary background with particular emphasis on the results needed here, which are not all prominent in the works on statistical mechanics. Even readers familiar with the subject may find it useful to glance at sections A2.3–A2.5 in this appendix. The question of nomenclature is a difficult one; there is little consistency between different authors in the present field and we have tried to adopt the best compromises. Appendix 1 lists the more important of the symbols used here and correlates them with those of other authors.

4.1 THE RRKM REACTION SCHEME

As indicated in Chapter 1, the reaction scheme used in the RRKM theory comprises the reactions shown in (4.1) and (4.2). This scheme is a

$$A + M \underset{k_2}{\overset{\delta k_{1(E^* \to E^* + \delta E^*)}}{\rightleftharpoons}} A^*_{(E^* \to E^* + \delta E^*)} + M \qquad (4.1)$$

$$A^*_{(E^*)} \xrightarrow{k_a(E^*)} A^+ \xrightarrow{k^+} \text{Products} \qquad (4.2)$$

more detailed version of the RRK mechanism, in which (4.2) was condensed to (4.3) [cf. (3.1)–(3.2)]. The superscript * in E^* has been introduced for reasons that will emerge later (section 4.3). As in any

$$A^*_{(E^*)} \xrightarrow{k_a(E^*)} \text{Products} \tag{4.3}$$

theory of this type the overall first-order rate-constant k_{uni} is given by (1.46), reproduced here as (4.4). There are essentially two new principles involved in the RRKM treatment.[1,2] Firstly, the energization rate-constant k_1 in (4.1) is evaluated as a function of energy by a quantum-statistical-mechanical treatment instead of the classical treatment used in

$$k_{uni} = -\frac{1}{[A]}\frac{d[A]}{dt} = \int_{E^*=E_0}^{\infty} \frac{k_a(E^*)\,dk_{1(E^* \to E^* + dE^*)}/k_2}{1 + k_a(E^*)/k_2[M]} \tag{4.4}$$

the HRRK and Slater theories. The de-energization rate-constant k_2 is considered as in those theories to be independent of energy, and is often equated to the collision number Z or to λZ where λ is a collisional deactivation efficiency. The assumption of constant k_2 might be relaxed eventually, but this type of calculation really requires a quite different approach (see Chapter 10). The second major feature of RRKM theory is the application of Absolute Rate Theory (ART) to the calculation of $k_a(E^*)$. For this purpose the overall reaction (4.3) is written in terms of two steps, (4.2), in which a careful distinction has been made between the *energized molecule* A* (sometimes called the active molecule) and the *activated complex* A⁺ (occasionally called the activated molecule).

The *energized molecule* A* is basically an A molecule, but is characterized loosely by having enough energy to react; a more precise definition will be possible in section 4.3. The energy distribution will not usually be such that reaction occurs immediately, however; there will be numerous quantum states of the energized molecule in a given small energy range, and only a few of these will correspond to energy distributions with which the molecule can actually undergo conversion to products. Moreover, the energized molecules will not react instantaneously even when one of these relatively rare quantum states is reached, since the vibrational modes involved in the reaction will not in general be correctly phased at first. The energized molecules thus have lifetimes to decomposition which are much greater than the periods of their vibrations (ca 10^{-13} s). The actual lifetimes to de-energization or decomposition depend on the values of $k_2[M]$ and $k_a(E^*)$ respectively, but are typically in the range 10^{-9} to 10^{-4} s.

The *activated complex* A⁺ is basically a species which is recognizable as being intermediate between reactant and products, and is characterized by having a configuration corresponding to the top of an energy barrier between reactant and products.† As in one version of ART[10] the conversion of A* into products is considered in terms of translational motion in the reaction coordinate. The energy profile along the reaction coordinate involves a potential energy barrier between reactant and products, of height E_0 (the critical energy requirement), and this barrier must be surmounted for reaction to occur. The activated complex is a molecule for which the extension of the reaction coordinate lies in an arbitrarily small range δ at the top of the barrier (see, however, the footnote in section 4.5.1). The activated complex is thus unstable to movement in either direction along the reaction coordinate and, in contrast to an energized molecule, has no measurable life. There will usually be more than one quantum state of A+ which can be formed from a given A*, because of the different possible distributions of the energy between the reaction coordinate and the vibrational and rotational degrees of freedom of the complex. Thus the rate-constant $k_a(E^*)$ (equation 4.2) will be evaluated, as in ART, as the sum of a set of contributions from the various possible activated complexes.

4.2 CLASSIFICATION OF ENERGIES AND DEGREES OF FREEDOM

The RRKM theory uses statistical mechanics for calculating the equilibrium concentrations of A* and A+, and is thus concerned very much with evaluating the number of ways of distributing a given amount of energy between the various degrees of freedom of a molecule. Any *fixed energy*, which cannot be redistributed, is clearly of no interest from this point of view; for example, the zero-point energy of the molecular vibrations is always present and always the same, and hence is said to be fixed. Overall translation of a molecule has no effect on the rate of its unimolecular reactions, and translational energy of the molecule as a whole is therefore fixed. In contrast, some of the energy content of the molecule (the *non-fixed energy*) is not fixed by any basic principle and is considered to be free to move around the molecule. In particular, the vibrational energy of the molecule (apart from the fixed zero-point energy) is assumed to be subject to rapid statistical redistribution. This assumption is the same as that of the RRK theories and is diametrically opposed to that of the Slater theory.

† Bunker has criticized the use of the term *activated complex* in this context[8] and suggested an alternative definition[9] of the *critical molecular configuration* as having a value of the reaction coordinate such that the density of internal quantum states is minimized; see also section 6.2.2.

Rotational energy poses a more difficult problem; it might well contribute to a reaction (for example, centrifugal force might assist a bond-breaking reaction), but there might be limitations due to the requirement of conservation of the angular momentum. If, as a result, a rotational degree of freedom stays in the same quantum state during the reaction, the mode is said to be *adiabatic*. There is no *random* exchange of energy between an adiabatic mode and the other degrees of freedom, but if the moment of inertia changes the rotational energy must also change, and this will affect the rate of reaction (see section 4.10).

The RRKM theory as normally applied assumes that all the non-adiabatic degrees of freedom of the molecule are *active*, i.e. can contribute their non-fixed energy to the reaction. Marcus[1] has also made provision for some of the modes to be *inactive* although not adiabatic; such degrees of freedom exchange energy between themselves but cannot contribute it towards the energy required for surmounting the potential energy barrier (see also section 4.12.1). This sort of model has not been found necessary in practice, and we treat here only the case where all modes are either active or adiabatic. A potentially more useful theory might permit energy transfer but only at a limited rate, but this is of course more difficult to cater for and only the beginnings of such a theory have yet emerged.[11]

4.3 TERMINOLOGY FOR ENERGIES

In order to handle the statistical mechanics used in the RRKM theory it is important to understand clearly the significance of the various energy terms used, and we are now in a position to discuss these in detail. As an aid to the discussion, Figure 4.1 illustrates the energetics of a simple unimolecular reaction and some of the relationships between the terms to be used.

An energized molecule, A*, can now be defined precisely as a molecule which contains in its active degrees of freedom a non-fixed energy E^* greater than a critical value E_0 below which classical reaction cannot occur. This critical energy may alternatively be defined as the difference between the ground-state energies of A^+ and A; it is closely related to the Arrhenius activation energy of the reaction, but may differ from it by a few multiples of kT. The energy E^* may comprise both vibrational and rotational non-fixed energy, denoted by E_v^* and E_r^* respectively, and the sum of these is clearly E^*, which will sometimes be subscripted as follows for emphasis:

$$E^* \equiv E_{vr}^* = E_v^* + E_r^*$$

The total non-fixed energy of a given activated complex A^+ is denoted by E^+. This quantity is closely related to the non-fixed energy E^* of the energized molecule A^* from which the activated complex was formed. In this process an amount of energy E_0 has been used to surmount the energy

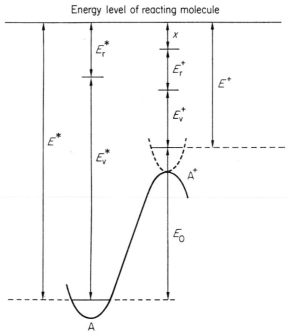

Energy level of reacting molecule

Figure 4.1 Illustration of energy terminology for a unimolecular reaction; adiabatic and inactive degrees of freedom are excluded

barrier and has thus become fixed; any redistribution of this energy corresponds to motion along the reaction coordinate and the molecule would then no longer be an activated complex. Thus E^+ and E^* for a corresponding complex and energized molecule are related simply by

$$E^+ = E^* - E_0$$

The energy E^+ may, like E^*, contain both vibrational and rotational non-fixed energy, denoted by E_v^+ and E_r^+ respectively, the sum of these being E_{vr}^+. It is also necessary to separate out for special treatment the energy x associated with the translational motion of A^+ in the reaction coordinate, and we therefore have

$$E^+ = E_{vr}^+ + x = E_v^+ + E_r^+ + x$$

The different possible distributions of the energy between vibration, rotation, and the translation along the reaction coordinate are an important factor in determining the number of quantum states corresponding to a given E^* or E^+, and hence the equilibrium concentrations of A* and A+.

4.4   EXPRESSION FOR $\delta k_{1(E^*\to E^*+\delta E^*)}/k_2$

The quantity $\delta k_{1(E^*\to E^*+\delta E^*)}/k_2$, which is required for the evaluation of (4.4), is the equilibrium constant for the reaction (4.1) in which the A molecules are thermally energized into the small range E^* to $E^*+\delta E^*$ (compare section 1.3.3). This equilibrium constant is given by statistical mechanics as the partition function ratio $Q(A^*_{(E^*\to E^*+\delta E^*)})/Q(A)$, both partition functions being calculated from the common energy zero, that of A (see Appendix 2, section A2.5). The function $Q(A)$, henceforth denoted Q_2, is simply the ordinary molecular partition function for all the active modes of A; these usually comprise all the vibrations and internal rotations of the molecule (see section 4.12.1). The function $Q(A^*_{(E^*\to E^*+\delta E^*)})$ is the partition function for those A* having non-fixed energy specifically in the range E^* to $E^*+\delta E^*$. If δE^* is small the exponential terms in the expression $Q = \sum g_i \exp(-E_i/kT)$ are all the same, simply $\exp(-E^*/kT)$, and the partition function can be expressed as follows:

$$Q(A^*_{(E^*\to E^*+\delta E^*)}) = \left(\sum_{E^*\to E^*+\delta E^*} g_i\right) \exp(-E^*/kT)$$

$$= N^*(E^*)\,\delta E^* \exp(-E^*/kT)$$

Division by $Q(A)\equiv Q_2$ then gives the required result (4.5); this expression is in fact the quantum Boltzmann distribution function $K(E)\,\delta E$ giving the thermal equilibrium proportion of molecules in the given energy range.

$$\frac{\delta k_{1(E^*\to E^*+\delta E^*)}}{k_2} = \frac{N^*(E^*)\exp(-E^*/kT)}{Q_2}\,\delta E^* \qquad (4.5)$$

In these equations $\sum g_i$, the number of quantum states of A* in the energy range E^* to $E^*+\delta E^*$, has been replaced by $N^*(E^*)\,\delta E^*$. Thus $N^*(E^*)$ is the *density of quantum states* or the *number of quantum states per unit range of energy* at energies close to E^*, although the physical significance of $N^*(E^*)\,\delta E^*$ is probably easier to grasp than that of $N^*(E^*)$ itself. A further discussion of state densities and related quantities is given in section 4.11. The replacement of $\sum g_i$ by $N^*(E^*)\,\delta E^*$ involves the implicit assumption that the energy levels are closely spaced relative to the width δE^* of any energy range of interest. Only under these conditions is it adequate to replace the essentially stepwise increase in the number of energy levels by a continuous distribution function. The density of energy levels at the energies of interest for the energized molecule is, in fact, usually very high. For example, cyclopropane at energies E^* near the critical energy ($E_0 \equiv 63 \text{ kcal mol}^{-1}$) has a vibrational state density of about 2×10^9 states per kcal energy range, so that a continuous distribution is quite adequate even for energy ranges δE^* as low as $10^{-4} \text{ cal mol}^{-1}$. Incidentally, this treatment does not necessarily imply that the vibrations have to be treated on a classical basis; $N^*(E^*)$ can be evaluated on a quantum-mechanical basis, giving values which can be quite different from those obtained from a classical approximation (see section 5.4). Rice[12] has argued further in favour of the continuous distribution treatment, on the basis that the energized molecules have a limited mean life $\tau \approx 1/k_a(E^*)$ and that the Heisenberg Uncertainty Principle thus requires the energy levels themselves to be broadened by an amount $\Delta E \approx h/\tau$, where h is Planck's constant. It so happens that this broadening is, as a matter of principle, of the same order of magnitude as the spacing of the energy levels, so that the broadened levels overlap to some extent and the stepwise variation of energy approaches a continuous variation. For example, for cyclopropane at $0.5 \text{ kcal mol}^{-1}$ above the threshold, $k_a(E^*) \approx 10^4 \text{ s}^{-1}$. Since $h = 0.96 \times 10^{-13} \text{ kcal mol}^{-1} \text{ s}$, the energy levels are blurred to the extent of $\Delta E \approx 10^{-9} \text{ kcal mol}^{-1}$, compared with the energy-level spacing of $1/N^*(E^*) \approx 10^{-9} \text{ kcal mol}^{-1}$ at this energy.

4.5   EXPRESSION FOR $k_a(E^*)$

An expression for $k_a(E^*)$, the rate-constant for the formation of activated complexes from energized molecules with non-fixed energy E^*, is obtained by a detailed application of the steady-state treatment to A^+ in the reactions (4.2). For any particular energy levels these are of the form:

$$A^* \xrightarrow{\ k_a\ } A^+ \xrightarrow{\ k^+\ } \text{Products}$$

and the steady-state treatment gives $k_a[A^*] = k^+[\overrightarrow{A^+}]$ and hence

$$k_a(E^*) = k^+ \left(\frac{[\overrightarrow{A^+}]}{[A^*]}\right)_{\text{steady state}}$$

$$= \tfrac{1}{2}k^+ \left(\frac{[A^+]}{[A^*]}\right)_{\text{eqm}} \tag{4.6}$$

In deducing (4.6) it has been assumed, as in ART, that the steady-state concentration $[\overrightarrow{A^+}]$ of complexes which are crossing the barrier in the forward direction during the reaction is the same as it would be if an equilibrium was set up between A^* and A^+. At equilibrium the concentration of forward-crossing complexes is half the total concentration of complexes, and since [A$^+$] here denotes the total concentration this accounts for the factor $\frac{1}{2}$ in (4.6). The validity of this 'equilibrium assumption' is discussed in section 4.12.3.

The non-fixed energy of an activated complex A$^+$ formed from an energized molecule A* of energy E^* is $E^+ = E^* - E_0$ (see section 4.3 and Figure 4.1) and this energy E^+ can be divided in different ways into energy of vibration and rotation (E_{vr}^+) and energy of translational motion in the reaction coordinate (x). Accordingly, the overall reaction (4.2) is broken down into contributions from the different complexes as shown in (4.7), and $k_a(E^*)$ is evaluated as the sum (4.8) of corresponding contributions

$$A^*(E^*) \xrightarrow{k_a(E_{vr}^+, x)} A_{(E_{vr}^+, x)}^+ \xrightarrow{k^+(x)} \text{Products} \qquad (4.7)$$

$$k_a(E^*) = \sum_{E_{vr}^+=0}^{E^+} \tfrac{1}{2}k^+(x)\left(\frac{[A_{(E_{vr}^+, x)}^+]}{[A_{(E^*)}^*]}\right)_{\text{eqm}} \qquad (4.8)$$

where

$$E^* = E_0 + E^+ \quad \text{and} \quad x = E^+ - E_{vr}^+$$

of the form (4.6). The sum covers all possible distributions of E^+ between E_{vr}^+ and x, from $E_{vr}^+ = 0$ (in which case all the non-fixed energy is translational energy in the reaction coordinate and $x = E^+$), up to the maximum possible value of E_{vr}^+ below E^+. The latter case corresponds to complexes having the minimum possible energy in the reaction coordinate (for the given E^+), but since E_{vr}^+ is quantized it may well not have a value corresponding precisely to E^+. The rate-constant k^+ depends only on x and is denoted $k^+(x)$, but the contribution to [A$^+$]/[A*] depends on E_{vr}^+, x and E^*, any two of which, together with E_0, serve to determine the third. The treatments of $k^+(x)$ and [A$^+$]/[A*] are detailed separately in the following sections.

4.5.1   Evaluation of $k^+(x)$

Marcus treated decomposition of the activated complex into products as the translation of a particle of mass μ in a one-dimensional box of length δ, the small region at the top of the energy barrier which is considered to define the activated complex.† The mass μ is often considered

† This aspect of the ART formulation has been criticized and an alternative derivation of the same result suggested by Slater.[13] The number of complexes crossing the barrier in unit time is given by their speed in the reaction coordinate divided by their 'distance apart along the reaction coordinate', thus avoiding the concept of an arbitrary length δ which has no physical significance. The 'trap' described by Slater is avoided in the type of derivation used here, which does not make the simplification of giving all complexes the mean velocity or mean life. There are also other treatments, again leading to the same result.[14]

to be a reduced mass for the atoms involved in the reaction coordinate, but its precise significance, like that of δ, need not be considered in detail because both cancel out in the final expression. If the energy in the reaction coordinate is x, the speed of translation is $(2x/\mu)^{\frac{1}{2}}$ and the time taken for the mass μ to cross the box is $\delta/(2x/\mu)^{\frac{1}{2}}$. The rate-constant for crossing of the barrier is therefore

$$k^+(x) = (2x/\mu)^{\frac{1}{2}}/\delta = (2x/\mu\delta^2)^{\frac{1}{2}} \tag{4.9}$$

4.5.2   Evaluation of $([A^+]/[A^*])_{\text{eqm}}$

As previously indicated, the RRKM theory uses the equilibrium ratio of concentrations of A^+ and A^*. This is calculated from statistical mechanics as the ratio $Q(A^+)/Q(A^*)$ of the partition functions of the activated complex and energized molecule, using energies reckoned from a common energy zero, that of the A molecule. Since the two species under consideration both have a total energy in the small range $E^* \rightarrow E^* + \delta E^*$, each partition function is of the form $(\sum g_i) \exp(-E^*/kT)$, where $\sum g_i$ is the number of quantum states in this small energy range, and $Q(A^+)/Q(A^*)$ reduces simply to $\sum g_i^+/\sum g_i^*$. Although the A^+ and the A^* have the same *total* energy, the former has much less *non-fixed* energy, and there are thus many less quantum states for A^+ in the given energy range and $[A^+]/[A^*]$ will be small, as is physically reasonable. As before (section 4.4) $\sum g_i^*$ can be replaced by a continuous distribution function $N^*(E^*) \, \delta E^*$, and a similar treatment is valid *at this stage* for the activated complex, since it contains a translational degree of freedom (the reaction coordinate). Translational motion usually has an extremely close spacing of energy levels (see Appendix 2, section A2.2) and the energy can reasonably be treated as continuous rather than quantized. The number of quantum states of the activated complex in the range of total energy $E^* \rightarrow E^* + \delta E^*$ could be denoted by $N^+(E^*) \, \delta E^*$, but is more commonly written $N^+(E^+) \, \delta E^*$ or $N^+(E^+) \, \delta E^+$ since E^* and δE^* serve to define $E^+ (= E^* - E_0)$ and $\delta E^+ (= \delta E^*)$, and the latter quantities are more fundamental to the behaviour of the complex. It is easier to think of 'a complex with non-fixed energy E^+' than of 'a complex which was formed from an energized molecule with non-fixed energy E^*'. Thus the concentration ratio for the small energy range of interest reduces to (4.10).

$$([A^+]/[A^*])_{\text{eqm}} = \sum g_i^+/\sum g_i^* = N^+(E^+)/N^*(E^*) \tag{4.10}$$

In order to evaluate the concentration ratio in (4.8) we need the density of states $N^+(E^+)$ for a specified division of E^+ into the vibrational rotational energy E_{vr}^+ and the translational energy x in the reaction coordinate. This will be denoted by $N^+(E_{\text{vr}}^+, x)$ to avoid confusion with the total state density $N^+(E^+)$. Although the translational energy x can be treated in terms of a continuous distribution, the vibrational and rotational energies of the activated complex will both be treated as quantized for the time being. The quantum treatment is certainly essential for the vibrational degrees of freedom of the activated complex, because the non-fixed energy of A^+ is in general relatively small, so that the vibrations of A^+ are much less highly excited than those of A^*. A continuous treatment is only reasonable at high energies where the density of quantum states is high; see sections 4.11 and 5.4. For example, for cyclopropane with $E^* = 70 \, \text{kcal mol}^{-1}$ and $E^+ = 5 \, \text{kcal mol}^{-1}$, A^* has 4×10^9 vibrational quantum states per kcal energy range, while A^+ has only ca 10^2; a continuous function is much less valid for A^+ than for A^*.

Thus in order to split $N^+(E_{vr}^+, x)$ into contributions from E_{vr}^+ and x we define:

$P(E_{vr}^+) = $ Number of vibrational–rotational quantum states of A^+ with vibrational–rotational non-fixed energy *equal to* E_{vr}^+ (precisely),

$N_{rc}^+(x) \, \delta x = $ Number of translational quantum states of A^+ with energy in the reaction coordinate *in the range* $x \to x + \delta x$.

The overall number of states in an energy range δx is given by the product of these degeneracies, and is $N^+(E_{vr}^+, x) \, \delta x$, whence

$$N^+(E_{vr}^+, x) = P(E_{vr}^+) \, N_{rc}^+(x) \qquad (4.11)$$

The required concentration ratio is thus obtained by substituting $N^+(E_{vr}^+, x)$ from (4.11) for $N^+(E^+)$ in (4.10), the result being:

$$\left(\frac{[A_{(E_{vr}^+, x)}^+]}{[A_{(E^*)}^*]} \right)_{\text{eqm}} = \frac{P(E_{vr}^+) \, N_{rc}^+(x)}{N^*(E^*)} \qquad (4.12)$$

4.5.3   Expression for $N_{rc}^+(x)$

This is derived in a manner similar to that used in the more sophisticated versions of ART.[15] The wave-mechanical treatment[16] of the translation of a particle of mass μ in a box of length δ produces the result that the energy is quantized, the energy x of the nth level being:

$$x = n^2 h^2 / 8\mu\delta^2$$

where h is Planck's constant. Thus the number of quantum states with energy up to and including x is $n = (8\mu\delta^2 x/h^2)^{\frac{1}{2}}$. The number of states in the energy range $x \to x + \delta x$ (where δx, although small, is sufficiently large compared with the energy level spacing) is $\delta n = (dn/dx) \, \delta x$, and this number of states is equal to $N_{rc}^+(x) \, \delta x$, whence

$$N_{rc}^+(x) = dn/dx = (2\mu\delta^2/h^2 x)^{\frac{1}{2}} \qquad (4.13)$$

4.5.4   Result for $k_a(E^*)$

The expression (4.8) for $k_a(E^*)$ can now be evaluated using (4.9), (4.12) and (4.13) to form the summand for a given energy distribution, and the result obtained is (4.15).

$$k_a(E^*) = \sum_{E_{vr}^+=0}^{E^+} \frac{1}{2} \left(\frac{2x}{\mu\delta^2}\right)^{\frac{1}{2}} \frac{P(E_{vr}^+)(2\mu\delta^2/h^2 x)^{\frac{1}{2}}}{N^*(E^*)}$$

$$= \sum_{E_{vr}^+=0}^{E^+} \frac{P(E_{vr}^+)}{hN^*(E^*)} \tag{4.14}$$

$$= \frac{1}{hN^*(E^*)} \sum_{E_{vr}^+=0}^{E^+} P(E_{vr}^+) \tag{4.15}$$

It will be noted that when the contribution to k_a from complexes with a specified distribution of E^+ between E_{vr}^+ and x is evaluated [i.e. the summand in (4.14)] the explicit characteristics of the reaction coordinate (μ and δ) disappear. The value of the contribution thus depends only on the properties of the energy levels in the energized molecule and in the ordinary degrees of freedom of the activated complex (i.e. excluding the reaction coordinate).

The sum in (4.15), often written less explicitly as $\sum P(E_{vr}^+)$, is the sum of the numbers of vibrational–rotational quantum states at all the quantized energy levels of energy less than or equal to E^+; i.e. it is simply the total number of vibrational–rotational quantum states of the activated complex with energies $\leqslant E^+$. If all the energy levels of the complex were plotted on an energy diagram, $\sum P(E_{vr}^+)$ could be evaluated by simply counting the number of levels at energies up to and including E^+. The physical significance of this quantity is discussed further in section 4.11 and its numerical evaluation is dealt with in Chapter 5.

4.5.5   Two modifications, and final result for $k_a(E^*)$

It will be profitable to introduce at this stage two minor modifications, concerned with adiabatic rotations and statistical factors respectively.

It has already been pointed out that adiabatic rotations, which stay in the same quantum state while the energized molecule becomes an activated complex, nevertheless suffer an energy change because of the different moments of inertia of A* and A+. Usually energy is released into the other degrees of freedom of the molecule (see section 4.10), and this obviously has an effect on the rate-constant $k_a(E^*)$. The ART treatment of the reaction at high pressures (section 4.7) shows that in this limiting case the correct k_∞ is obtained if the expression (4.15) for $k_a(E^*)$ is multiplied by the factor Q_1^+/Q_1, where Q_1^+ and Q_1 are the partition functions for the adiabatic rotations in the activated complex and the A molecule respectively. The appearance of this factor may be traced to the calculation of $[A^+]/[A^*]$ (section 4.5.2). Marcus suggested originally[1] that the same correction is reasonably accurate even at low pressures, although later work by Marcus[26, 28] and others has shown that a more sophisticated treatment is really required. Such treatments are discussed in section 4.10, but the crude correction by the factor Q_1^+/Q_1 will be included for the time being.

The second modification concerns the possibility that a reaction can proceed by several distinct paths which are kinetically equivalent, i.e. equivalent with regard to the energetics and rate of reaction. The number of such paths is termed the *statistical factor L^+*. A simple case would be the dissociation of H_2O to $OH + H$, for which $L^+ = 2$ since either of the two identical OH bonds can be broken. The problem is discussed in detail in section 4.9, from which it emerges that the correct rate-constant $k_a(E^*)$ for disappearance of energized molecules by all the paths together is obtained by including a factor L^+ in (4.15). When L^+ is defined as in section 4.9, symmetry numbers must be omitted from the rotational partition functions.

The final equation for $k_a(E^*)$ is therefore (4.16), and unless a better treatment of adiabatic rotations is required (4.16) is the expression generally used in the RRKM formulation.

$$
\left.
\begin{aligned}
k_a(E^*) &= L^+ \frac{Q_1^+}{Q_1} \frac{1}{hN^*(E^*)} \sum_{E_{vr}^+=0}^{E^+} P(E_{vr}^+) \\
&\equiv L^+ \frac{Q_1^+}{Q_1} \frac{\sum P(E_{vr}^+)}{hN^*(E^*)}
\end{aligned}
\right\}
\tag{4.16}
$$

4.6 RRKM EXPRESSION FOR k_{uni}

As indicated in section 4.1 the overall first-order rate-constant k_{uni} is obtained by substituting the expressions (4.5) and (4.16) into (4.4), whence

$$
k_{uni} = \frac{L^+ Q_1^+}{hQ_1 Q_2} \int_{E^*=E_0}^{\infty} \frac{\{\sum P(E_{vr}^+)\} \exp(-E^*/kT) \, dE^*}{1 + k_a(E^*)/k_2[M]}
$$

Or, since $E^* = E_0 + E^+$ and $dE^* = dE^+$, we obtain the final result:

$$
k_{uni} = \frac{L^+ Q_1^+ \exp(-E_0/kT)}{hQ_1 Q_2} \int_{E^+=0}^{\infty} \frac{\left\{ \sum_{E_{vr}^+=0}^{E^+} P(E_{vr}^+) \right\} \exp(-E^+/kT) \, dE^+}{1 + k_a(E_0 + E^+)/k_2[M]}
\tag{4.17}
$$

in which $k_a(E_0 + E^+)$ is given by (4.16), rewritten in the equivalent form (4.18):

$$k_a(E_0 + E^+) = \frac{L^+ Q_1^+}{hQ_1 \, N^*(E_0 + E^+)} \sum_{E_{vr}^+=0}^{E^+} P(E_{vr}^+) \qquad (4.18)$$

Since the result of the summation in (4.17) and (4.18) is simply a function of E^+ (see section 4.11) the whole integrand is a function of E^+ and can be integrated numerically to give k_{uni} if $N^*(E_0 + E^+)$ and $\sum P(E_{vr}^+)$ are known as functions of energy, i.e. if the distributions of the vibrational–rotational energy levels of the reactant and the activated complex are known. The physical significance of these quantities is discussed in detail and further illustrated in section 4.11. Their exact evaluation is usually laborious in the extreme, and application of the RRKM theory depends very much on the availability of simple but reasonably accurate approximations which will be discussed in Chapter 5. The basic assumptions which have been made in the derivation of (4.17) are reviewed and their validity discussed in section 4.12.

Apart from the approximate treatment of adiabatic rotations, (4.17) is to be regarded as the fundamental RRKM result for k_{uni}. If adiabatic rotations are to be ignored (by putting $Q_1^+/Q_1 = 1$) the result is then accurate within the framework of the basic assumptions discussed in section 4.12. Reference to section 4.10 is desirable if adiabatic rotations are to be included; it emerges that (4.17) and (4.18) give erroneous results in the fall-off and low-pressure regions. For low values of Q_1^+/Q_1 a better result is obtained by including this factor in (4.17) but not in (4.18),[28] but in cases where the effect is really worth considering further modifications are necessary.

It will be noted that in (4.17) vibrations and rotations have both been treated as quantized, in contrast to the original formulations. In Chapter 5 a classical treatment will be applied to the active rotations, following Marcus and leading to his result (5.20), but this simplification is not an essential part of the basic theory. The integration over a continuously varying E^+ arises, despite the quantum treatment and the consequent stepwise variation of $\sum P(E_{vr}^+)$ and $k_a(E^*)$, from the essentially continuous energies of the translational degree of freedom in A^+ (the reaction coordinate) and the very high density of energy levels of A^* at the energies of interest, already referred to in section 4.4.

4.7   THE HIGH-PRESSURE LIMIT

It is interesting to consider the result obtained from the RRKM theory for a unimolecular reaction at high pressures, and to compare this with the predictions of other theories. In particular the RRKM theory should give a similar result to the Absolute Rate Theory, since both are based on a statistical-mechanical approach. The high-pressure limit is easily obtained from (4.17) by putting $[M] \to \infty$, when the pseudo-first-order rate-constant k_{uni} becomes the genuine pressure-independent first-order rate-constant k_∞, given by (4.19).

$$k_\infty = \frac{L^+ Q_1^+}{hQ_1 Q_2} \exp(-E_0/kT) \int_{E^+=0}^\infty \left\{ \exp(-E^+/kT) \sum_{E_{vr}^+=0}^{E^+} P(E_{vr}^+) \right\} dE^+ \quad (4.19)$$

It will be noted that $N^*(E^*)$ does not appear in this result, except implicitly in the calculation of Q_2, the partition function for the active degrees of freedom of the reactant molecule A. The integral in (4.19) can be evaluated by reversing the order of summation and integration, with careful attention to the limits involved. It is necessary to consider all possible activated complexes, and this is achieved in (4.19) by integrating from $E^+ = 0$ to ∞ (thus covering all possible total energies of the complex), and for each value of E^+ summing the contributions from the different complexes having the different possible divisions of this energy E^+ between E_{vr}^+ and x, i.e. all the complexes with $0 \leqslant E_{vr}^+ \leqslant E^+$. In the reversed expression (4.20) the same effect is achieved by summing over all possible

$$k_\infty = \frac{L^+ Q_1^+}{hQ_1 Q_2} \exp(-E_0/kT) \sum_{E_{vr}^+=0}^\infty \left\{ P(E_{vr}^+) \int_{E^+=E_{vr}^+}^\infty \exp(-E^+/kT) dE^+ \right\} \quad (4.20)$$

values of E_{vr}^+ (from 0 to ∞), and for each value of E_{vr}^+ considering all possible x (from 0 to ∞) so that E^+ varies continuously from E_{vr}^+ to ∞, these being the limits for the inner integration. In (4.20) $P(E_{vr}^+)$ stays constant while the integral is evaluated as a simple function of E_{vr}^+, as follows:

$$\int_{E^+=E_{vr}^+}^\infty \exp(-E^+/kT) dE^+ = \left[-kT \exp(-E^+/kT) \right]_{E^+=E_{vr}^+}^\infty$$

$$= kT \exp(-E_{vr}^+/kT)$$

Summation over all possible values of E_{vr}^+ then gives

$$k_\infty = \frac{L^+ Q_1^+}{hQ_1 Q_2} kT \exp(-E_0/kT) \sum_{E_{vr}^+=0}^\infty [P(E_{vr}^+) \exp(-E_{vr}^+/kT)]$$

A little consideration shows that the sum in this expression is simply the partition function Q_2^+ for the active vibrations and rotations in the activated complex (cf. Appendix 2, section A2.1). This is very similar to the partition function Q_2 for the active vibrations and rotations in the ordinary molecule A, except that the frequencies and moments of inertia will have changed and that the activated complex has one degree of freedom fewer, since the motion in the reaction coordinate has been singled out for special treatment. The high-pressure limit from the RRKM theory is thus (4.21), in which Q and Q^+ are the complete vibrational–

$$k_\infty = L^{\ddagger} \frac{kT}{h} \frac{Q^+}{Q} \exp(-E_0/kT) \tag{4.21}$$

rotational partition functions for the reactant and the activated complex $(Q = Q_1 Q_2$ and $Q^+ = Q_1^+ Q_2^+)$. Except for the omission of a transmission coefficient this result is identical with that obtained from ART for a unimolecular reaction, which is reasonable in view of the similarity of the treatments involved. ART[10] calculates the rate on the assumption that the activated complexes A^+ are in thermal equilibrium with the reactant molecules A, i.e. that the thermal Boltzmann distribution is maintained at all energies, which is true at sufficiently high pressures. The ART scheme is thus (4.22). The RRKM theory, in the general case, admits

$$\left. \begin{array}{c} M + A \xrightleftharpoons{\text{eqm}} A^+ + M \\[2mm] A^+ \longrightarrow \text{Products} \end{array} \right\} \tag{4.22}$$

equilibrium between A^+ and A^*, but not between A^* and A in the reactions (4.1) and (4.2). At high pressures, however, A^* and A are also in equilibrium so that the model becomes the same as that treated in ART and the results naturally coincide.

It may be noted that the agreement between the RRKM and ART results for k_∞ is not complete unless the RRKM formulation includes a correction which reduces to Q_1^+/Q_1 at high pressures. The idea that a constant factor of Q_1^+/Q_1 will give approximately the correct result for k_{uni} at lower pressures has only limited validity—see section 4.10.

4.8   THE LOW-PRESSURE LIMIT

In the limit of very low pressures the first-order rate-constant from (4.17) becomes proportional to the pressure; the second-order rate-constant k_{bim} is then given by (4.23), in which Q_2^* is the partition function for

$$\begin{aligned}
k_{bim} &= \lim_{[M] \to 0} \left(\frac{k_{uni}}{[M]} \right) \\[2mm]
&= \frac{k_2}{Q_2} \exp(-E_0/kT) \int_{E^+=0}^{\infty} N^*(E_0 + E^+) \exp(-E^+/kT) \, dE^+ \\[2mm]
&= \frac{k_2}{Q_2} \int_{E^*=E_0}^{\infty} N^*(E^*) \exp(-E^*/kT) \, dE^* = \frac{k_2 Q_2^*}{Q_2} \tag{4.23}
\end{aligned}$$

energized molecules (i.e. specifically those A molecules which have non-fixed energy greater than E_0) *using the ground state of A for the zero of energy.* Thus Q_2^* is a part of the series

$$g_1 \exp\left(-E_1/kT\right) + g_2 \exp\left(-E_2/kT\right) + \ldots$$

which defines Q_2 (cf. Appendix 2), but includes only the terms for which $E_i \geqslant E_0$.

An alternative derivation of (4.23) starts by taking the low-pressure limit of (4.4), whence

$$k_{\text{bim}} = \int_{E^*=E_0}^{\infty} \mathrm{d}k_1$$

Insertion of the RRKM expression for δk_1 from (4.5) leads as above to (4.23). This is equivalent to noting that at sufficiently low pressures all the energized molecules react; the rate of reaction is thus simply the rate of energization, and the rate of formation of A^+ from A^* is no longer relevant. An expression for the rate of energization may be obtained as before by considering the equilibrium $M + A \rightleftharpoons M + A^*$; at equilibrium the rate of energization is equal to the rate of de-energization, $k_2[M][A^*]$, and the concentration of energized molecules is given by $[A^*]/[A] = Q_2^*/Q_2$. Thus the rate of energization is $(k_2 Q_2^*/Q_2)[M][A]$, and this is assumed to be correct even when A^* are removed by reaction (see section 4.12.3). There are no statistical factors to be included; this is because they occur in the rate-constant k_a for formation of A^+ from A^*, and this no longer affects the rate of reaction. Similarly there is no factor involving the adiabatic rotations in (4.23); this is a fault of the original Marcus treatment which is corrected in the later versions discussed in section 4.10. Apart from this reservation, however, we have the interesting result that the low-pressure rate-constant depends only on the properties of the reactant molecules and the height of the energy barrier, and in no way depends on the detailed properties of the activated complex.

Equation (4.23) may alternatively be written in terms of the partition function $Q_2^{*\prime}$ for energized molecules *relative to the ground state of* A^+ *as the zero of energy*; this level is in fact the 'ground state' for energized molecules, being the lowest energy that an A^* can have, and in some ways forms a more natural choice of energy zero. The two partition functions are related simply by the equation $Q_2^{*\prime} = Q_2^* \exp(E_0/kT)$ (see Appendix 2, section A2.1), and k_{bim} is therefore given by (4.24). The terms

$$k_{\text{bim}} = \frac{Q_2^{*\prime}}{Q_2} k_2 \exp\left(-E_0/kT\right) \tag{4.24}$$

$k_2 \exp(-E_0/kT)$ in this equation correspond to Lindemann's k_1, and the

4

ratio $Q_2^{*\prime}/Q_2$ is thus the quantum-statistical-mechanical equivalent of the term $(E_0/kT)^{s-1}/(s-1)!$ in Hinshelwood's k_1 [equation (1.29)]. The density of quantum states increases rapidly with energy (see section 4.11), and $Q_2^{*\prime}$ is therefore much greater than Q_2. Thus (4.24), like the Hinshelwood equation, can give rates of energization which are many times greater than those predicted on the simple collision theory picture.

It has already been seen that the Arrhenius activation energy of a unimolecular reaction varies with pressure. The RRKM theory does not lead to any simple equation for this variation, but the theoretical activation energy E_{Arr} at any pressure may be obtained in the usual way [equation (1.6)] from the first-order rate-constants calculated at a series of temperatures. The marked variation which can occur is illustrated by some calculated results for the isomerization of 1,1-dichlorocyclopropane (see section 6.4); with a critical energy of $E_0 = 55 \cdot 5 \ kcal \, mol^{-1}$, E_{Arr} is calculated to be $57 \cdot 7 \ kcal \, mol^{-1}$ at 1000 Torr and $48 \cdot 6 \ kcal \, mol^{-1}$ at 10^{-5} Torr, these being close to the high- and low-pressure limits of E_{Arr}. The extent of the variation increases with the complexity of the molecule; for isomerization of the relatively small methyl isocyanide molecule (see section 7.2) $E_0 = 37 \cdot 9 \ kcal \, mol^{-1}$, $E_\infty = 38 \cdot 4 \ kcal \, mol^{-1}$ and $E_{bim} = 36 \cdot 2 \ kcal \, mol^{-1}$, the variation being in good agreement with experiment.

The basic reason for the variation of E_{Arr} with pressure is the change in the energy distribution of the reacting molecules, illustrated in Figure 4.2 for the isomerization of 1,1-dichlorocyclopropane. This Figure is constructed from the data summarized in Table 6.7. At sufficiently low pressures, all the molecules which become energized react. The energy distribution is that associated with k_1, which decreases rapidly as the energy increases, thus favouring the reaction of molecules with energies near the critical energy. At high pressures there is a competition between reaction and collisional de-energization. The energized molecules with energies near the critical energy have long lifetimes before reaction, and are thus more likely to be de-energized than the rapidly reacting molecules with higher energies. The more highly energized molecules thus contribute more heavily to the overall rate of reaction, and the activation energy is correspondingly higher.

4.9 STATISTICAL FACTORS

The straightforward RRKM formulation (4.15) of $k_a(E^*)$ refers basically to the rate of reaction by a single reaction path from the reactant molecule to the products. It often happens that there are several paths

which are physically distinct but nevertheless completely equivalent so far
as the rate calculation is concerned. Such reaction paths involve activated
complexes which are geometrical or optical isomers of each other, and if
diagrams are drawn showing the movements of the various atoms during

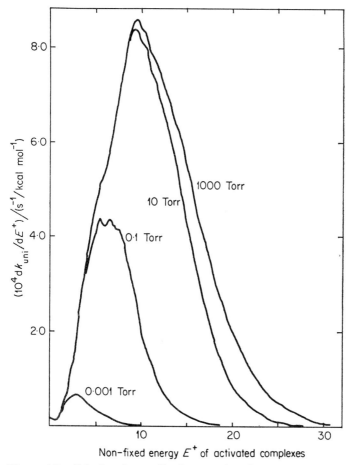

Figure 4.2 Calculated contributions to k_{uni} from 1,1-dichloro-
cyclopropane complexes with various energies as a function of
pressure (see section 6.4)

the reaction, the diagrams for the different paths will also be geometrically
or optically isomeric. In this case the calculated rate must be increased by
an appropriate factor known as the *statistical factor* or *reaction path
degeneracy*, denoted here by $L^{\ddagger}$. The terms are not always used with
exactly the same meaning as here, but any difference will be clear from the

context. The factor $L^{\pm}$ appears basically in the rate-constant $k_a(E^*)$ for formation of activated complexes from energized molecules [e.g. (4.16)], and hence appears in the final equation for k_{uni} both as part of k_a in the denominator and in front of the whole expression [e.g. (4.17)]. In Absolute Rate Theory the factor is generally derived from the ratio σ/σ^+ of the symmetry numbers of the rotations (both internal and overall) of A and A^+. Thus the usual formulation can be written as:

$$k_\infty = \frac{\sigma}{\sigma^+} \frac{kT}{h} \frac{Q^+}{Q} \exp(-E_0/kT)$$

(in which the partition functions do not include the symmetry numbers), and the statistical factor is found to be σ/σ^+.

It has been recognized for some time that this procedure can be in error, however, in connection with both equilibrium[17,18] and rate[18-20] studies, and a direct count of the number of reaction paths has generally been preferred. In what follows we have particularized the treatments of various workers to refer specifically to unimolecular reactions. Bishop and Laidler[20,21] defined a statistical factor $l^{\pm}$ for the formation of activated complexes as the number of different complexes that can be formed if all identical atoms in A are numbered. With this definition the correct rate of reaction was shown to be obtained by omitting the symmetry numbers from the partition functions and multiplying the rate expression by $l^{\pm}$. This remains a simple and accurate way of handling the problem, the $l^{\pm}$ of Bishop and Laidler being identical with our statistical factor $L^{\pm}$. Elliott and Frey[22] similarly defined the reaction path degeneracy $L^{\pm}$ as 'the number of different operations producing all equivalent activated complexes and comprising equivalent motion of different sets of atoms or different motion of equivalent sets of atoms'.

As an example, consider the isomerization of 1,1-dichlorocyclopropane through a complex of the type shown in Figure 4.3. We number the atoms in the molecule A and count four equivalent but distinct complexes I–IV; the statistical factor for this calculation is thus 4. Note that I is distinct from IV (and II from III) because of the numbering, although physically superimposable by rotation, and that I and IV are enantiomorphic with II and III, i.e. they are non-superimposable mirror-images of II and III.

Schlag and Haller[23] have given an alternative and carefully defined direct-count procedure for determining the statistical factor. The atoms are again numbered, and a count is made of the number of complexes *which would be superimposable if the numbering were removed* (and are not related by simple rotation with the numbering present). This number is then multiplied by two if the reactant is symmetric and the complex asymmetric, and divided by two if the reverse is true. In the above example there are only two superimposable complexes (I and IV or II and III), but this number is multiplied by two since the reactant is symmetric and the complexes asymmetric, producing a statistical factor of 4 as before.

It is also possible to formulate a correct answer in terms of symmetry numbers, but an additional factor may be necessary. Bishop and Laidler[20]

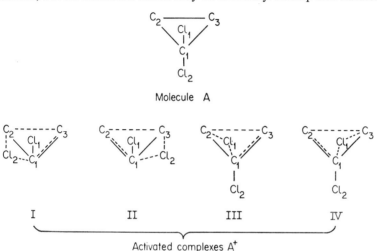

Figure 4.3   Activated complexes for isomerization of 1-1,dichloro-cyclopropane

showed that $l^{\pm} = r^{\pm} \sigma / \sigma^{+}$, where $r^{\pm}$ is the statistical factor for the return of activated complexes to reactant molecules, defined in a similar way to l^{+}, and σ and σ^{+} are the symmetry numbers, taken as $\frac{1}{2}$ for asymmetric molecules.[18] In the above example $\sigma = 2$, $\sigma^{+} = \frac{1}{2}$ and $r^{\pm} = 1$ (since each complex I–IV can return to reactant by only one path); hence the statistical factor is again 4. The application of this approach has been simplified by the realization[24] that an activated complex can only have $r^{\pm} = 1$. It is not possible to have a species in a configuration corresponding to the top of an energy barrier in more than one degree of freedom at a time, so that an activated complex can have only one route for returning to reactant and only one route for going on to product. If this appears not to be the case then the species under consideration is not an activated complex according to the normal definition. For example, a plausible-looking complex (V) for cyclopropane isomerization is symmetrical with a C—C bond stretched to a critical length; H migration will occur only after the barrier is passed and will occur by four distinct paths. In fact[24] there *must* be four equivalent routes of lower potential energy, proceeding via four distinct complexes through each of which there is only one path from reactant to products; complexes similar to those in Figure 4.3 would be satisfactory in this respect. Complications have been discussed[25] for the case where reaction proceeds via a single symmetric complex but the

$$
\begin{array}{c}
\text{H} \\
\text{H} \quad \text{H} \quad \text{H} \\
\text{H} \quad\quad\quad \text{H} \\
(V)
\end{array}
$$

product 'valley' on the potential energy surface branches to give more than one possible set of identical products. The real difficulty in these cases seems to be that the systems require a more detailed mechanistic scheme than the simple ART formulation ($A^+ \rightarrow$ products).

Provided that these complications are avoided, however, and the complex is chosen so that it can form products or return to reactants by one route only, then the statistical factor is given by $L^{+} = \sigma/\sigma^{+}$ if the symmetry numbers are taken as $\frac{1}{2}$ for asymmetric molecules or $L^{+} = \alpha\sigma/\sigma^{+}$ where α is the number (1 or 2) of optical isomers of the complex and the symmetry numbers are now taken as 1 for asymmetric species.

This last equation is effectively the formulation used by Marcus,[26, 27] the symmetry numbers there being included in the partition functions. In that formulation the rate is also summed for a number of 'geometrical isomers' of the reaction path, but these are really different reactions with different potential energy barriers and different properties for the activated complex, and would not normally be dealt with together as parts of one calculation. For example, Wieder and Marcus[6, 28] drew eight activated complexes for the isomerization of methylcyclopropane to but-1-ene and

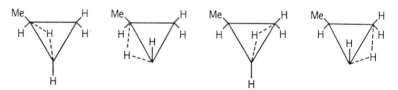

Figure 4.4   Activated complexes[6] for isomerization of methylcyclopropane

but-2-ene, and obtained a statistical factor of 8 since $\sigma = 1$, $\sigma^{+} = 1$, and the number of 'optical isomeric forms' of the complex = 8. The complexes are those shown in Figure 4.4 together with their mirror-images (not shown). On careful inspection it is clear that the structures in Figure 4.4 are not *equivalent* complexes, but involve migration of two differently situated hydrogen atoms (*cis-* or *trans-* to CH_3) to two different sites on the ring ($CH_3 \cdot CH<$ or $CH_2<$). The complexes will have different properties and the potential energy curves through them will be different. For each of the four distinct reactions the statistical factor is 2. For example, for the first complex in Figure 4.4 there are $L^{\ddagger} = 2$ *cis*-hydrogen atoms which can migrate to the $CH_3 \cdot CH$ group; alternatively $L^{\ddagger} = r^{\ddagger}\sigma/\sigma^{+} = 1 \times 1/\frac{1}{2} = 2$;   or   finally $L^{\ddagger} = \alpha\sigma/\sigma^{+} = 2 \times 1/1 = 2$. For computational purposes it may be useful to assume that the disappearance of reactant proceeds with the same rate-constant k by each of the four reactions, but this should be taken into account by putting

$$k_{\text{uni}} = \sum_{g=1}^{4} (L^{\ddagger}k_g) = 4(2k) = 8k$$

rather than by calling the eight complexes 'optical isomers'. There can only be one or two geometrically equivalent optical isomers of a structure, since reflection of a mirror-image in a mirror is bound to regenerate the original structure and cannot possibly generate a third structure.

   Another case which has caused some confusion in the past is the dehydrochlorination of ethyl chloride, and this is discussed in section 7.2.4.

4.10   IMPROVED TREATMENT OF ADIABATIC ROTATIONS

   In previous sections the adiabatic degrees of freedom (usually the overall rotations of the molecule) have been largely ignored and taken into account only by the semi-empirical correction factor Q_1^{+}/Q_1. In the present section more sophisticated treatments are discussed. The whole subject has been reviewed recently by Waage and Rabinovitch.[29]

   An adiabatic rotation is one for which the angular momentum stays constant during the conversion of the energized molecule to an activated complex, i.e. the rotation stays in the same quantum state throughout this process. Since the energy of the rotation is given by $E_J = (h^2/8\pi^2 I)J(J+1)$, the energy will change as the geometry of the molecule and hence the moment of inertia I changes. In most cases where such effects are worth considering, $I^{+} > I$ so that $E_J > E_J^{+}$ and the adiabatic rotations release energy into the other (active) degrees of freedom of the molecule, thus increasing the multiplicity of available quantum states of the complex and increasing the specific rate-constant k_a. An alternative interpretation is that this 'centrifugal effect' allows part of the adiabatic rotational energy to be used for overcoming the potential energy barrier, thus effectively reducing E_0. In bond fission reactions (such as the dissociation of ethane into methyl radicals) the moments of inertia can change substantially and the effect can be quite marked, amounting to an effective reduction of E_0 by more than kT and an increase in k_∞ by more than a factor of e. The effect was first treated by Rice and Gershinowitz,[30] who considered only the high-pressure (equilibrium) limit and derived the correction factor Q_1^+/Q_1 for a simple bond fission reaction. Bunker and Pattengill[9] have extended a similar treatment to the fall-off region for the specific case of a classical triatomic molecule and found improved agreement with the results of Monte-Carlo 'experiments' on the same models (see also section 6.2.2).

4.10.1 Basic treatment

The general treatment described below is basically that due to Marcus,[26, 31] but reaction path degeneracy has been accounted for by the inclusion of a statistical factor L^+ rather than a symmetry number ratio (see section 4.9). In addition the case dealt with is specifically the usual one where there is a single doubly-degenerate adiabatic rotational degree of freedom.† This corresponds to the model often used for a unimolecular bond fission reaction, as discussed further in section 6.2.2.

The basic effect of the change in I is best seen in the modified energy diagram of Figure 4.5. This is based on Figure 4.1, but now shows in addition the energies E_J and E_J^+ of the adiabatic rotations in A* and A$^+$ respectively in their Jth energy level.‡ The diagram shows that the energy in the active degrees of freedom of A*, which was previously $E_0 + E^+$, is now

$$E^*_{\text{active}} = E_0 + E^+ + E_J^+ - E_J = E_0 + E^+ + \Delta E_J$$

where

$$\Delta E_J = E_J^+ - E_J$$

The symbol E^*_{active} is used here to avoid confusion with the use[26, 31] of E^* for the quantity $E_0 + E^+$; thus $E^*_{\text{active}} = E^* + \Delta E_J$. The integral corresponding to (4.17) can now be reformulated, in the first place with reference to a particular rotational state with quantum number J. The basic equation is still (4.4) [i.e. (1.46)], but a consideration of sections 4.4 and 4.5 shows

---

† Marcus has suggested[26] that an extension to other cases can be made by multiplying E_J etc. in what follows by $l/2$, where l is the number of adiabatic rotational degrees of freedom. In particular, the expression (4.29) for $\langle \Delta E_J \rangle$ is multiplied by $l/2$ and the same factor thus appears in the resulting expressions (4.30) and (4.31).

‡ In ref. 26, J was used to denote 'the totality of quantum numbers that are approximately conserved', i.e. in the present case it would include the K quantum number (see section 5.2.4). Thus where Marcus had $\sum_J (...)$ we must have, for a doubly degenerate rotation, $\sum_J (2J + 1) (...)$.

that the equations (4.5) and (4.15) for $\delta k_1/k_2$ and k_a should be modified as follows:

$$\left(\frac{\delta k_1}{k_2}\right)_J = \frac{N^*(E^*_{\text{active}})\exp\left[-(E_0+E^++E_J^+)/kT\right]\delta E^*_{\text{active}}}{Q}$$

$$k_{EJ}(E^*_{\text{active}}) = \frac{L^{\ddagger}}{hN^*(E^*_{\text{active}})}\sum_{E_{\text{vr}}^+=0}^{E^+} P(E_{\text{vr}}^+)$$

In the equation for $(\delta k_1/k_2)_J$, Q is the partition function for all the degrees of freedom (both active and adiabatic) of A, and $(E_0+E^++E_J^+)$ is the

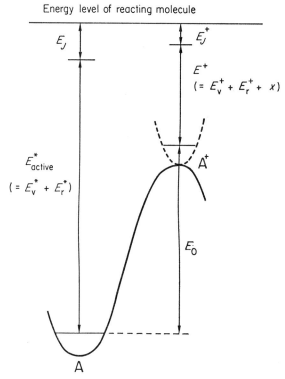

Energy level of reacting molecule

Figure 4.5 Energy diagram for a unimolecular reaction with adiabatic rotations included

total energy of the A* in question relative to the ground state of A (cf. Appendix 2, sections A2.3 and A2.5). The rate-constant for conversion of energized molecules to activated complexes is now a function of J as well as the total energy, and is therefore denoted by k_{EJ}. The integration corresponding to (4.17) is still from $E^+ = 0$ to ∞, and the contribution to

k_{uni} from each of the $(2J+1)$ states with quantum number J is therefore given by (4.25).

$$(k_{uni})_J =$$

$$\frac{L^{\ddagger} \exp(-E_0/kT) \exp(-E_J^+/kT)}{hQ} \int_{E^+=0}^{\infty} \frac{\{\sum P(E_{vr}^+)\} \exp(-E^+/kT) \, dE^+}{1 + k_{EJ}(E_{active}^*)/k_2[M]}$$

$$(4.25)$$

The term $\exp(-E_J^+/kT)$ appears outside the integral sign, being a constant for this particular state, and from the above considerations $dE_{active}^* = dE^+$ for a specified state. The total k_{uni} is obtained by summing the contributions from all the adiabatic rotational quantum states. In the present case there are $(2J+1)$ states for each value of J, and k_{uni} is therefore given by (4.26).

$$k_{uni} = \frac{L^{\ddagger} \exp(-E_0/kT)}{hQ}$$

$$\times \sum_{J=0}^{\infty} (2J+1) \exp(-E_J^+/kT) \int_{E^+=0}^{\infty} \frac{\{\sum P(E_{vr}^+)\} \exp(-E^+/kT) \, dE^+}{1 + k_{EJ}(E_0+E^++\Delta E_J)/k_2[M]}$$

$$(4.26)$$

4.10.2 First approximation

The exact evaluation of (4.26) can only be achieved by a laborious double numerical integration, since the inner integral involves ΔE_J which is strictly a function of J. The expression can be greatly simplified, however, by replacing ΔE_J by its mean value $\langle \Delta E_J \rangle$ for all the rotational states of interest. This step is justified by the statement that k_{EJ} is insensitive to fluctuations of ΔE_J about this mean, and leads to the replacement of k_{EJ} by $k_a(E_0+E^++\langle \Delta E_J \rangle)$, which is no longer a function of J. The appropriate expression for $\langle \Delta E_J \rangle$ is obtained by noting that the rotational energies of A* and A+ in a given level J are related by $E_J/E_J^+ = I^+/I$, whence $\Delta E_J = E_J^+ - E_J = (1 - I^+/I) E_J^+$ and $\langle \Delta E_J \rangle$ is given by

$$\langle \Delta E_J \rangle = (1 - I^+/I) \langle E_J^+ \rangle \qquad (4.27)$$

An expression for $\langle E_J^+ \rangle$ is obtained by averaging E_J^+ with a weighting factor which is proportional to the number of molecules in the given quantum state which undergo reaction, i.e. to the right-hand side of (4.25). At high pressures this is simply $\exp(-E_J^+/kT)$ multiplied by a term independent of J, and at low pressures it is effectively the same since k_{EJ} varies much less rapidly with J than does $\exp(-E_J^+/kT)$. Since in the present case there are $(2J+1)$ rotational states for each value of J, the expression for $\langle E_J^+ \rangle$ becomes

$$\langle E_J^+ \rangle = \int_{J=0}^{\infty} E_J^+(2J+1)\exp(-E_J^+/kT)\,dJ \bigg/ \int_{J=0}^{\infty} (2J+1)\exp(-E_J^+/kT)\,dJ$$

Putting $x = (E_J^+/kT) = h^2 J(J+1)/8\pi^2 I^+ kT$, whence

$$dx = (h^2/8\pi^2 I^+ kT)(2J+1)\,dJ,$$

the simple result (4.28) is obtained [the integrals in (4.28) being standard forms and both equal to unity—the first is $\Gamma(2)$, see Appendix 7].

$$\langle E_J^+ \rangle = kT \int_{x=0}^{\infty} x\exp(-x)\,dx \bigg/ \int_{x=0}^{\infty} \exp(-x)\,dx = kT \qquad (4.28)$$

Thus from (4.27) and (4.28) the value of $\langle \Delta E_J \rangle$ to be inserted into (4.26) is given by (4.29). Alternatively it may be noted that the average rotational

$$\langle \Delta E_J \rangle = (1 - I^+/I)kT \qquad (4.29)$$

energy of the energized molecules [corresponding to (4.27) for the activated complexes] is $\langle E_J \rangle = (I^+/I)\langle E_J^+ \rangle = (I^+/I)kT$, and therefore

$$\langle \Delta E_J \rangle = \langle E_J^+ \rangle - \langle E_J \rangle = (1 - I^+/I)kT$$

It is interesting to note that $\langle E_J \rangle$ can be considerably in excess of the average rotational energy of all the molecules, which is of course kT. Thus the molecules undergoing reaction are rotationally 'hot'; the centrifugal effect can increase the rate so much that activated complexes are formed with high preference from the energized molecules having rotational energy considerably above the average of the thermal distribution.[32]

When ΔE_J in (4.26) has thus been replaced by $\langle \Delta E_J \rangle$ the only function of J remaining is $(2J+1)\exp(-E_J^+/kT)$. The equation can therefore be rewritten as

$$k_{\mathrm{uni}} = C \sum_{J=0}^{\infty} [(2J+1)\exp(-E_J^+/kT)]$$

$$= C \int_{J=0}^{\infty} (2J+1)\exp(-E_J^+/kT)\,dJ = CQ_1^+$$

in which C comprises the remaining terms in (4.26) and is independent of J, and Q_1^+ is the classical partition function for this two-dimensional rotation.

The final equations for k_{uni} are therefore (4.30) and (4.31), which differ

$$k_{uni} = \frac{L^{\ddagger} Q_1^+ \exp(-E_0/kT)}{hQ} \int_{E^+=0}^{\infty} \frac{\left\{ \sum\limits_{E_{vr}^+=0}^{E^+} P(E_{vr}^+) \right\} \exp(-E^+/kT) \, dE^+}{1 + k_a(E_0 + E^+ + \langle \Delta E_J \rangle)/k_2[M]} \tag{4.30}$$

$$k_a(E_0 + E^+ + \langle \Delta E_J \rangle) = \frac{L^{\ddagger}}{hN^*(E_0 + E^+ - (I^+/I-1)kT)} \sum_{E_{vr}^+=0}^{E^+} P(E_{vr}^+) \tag{4.31}$$

from (4.17) and (4.18) only in the omission of the factor Q_1^+/Q_1 in the expression for k_a and in the modified energy at which k_a is evaluated. The high-pressure limit is easily derived by the technique of section 4.7; the result is (4.32), which now includes the factor Q_1^+/Q_1 automatically and therefore agrees identically with the Absolute Rate Theory expression.

$$k_\infty = L^{\ddagger} \frac{kT}{h} \frac{Q^+}{Q} \exp(-E_0/kT) \tag{4.32}$$

where

$$Q^+ = Q_1^+ Q_2^+ \quad \text{and} \quad Q = Q_1 Q_2$$

4.10.3   Further approximations

The numerical integration of (4.30) is considerably simpler than that of (4.26), and represents a reasonable approach to the practical evaluation of centrifugal effects. A useful further simplification may be obtained, however, by writing (4.31) in the form

$$k_a(E_0 + E^+ + \langle \Delta E_J \rangle) = \frac{1}{F} k_a(E_0 + E^+)$$

where

$$F = N^*(E_0 + E^+ + \langle \Delta E_J \rangle) \Big/ N^*(E_0 + E^+) \tag{4.33}$$

The factor F (which is less than unity) is actually a weak function of E^+, but to a reasonable approximation may be considered constant and given, for example, the value in (4.34). Numerical values for F can be obtained

$$F = N^*(E_0 + \langle \Delta E_J \rangle)/N^*(E_0) \tag{4.34}$$

using various expressions for the state densities, e.g. the Whitten–Rabinovitch equations developed in section 5.4. The high-pressure limit is still given by (4.32), and the low-pressure second-order rate-constant is found to be (4.35) [cf. (4.23)]. Comparison of these equations with the

$$k_{\text{bim}} = \left(F \frac{Q_1^+}{Q_1} \right) \left(k_2 \frac{Q_2^*}{Q_2} \right) \qquad (4.35)$$

corresponding results obtained when centrifugal effects are ignored† leads to the expressions (4.36) and (4.37) for the factors f_∞ and f_0 by which centrifugal effects increase the reaction rate at high and low pressures respectively. For the intermediate pressure range k_{uni} can now be written

$$f_\infty = Q_1^+/Q_1 \qquad (4.36)$$

$$f_0 = FQ_1^+/Q_1 \qquad (4.37)$$

as (4.38), which shows that the effect on the integral is effectively to

$$k_{\text{uni}} = \frac{L^{\pm} Q_1^+ \exp(-E_0/kT)}{hQ} \int_{E^+=0}^{\infty} \frac{\sum P(E_{\text{vr}}^+) \exp(-E^+/kT) \, dE^+}{1 + k_a(E_0+E^+)/Fk_2[M]} \qquad (4.38)$$

replace [M] by $F[M]$. The net effect on k_{uni} is thus to increase k_∞ by the factor Q_1^+/Q_1 and to shift the fall-off curve of $\log k_{\text{uni}}/k_\infty$ vs $\log p$ to higher pressures by an effectively constant increment $-\log F$ along the $\log p$ axis.

Other workers have approached the simplification of (4.26) in different ways but cast the results in the same form indicated by (4.32) and (4.35)–(4.38). In particular, Hay and Belford[33] derived (4.39), and Waage and Rabinovitch[34] derived (4.40) as expressions for F to be used in these equations. For comparison, the Whitten–Rabinovitch expression for $N^*(E^*)$ (section 5.4) leads to (4.41) for $F_{34} = N^*(E_0 + \langle \Delta E_J \rangle)/N^*(E_0)$. In

$$F_{\text{HB}} = \exp\left[-(s-1)(I^+/I-1)kT/(E_0+E_z+kT)\right] \qquad (4.39)$$

$$F_{\text{WR}} = [1+(s-1)(I^+/I-1)kT/(E_0+aE_z)]^{-1} \qquad (4.40)$$

$$F_{34} = [1-(I^+/I-1)kT/(E_0+aE_z)]^{s-1} \qquad (4.41)$$

these equations s is the number of active vibrational degrees of freedom (replaced by $s+\frac{1}{2}r$ if there are in addition r active rotational degrees of freedom), E_z is the zero-point energy of the molecular vibrations and a is a parameter having a value between 0 and 1 which can be calculated from the properties of the molecule (see section 5.4).

Waage and Rabinovitch[29, 34] have compared the results obtained by using various expressions for F in (4.37) and (4.38) with the results of exact calculations using (4.26). All the treatments give the correct $f_\infty = Q_1^+/Q_1$, and the maximum errors occur at the low-pressure limit, for which some results are given in Table 4.1. It is clear that the three treatments

---

† Some workers have compared equations like (4.30) and (4.35) with expressions embodying a cruder allowance for centrifugal effects rather than no allowance at all. Apparent contradictions have arisen from a lack of clarity in this respect.

based on (4.37) are all adequate up to $I^+/I \approx 3$, but that only F_{WR} [equation (4.40)] gives a good result for high values of I^+/I. A wider set of comparisons shows [29, 34] that for I^+/I up to 10 the error in f_0 obtained from $F_{WR} Q_1^+/Q_1$ is no more than a few per cent whatever the size of the molecule involved. Table 4.1 also includes the prediction of f_0 by the

Table 4.1 Comparison of calculated low-pressure centrifugal factors f_0 for ethane decomposition[a]

I^+/I	Exact f_0 [from (4.26)]	f_0 from (4.23)[b]	Ref. 28[c]	$F_{34} \dfrac{Q_1^+}{Q_1}$	$F_{HB} \dfrac{Q_1^+}{Q_1}$	$F_{WR} \dfrac{Q_1^+}{Q_1}$
1	1	1	1	1	1	1
3	2·26	1	3	2·16	2·19	2·27
6	3·28	1	6	2·60	2·72	3·34
10	3·99	1	10	2·14	2·41	4·10

[a] For details of model see refs. 29 and 34.
[b] (4.23) arises from the original Marcus treatment adopted in (4.17) and (4.18).
[c] The formulation of ref. 28 leads to $f_0 = Q_1^+/Q_1$.

earlier cruder treatments. Inclusion of the factor Q_1^+/Q_1 in both k_{uni} and k_a [(4.17) and (4.18)] leads to $f_0 = 1$. A somewhat better result is in fact obtained by inserting the Q_1^+/Q_1 in (4.17) but omitting it from (4.18),[28] but the use of (4.37) and (4.40) is clearly desirable and not appreciably more difficult. For use in the fall-off region a modified factor F_0 can be defined as that which gives agreement with (4.26) at low pressures, but F_0 is found to be close to F_{WR} and the necessary numerical integration of (4.26) is not likely to be worth while in most cases.

In some treatments[29, 35] I^+ is regarded in the first instance as a function of J. The denominator terms corresponding to k_{EJ} in (4.26) can be treated as a function of J by using a specific expression for $N^*(E^*_{active})$ such as the Whitten–Rabinovitch expression, and the integration carried out numerically. Waage and Rabinovitch have again suggested a simpler treatment in terms of a modified F-factor, and found this to be a reasonable approximation in practice.[29]

4.10.4 Conclusions

To conclude, an accurate treatment of centrifugal effects for practical cases is obtained as follows. The high-pressure unimolecular rate-constant k_∞ is given by (4.32), i.e. the high-pressure centrifugal factor is given by $f_\infty = Q_1^+/Q_1$. The low-pressure rate-constant k_{bim} is given by (4.35) with F equated to F_{WR} from (4.40), i.e. $f_0 = F_{WR} Q_1^+/Q_1$. The general-pressure rate-constant is given by (4.38), so that k_{uni}/k_∞ has the same value at a concentration $[M]/F$ as it would have at concentration $[M]$ in the absence of centrifugal effects. Since $F < 1$ the fall-off curve is shifted to higher pressures by an increment $-\log F$ along the $\log p$ axis.

4.11   THE QUANTITIES $\sum\limits_{E_{vr}^+=0}^{E^+} P(E_{vr}^+)$ AND $N^*(E^*)$

These quantities are so fundamental to any application of the RRKM theory that it seems worth while to digress a little to illustrate and clarify further their precise significance. The discussion will not be in any way specific to the properties of an activated complex or an energized molecule, and it will be simpler for the present purpose to drop the superscripts $+$ and $*$ and to refer to the sum and density of states for any general molecule. Further, we shall consider by way of example the more common situation in which there are no active rotations, and will therefore discuss the quantities $\sum\limits_{E_v=0}^{E} P(E_v)$ and $N(E)$. The sum, often written less explicitly as $\sum P(E_v)$, will be used mainly when the quantized nature of the molecular vibrations must be recognized, i.e. when the energies of interest are relatively low. The function $N(E)$ is related basically to a continuously varying energy content and is more appropriate at high energies.

4.11.1   Sum of states, $\sum\limits_{E_v=0}^{E} P(E_v)$

Consider first the sum $\sum P(E_v)$, which is particularly relevant to the quantized model of the activated complex employed in the RRKM theory. The degeneracy $P(E_v)$ is the number of vibrational quantum states with a vibrational energy of E_v, and $\sum_{E_v=0}^{E} P(E_v)$ is thus the total number of states with energy not exceeding E. Consider first the case of a single simple harmonic oscillator of frequency v, representing perhaps a diatomic molecule. Such an oscillator has one quantum state at each energy level 0, $hv, 2hv, ..., vhv, ...$ relative to the ground state as zero energy. The fixed zero-point energy $\tfrac{1}{2}hv$ is not included here. Thus $P(E_v) = 1$ at each level, and a plot of $P(E_v)$ against energy appears as shown in the lower part of Figure 4.6(a); this is the conventional energy level diagram plotted on its side. The sum $\sum P(E_v)$ is then obtained by counting the number of quantum states up to and including energy E; from $E/hv = 0$ to $0\cdot\dot{9}$ there is only one such state, $v = 0$; from $E/hv = 1$ to $1\cdot\dot{9}$ the *total* is two, $v = 0$ and $v = 1$; and so on. Thus $\sum P(E_v)$ is built up as a stepwise function which has a constant value throughout the energy range between any two of

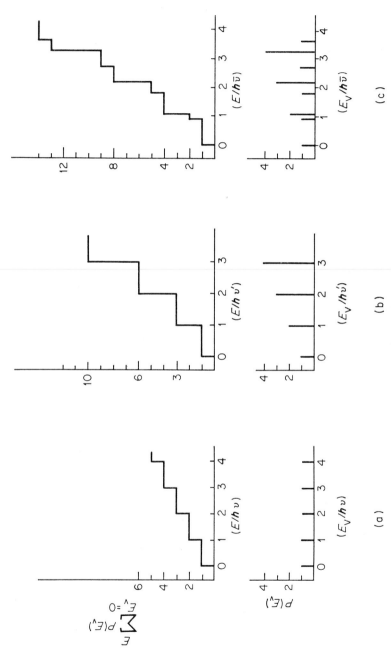

Figure 4.6 Distribution of vibrational quantum states for (a) a single oscillator of frequency v, (b) a doubly degenerate oscillator of frequency v', and (c) a system of three oscillators comprising (a) and (b)

the quantized energy levels; this is shown in the upper part of Figure 4.6(*a*).

Next consider the case of a doubly degenerate oscillator, i.e. two vibrations A and B with the same frequency v' ($= 1 \cdot 2v$ for the sake of illustration). The energy levels are simply vhv', since quanta of energy hv' may be fed into either or both of the vibrations, but now there may be more than one quantum state at a given energy level. There are in fact $v+1$ distinct quantum states of energy vhv', as may readily be verified by inspection in a few simple cases. For $v = 1$ the quantum can be in either oscillator, giving two physically distinct states of the system. For $v = 2$, A and B can contain 2 and 0, 1 and 1, or 0 and 2 quanta respectively, giving three quantum states, and so on; see also section 5.3. The corresponding graphs in Figure 4.6(*b*) show the resulting upward curvature of the plots of $\sum P(E_v)$ against E.

Finally, Figure 4.6(*c*) shows the irregular type of behaviour which arises when the two preceding cases are combined, so that the molecule under consideration consists of three oscillators, two of which have the same frequency. It is clear that diagrams of this type become rapidly more complex as the number of atoms in the molecule increases, and this is well illustrated in the typical curves shown in Figures 4.7 and 4.8. Other notable features are the rapid increase in $\sum P(E_v)$ with energy (a logarithmic plot is needed to display the results over a reasonable range of energies in Figure 4.8), and the greater numbers of states available to molecules with a greater number of vibrations.

It will be noted that for a given system $\sum_{E_v=0}^{E} P(E_v)$ depends only on the total energy E, a relationship which is by no means apparent in the common abbreviation $\sum P(E_v)$. The alternative nomenclature $W(E)$ is advantageous in this respect and will be used in parts of this book. For any general system

$$W(E) \equiv \sum_{E_n=0}^{E} P(E_n) \qquad (4.43)$$

where E_n and $P(E_n)$ are the energy levels and corresponding degeneracies for the particular degrees of freedom under consideration.

4.11.2   Density of quantum states, $N(E)$

It is clear from Figures 4.7 and 4.8 that if attention is focused on highly excited molecules, i.e. those with a large non-fixed energy, the stepwise variation in the number of states becomes much less apparent. An extension of Figure 4.6 to higher energies, shown in Figure 4.9, similarly yields plots of $\sum P(E_v)$ which are much better approximated by a smooth curve than the same plots at low energies. For the single oscillator in

Figure 4.9(a), for example, the plot is well approximated by the straight line $\sum P(E_v) = (E/h\nu)$. For the doubly degenerate oscillator of Figure 4.9(b) the plot is approximately represented by the smooth curve

$$\sum P(E_v) = \tfrac{1}{2}(E/h\nu')^2$$

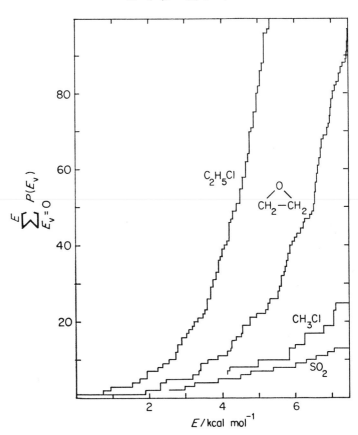

Figure 4.7 $\sum P(E_v)$ as a function of energy for some molecules at low energies; the frequency assignments are detailed in ref. 51 for ethyl chloride and in section 5.3 for the other molecules

and so on. These approximate representations are likely to be perfectly adequate at the high energies relevant to the energized molecules considered in RRKM theory. These molecules will generally contain at least 20–30 quanta of vibrational energy, and the curves also tend to be smoother for realistic models having several different vibration frequencies (see, for example, Figures 4.7 and 4.8).

Under these circumstances it may be useful to formulate results in terms of $N(E)$, the *density of quantum states* or the *number of quantum states per unit energy range* at energy E, which is found to be simply the

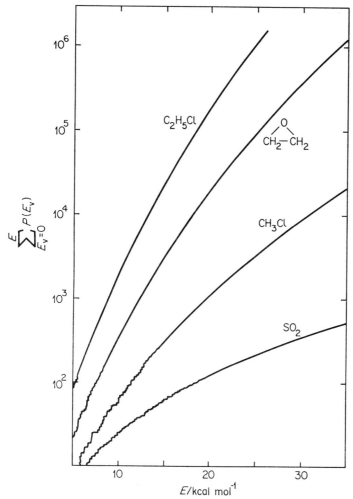

Figure 4.8   Extension of Figure 4.7 to higher energies; note the logarithmic plot required

gradient of the plot of $\sum P(E_v)$ against E. This important result will be required when the numerical evaluation of state densities is considered in Chapter 5, and may be verified for the general case as follows. The number of quantum states of any system in a small range E to $E + \delta E$ is the number

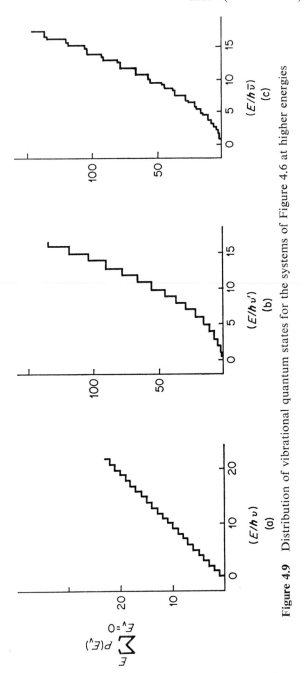

Figure 4.9 Distribution of vibrational quantum states for the systems of Figure 4.6 at higher energies

of states at all energies up to $E + \delta E$ minus the number of states at all energies up to E, i.e.

$$\sum_{E_n=0}^{E+\delta E} P(E_n) - \sum_{E_n=0}^{E} P(E_n) \qquad \text{or} \qquad W(E + \delta E) - W(E)$$

Provided δE is sufficiently small this number of states is also given by $N(E)\,\delta E$, and $N(E)$ is therefore given by (4.44).

$$N(E) = \lim_{\delta E \to 0} \left(\frac{W(E + \delta E) - W(E)}{\delta E} \right)$$

$$= \frac{\mathrm{d}}{\mathrm{d}E} W(E) \equiv \frac{\mathrm{d}}{\mathrm{d}E} \sum_{E_n=0}^{E} P(E_n) \qquad (4.44)$$

This result is most useful under conditions where the density of states is high and the $\sum P(E_n)$ plot is essentially a smooth curve; the density of states is then a continuous function and it is normally only under these conditions that a result is formulated in terms of $N(E)$. The result is still valid when $\sum P(E_n)$ is a coarsely stepped function, however; in this case $N(E)$ is infinite at energies corresponding to the quantized energy levels and zero at intermediate energies, and may be formulated in terms of the Dirac δ-function.[36]

Logarithmic plots of $N(E)$ against energy for some typical vibrational systems are shown in Figure 4.10. A striking feature of these plots is the very high density of quantum states which can be found at quite moderate energies, and this amply justifies the use of a continuous energy distribution function for the energized molecules considered in the RRKM theory. The broken lines at low energy are smoothed versions of the actual spiked variations discussed above; the behaviour at low energies is discussed in more detail in section 5.4.3.

4.12   ASSUMPTIONS OF THE BASIC RRKM THEORY

Before proceeding to discuss the application of the RRKM theory it is appropriate to list and examine the fundamental assumptions and approximations which were essential to the derivation of (4.17). In the following sections five points are discussed, although it will soon become apparent that these are not altogether separable.

4.12.1   Free exchange of energy between oscillators

The assumption that the non-fixed energy of the active vibrations and rotations is subject to rapid statistical redistribution means that every sufficiently energetic molecule will eventually be converted into products unless deactivated by collision. This is a much less restrictive assumption than that made in Slater's theory, in which there must be not only sufficient

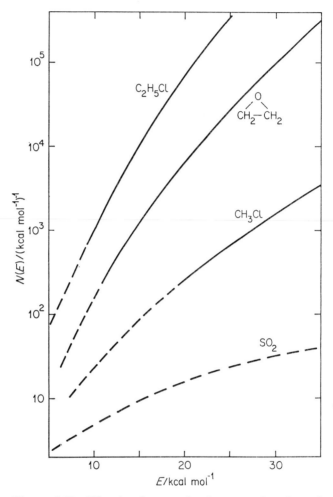

Figure 4.10 Vibrational state density as a function of energy for the molecules referred to in Figures 4.7 and 4.8 (note logarithmic plot)

energy but also a suitable distribution of the energy before a molecule is able to react. In phase-space terminology, the RRKM theory assumes that the phase-space of the molecules concerned is metrically indecomposable, i.e. cannot be subdivided into smaller regions such that the representative point cannot move from one region to another by any feasible trajectory.[3] If the phase-space were decomposable this would imply that not all states of the same energy were freely interchangeable, and purely statistical considerations would not be adequate to describe the model.

Marcus has in fact made provision for some of the degrees of freedom to be completely *inactive*.[1] This means that the energy in these degrees of freedom cannot flow into the reaction coordinate; an energized molecule A* is one with non-fixed energy greater than E_0 *in the active modes*, irrespective of that in the inactive modes. The energy in the inactive modes can, however, be redistributed at random among these modes, and thus does contribute to the number of quantum states at a given energy, and hence affects $N^*(E^*)$ and $k_a(E^*)$. A little progress has also been made towards a theory in which energy exchange between some of the oscillators occurs only at a limited rate.[11]

There is little evidence, however, to support any assumption other than that all modes are either active or adiabatic. In fact, there is positive evidence[37] for the assumption that energy is rapidly redistributed between all vibrational modes of the molecules involved in time intervals no greater than 10^{-11} s following excitation. Most of this evidence has been obtained by studies of the unimolecular reactions of chemically activated species, a subject discussed in detail in Chapter 8. The original work was that of Butler and Kistiakowsky[38] on the isomerization of vibrationally excited methylcyclopropane molecules formed by two different routes as shown in (4.45). The exothermicity of the forming reaction is in

$$CH_2\colon + CH_3CH\colon CH_2$$

(4.45)

each case sufficient to produce the methylcyclopropane with more than enough energy to isomerize to the various butenes, and this isomerization occurs unless the excited molecules are stabilized by collision. Comparison of the rates of formation of isomerization products and stabilized molecules then yields values for the isomerization rate-constant. The main argument from Butler and Kistiakowsky's work came from the observation that the butene fraction had the same composition no matter which way the excited molecules were formed. The energy distributions in the molecules at their moment of formation must be very different, and would be expected to lead to different relative rates of formation of the four butenes unless the energy is completely redistributed in the molecule before isomerization occurs. In addition, although the total rates of isomerization differ because of the differing excess energies of the excited molecules in the two cases, the rates are describable in terms of the same RRKM model involving statistical redistribution of the energy among all the vibrational and internal rotational degrees of freedom. This latter point has been very strikingly demonstrated for the decomposition of sec-butyl radicals formed by the addition of H (or D) atoms to but-1-ene or cis- or trans-but-2-ene (4.46).[39] The thermochemistry is more accurately known here, and despite obvious differences in the initial energy distributions these radicals

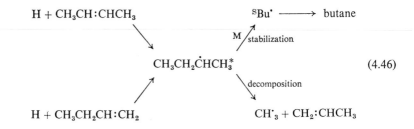

$$(4.46)$$

decompose at a rate which is determined only by the total excess energy and is described accurately by RRKM calculations (see Figure 8.4). A similar observation has been made for isomeric dichloroheptafluoro-sec-butyl radicals.[40] The decomposition of sec-butyl radicals formed from H atoms and cis-but-2-ene has also been studied over a very wide range of pressure (up to 200 atm) and hence a wide range of times between collisions.[41] The decomposition rate-constant k_a was found to be independent of pressure with collision lifetimes varying from 2×10^{-13} to 1×10^{-8} s, which would not be the case unless complete randomization of the vibrational energy occurred within even the shortest lifetimes.

Very recent experiments[42] have utilized the reaction scheme (4.47) in which an essentially symmetrical bicyclopropyl derivative is produced by a reaction involving only one side of the molecule. The chemically activated molecule decomposes by elimination of the carbene CF_2, and the molecule is labelled so that the products of elimination from the two rings are distinguishable by mass spectrometry. The product ratio I/II was invariant over the pressure range 0·8–310 Torr, indicating complete intramolecular energy relaxation in times down to ca 10^{-10} s. Work at higher pressures revealed a change in product ratio and enabled the rate-constant for energy exchange between the two halves of the molecule to be measured as ca 10^{12} s^{-1}.

$$(4.47)$$

4.12.2   Strong collisions

The assumption of strong collisions means that relatively large amounts of energy ($\gg kT$) are transferred in molecular collisions. Thus the basic RRKM model treats the processes of activation and deactivation as essentially single-step processes as opposed to ladder-climbing processes in which molecules acquire or lose their energy in a series of small steps. The latter model requires a much more detailed treatment (see Chapter 10) and may lead, for example, to a situation in which the limited rate of transfer of molecules from one energy level to another below the critical energy may affect the concentration of energized molecules and hence the reaction rate. The strong collision hypothesis is thus closely related to the equilibrium hypothesis, discussed in the next section. A strong collision is also assumed to be so violent, and to cause such a complete redistribution of the coordinates and momenta of the atoms in the molecule, that the state of the molecule after collision is in no way governed by its state before collision, and is merely a random choice from all the available states of the appropriate energy. It is thus not necessary to consider any of the dynamical detail of the molecular collision.

There is, in fact, good evidence that the strong-collision assumption is reasonably realistic for thermal reactions in the temperature range of conventional kinetic studies unless very small molecules are involved. Studies of the energizing efficiencies of various gases in the second-order region indicate a constant limiting efficiency for molecules above a certain moderate size (e.g. C_3 or C_4 and higher) and this limiting efficiency is presumed to be unity. Other studies indicate that these molecules transfer usually about 5 kcal mol^{-1} or more of energy per collision, and since the average excitation energies in thermal reactions are typically 5–15 kcal mol^{-1} above the critical energy it should not require many collisions to deactivate the majority of energized molecules. The studies leading to these conclusions are described in more detail in Chapter 10. A crude allowance for limited energy transfer on collision may be made by replacing $k_2[M]$ in (4.17) and similar equations by λZp, the collision rate-constant Zp multiplied by a constant collisional deactivation efficiency λ. This assumes that the number of collisions required to de-energize an energized molecule is a constant $(1/\lambda)$, and leads to a simple shift of the fall-off curve ($\log k_{uni}$ vs $\log p$) to higher pressures by an increment $\log \lambda$. It is an oversimplification, since the number of collisions needed to de-energize a given energized molecule must clearly depend on the excess energy E^+. Since the average excess energy varies with pressure (see section 4.8) the average λ must also vary with pressure. On the other hand, the more detailed treatments described in Chapter 10 show that this simple modification produces for many purposes a reasonable approximation to the fall-off curves calculated on a rigorous weak-collision basis with realistic parameters for the collision processes. The introduction of a constant λ thus remains a useful semi-empirical method of obtaining agreement between theory and experiment.

4.12.3   The 'equilibrium hypothesis'

The assumption is made in the RRKM theory, as in the Absolute Rate Theory, that the concentration of forward-crossing complexes is the same in the steady state as it would be at total equilibrium where no net reaction was occurring. This assumption has been defended on various grounds,[43, 44] the main argument being that energized molecules about to become activated complexes do not know that there are no molecules travelling in the opposite direction, and hence enter the critical section of the reaction coordinate at the same rate as they would in a true equilibrium situation. Perhaps more convincingly, non-equilibrium calculations show[45, 46] that the equilibrium hypothesis provides a good approximation except again for cases where E_0/kT is small (less than about 10) in which case the reaction is so rapid as to disturb the Boltzmann distribution of molecules with energies near the critical energy. The effect has been calculated[45] to give a maximum error of about 8%, this occurring at $E_0/kT \approx 5$.

4.12.4 Random lifetimes

The RRKM theory assumes[47] that the energized molecules A* have random lifetimes before their unimolecular dissociation, and this infers that the process $A^* \rightarrow A^+$ is governed by purely statistical considerations, and that there is no particular tendency for all A* to decompose soon after their formation, or to exist for some particular length of time before

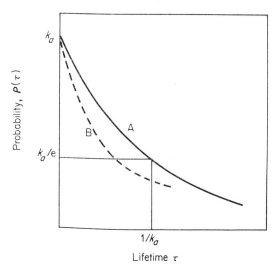

Figure 4.11 Distributions of lifetimes of energized molecules:
curve A, random lifetimes $[P(\tau) = k_a \exp(-k_a \tau)]$;
curve B, a non-random distribution

decomposing. Molecules with a particular energy have a distribution $P(\tau)$ of lifetimes τ which follows the exponential equation:

$$P(\tau) = k_a \exp(-k_a \tau)$$

In this distribution the proportionality constant and the exponential decay constant are both equal to the first-order rate-constant for conversion of the energized molecules into activated complexes. This random distribution is illustrated in Figure 4.11, curve A; it is very similar to the distribution of the number of times a coin can be tossed without a head appearing. It has been noted[47] that even if the distribution of lifetimes is non-random the zero-time value of $P(\tau)$ is still k_a, defined a little more specifically as the *high-pressure* rate-constant for conversion of A^* into A^+. Non-random lifetimes may thus correspond to $P(\tau)$ curves which descend at the wrong rate from the initial value, although they may still have the general appearance of exponential decay. Bunker[47] has illustrated some cases where there is abnormally high probability of dissociation in the period immediately following an energizing collision, and this behaviour is illustrated in Figure 4.11, curve B.

The random-lifetime assumption is closely related to the assumptions described in the previous two sections; since the collisional activation and deactivation are treated as random statistical processes, the dissociation must clearly be treated in the same way if a consistent theory is to emerge. The validity of this assumption has been discussed at length,[3, 9, 12, 47–50] and it appears to be accurate except perhaps[47, 48] for small molecules (diatomics and some triatomics) or for cases where E_0/kT is exceptionally low.

4.12.5   Continuous distribution function $N^*(E^*)$

The treatment of the energized molecule in terms of a continuously variable energy rather than of quantized energy levels has already been discussed (sections 4.4 and 4.11) and seems to be well justified, at least in the majority of cases. It is perhaps worth noting that a very different treatment would be required for the case where the energy levels of A were significantly separated.

References

1. R. A. Marcus, *J. Chem. Phys.*, **20**, 359 (1952).
2. R. A. Marcus and O. K. Rice, *J. Phys. and Colloid Chem.*, **55**, 894 (1951).
3. D. L. Bunker, *Theory of Elementary Gas Reaction Rates*, Pergamon, Oxford, 1966.
4. K. J. Laidler, *Theories of Chemical Reaction Rates*, McGraw–Hill, New York, 1969.
5. (a) E. E. Nikitin, *Theory of Thermally Induced Gas Reactions*, Indiana University Press, Bloomington and London, 1966. (b) O. K. Rice, *Statistical Mechanics Thermodynamics and Kinetics*, Freeman, San Francisco, 1967. (c) R. P. Wayne, 'Theory of Elementary Gas-phase Reactions', in *Comprehensive Chemical Kinetics*, vol. 2 (Ed. C. H. Bamford and C. F. H. Tipper), Elsevier, Amsterdam, 1969, p. 189.
6. G. M. Wieder, Ph.D. Thesis, Polytechnic Institute of Brooklyn, 1961.
7. B. S. Rabinovitch and D. W. Setser, *Advan. Photochem.*, **3**, 1 (1964).
8. Ref. 3, p. 54.
9. D. L. Bunker and M. Pattengill, *J. Chem. Phys.*, **48**, 772 (1968).
10. (a) S. Glasstone, K. J. Laidler and H. Eyring, *The Theory of Rate Processes*, McGraw–Hill, New York and London, 1941. (b) Ref. 4, Chap. 3. (c) R. H. Fowler and E. A. Guggenheim, *Statistical Thermodynamics*, Cambridge University Press, 1939, § 1208. (d) K. J. Laidler and J. C. Polanyi, *Progr. Reac. Kinetics*, **3**, 1 (1965).

11. E. K. Gill and K. J. Laidler, *Proc. Roy. Soc. (A)*, **250**, 121 (1959); M. Solc, *Mol. Phys.*, **11**, 579 (1969); **12**, 101 (1967); *Zeit. Physik. Chem. (Leipzig)*, **234**, 185 (1967); *Chem. Phys. Lett.*, **1**, 160 (1967); N. B. Slater, *Mol. Phys.*, **12**, 107 (1967).

12. O. K. Rice, *J. Phys. Chem.*, **65**, 1588 (1961).

13. N. B. Slater, *Theory of Unimolecular Reactions*, Methuen, London, 1959, p. 108.

14. Ref. 4, Chap. 3; J. D. Macomber and C. Colvin, *Internat. J. Chem. Kinetics*, **1**, 483 (1969); R. P. Bell, *Trans. Faraday Soc.*, **66**, 2770 (1970).

15. Ref. 10(c); V. N. Kondrat'ev, *Chemical Kinetics of Gas Reactions*, Pergamon, Great Britain, 1964, p. 174.

16. L. D. Landau and E. M. Lifshitz, *Quantum Mechanics*, Pergamon, London, 2nd ed., 1965; L. I. Schiff, *Quantum Mechanics*, McGraw–Hill, New York, 2nd ed., 1955.

17. R. P. Bell and E. Gelles, *Proc. Roy. Soc. (A)*, **210**, 310 (1952); D. R. Augood, *Nature*, **178**, 754 (1956).

18. V. Gold, *Trans. Faraday Soc.*, **60**, 739 (1964).

19. E. A. Guggenheim, *Trans. Faraday Soc.*, **50**, 574 (1954); D. Rapp and R. E. Weston, *J. Chem. Phys.*, **36**, 2807 (1962); S. W. Benson and P. S. Nangia, *J. Chem. Phys.*, **38**, 18 (1963); E. W. Schlag, *J. Chem. Phys.*, **38**, 2480 (1963).

20. D. M. Bishop and K. J. Laidler, *J. Chem. Phys.*, **42**, 1688 (1965); see also *Trans. Faraday Soc.*, **66**, 1685 (1970).

21. Ref. 4, p. 65.

22. C. S. Elliott and H. M. Frey, *Trans. Faraday Soc.*, **64**, 2352 (1968).

23. E. W. Schlag and G. L. Haller, *J. Chem. Phys.*, **42**, 584 (1965).

24. J. N. Murrell and K. J. Laidler, *Trans. Faraday Soc.*, **64**, 371 (1968).

25. J. N. Murrell and G. L. Pratt, *Trans. Faraday Soc.*, **66**, 1680 (1970); see also M. R. Wright and P. G. Wright, *J. Phys. Chem.*, **74**, 4394, 4398 (1970).

26. R. A. Marcus, *J. Chem. Phys.*, **43**, 2658 (1965); **52**, 1018 (1970).

27. R. A. Marcus, *J. Chem. Phys.*, **43**, 1598 (1965).

28. G. M. Wieder and R. A. Marcus, *J. Chem. Phys.*, **37**, 1835 (1962).

29. E. V. Waage and B. S. Rabinovitch, *Chem. Rev.*, **70**, 377 (1970).

30. O. K. Rice and H. Gershinowitz, *J. Chem. Phys.*, **2**, 853 (1934).

31. R. A. Marcus, rewritten by H. Heydtmann, 'Unimolecular Reaction Rate Theory', in *Chemische Elementarprozesse* (Ed. H. Hartmann), Springer–Verlag, Berlin, Heidelberg and New York, 1968, p. 109.

32. S. W. Benson, *J. Amer. Chem. Soc.*, **91**, 2152 (1969).

33. A. J. Hay and R. L. Belford, *J. Chem. Phys.*, **47**, 3944 (1967).

34. E. V. Waage and B. S. Rabinovitch, *J. Chem. Phys.*, **52**, 5581 (1970); **53**, 3389 (1970).

35. W. Forst, *J. Chem. Phys.*, **48**, 3665 (1968).

36. P. A. M. Dirac, *The Principles of Quantum Mechanics*, Oxford University Press, 4th ed., 1958; E. Thiele, *J. Chem. Phys.*, **39**, 3258 (1963).

37. D. L. Spicer and B. S. Rabinovitch, *Ann. Rev. Phys. Chem.*, **21**, 349 (1970).

38. J. N. Butler and G. B. Kistiakowsky, *J. Amer. Chem. Soc.*, **82**, 759 (1960).

39. B. S. Rabinovitch, R. F. Kubin and R. E. Harrington, *J. Chem. Phys.*, **38**, 405 (1963).

40. A. S. Rodgers, *J. Phys. Chem.*, **72**, 3400, 3407 (1968).

41. D. W. Placzek, B. S. Rabinovitch and F. H. Dorer, *J. Chem. Phys.*, **44**, 279 (1966); I. Oref, D. Schuetzle and B. S. Rabinovitch, *J. Chem. Phys.*, **54**, 575 (1971).

42. J. D. Rynbrandt and B. S. Rabinovitch, *J. Phys. Chem.*, **74**, 4175 (1970); *J. Chem. Phys.*, **54**, 2275 (1971).

43. (*a*) K. J. Laidler, *Chemical Kinetics*, McGraw–Hill, New York, 2nd ed., 1965, Chap. 3. (*b*) Ref. 4, Chap. 3. (*c*) D. M. Bishop and K. J. Laidler, *J. Chem. Phys.*, **42**, 1688 (1965). (*d*) Ref. 10(*d*). (*e*) K. J. Laidler and A. Tweedale, *Advan. Chem. Phys.*, **21**, 113 (1971).

44. R. A. Marcus, 'Remarks on the Generalization of Activated Complex Theory', in ref. 31, p. 23.

45. B. Morris and R. D. Present, *J. Chem. Phys.*, **51**, 4862 (1969).

46. R. D. Present, *J. Chem. Phys.*, **31**, 747 (1959); E. W. Montroll and K. E. Shuler, *Advan. Chem. Phys.*, **1**, 361 (1958); K. E. Shuler, *J. Chem. Phys.*, **31**, 1375 (1959); H. Eyring, T. S. Ree, T. Ree and F. M. Wanlass, *Chem. Soc. Special Pub. No. 16*, 1962, p. 3; Ref. 4, Chap. 8.

47. D. L. Bunker, *J. Chem. Phys.*, **40**, 1946 (1964).

48. D. L. Bunker, *J. Chem. Phys.*, **48**, 772 (1968).

49. E. Thiele, *J. Chem. Phys.*, **36**, 1466 (1962); **38**, 1959 (1963).

50. W. Forst and P. St. Laurent, *Can. J. Chem.*, **45**, 3169 (1967).

51. L. W. Daasch, C. Y. Liang and J. R. Nielsen, *J. Chem. Phys.*, **22**, 1293 (1954).

5 The Evaluation of Sums and Densities of Molecular Quantum States

Application of the RRKM theory depends critically on the ability to evaluate the quantities $N^*(E_{vr}^*)$ for a molecule A and $\sum P(E_{vr}^+)$ for a postulated model of the corresponding activated complex A^+. The significance of these quantities has already been discussed in some detail (section 4.11) and it will be realized that their evaluation depends in one way or another on an assessment of the distribution of vibrational–rotational quantum states at various energy levels. For the purpose of the present chapter the presence of the superscripts * and + is of little relevance, and we therefore discuss more generally the calculation of the density $N(E_{vr})$ of vibrational–rotational states at a non-fixed vibrational–rotational energy E_{vr}, and the sum of states up to this energy, $W(E_{vr})$, where

$$W(E_{vr}) \equiv \sum_{E_{vr}=0}^{E_{vr}} P(E_{vr}) \quad [\equiv \sum P(E_{vr})] \tag{5.1}$$

In this chapter the symbols E, E_{vr}, etc. are used to denote the total energies at which $N(E)$, $W(E)$, etc. are to be evaluated, while E, E_{vr}, etc. denote specific energy levels referred to in the derivations; the distinction is necessary to avoid ambiguity. The numbers of states of various degrees of freedom with energies up to E_{vr}, E_v, etc. will be denoted as in (5.1) by either the expressions $W(E_{vr})$, $W(E_v)$, etc., or by the abbreviations $\sum P(E_{vr})$, $\sum P(E_v)$, etc., according to the emphasis required.

In section 5.1 the distributions of the vibrational and rotational states are separated for individual treatment. If there are no rotational degrees of freedom in the model under discussion, $N(E_{vr})$ and $W(E_{vr})$ are simply replaced by $N(E_v)$ and $W(E_v)$ and one proceeds directly to section 5.3. Section 5.2 describes the treatment of rotational states on the basis of a classical approximation, this being a simplification which is made in virtually all work on the present subject. It is noted in passing that this treatment converts the general equation (4.17) for k_{uni} into the expression (5.20) derived by Marcus.[1] The next three sections (5.3–5.5) deal mainly with the distributions of vibrational states, and describe three basic methods which are available for the calculation of $N(E_v)$ and $W(E_v)$. The results can be used in conjunction with those of sections 5.1 and 5.2 to deal with vibrational–rotational systems; there are also more direct combined treatments which are mentioned as appropriate in sections 5.4–5.5. Finally, section 5.6 makes numerical comparisons between various expressions for $N(E)$ and $W(E)$, and conclusions are reached about the accuracies and practical utilities of the various approaches.

5.1 SEPARATION OF VIBRATIONAL AND ROTATIONAL DEGREES OF FREEDOM

Consider first the sum $W(E_{vr})$. As shown in (5.1), this is the total number of vibrational–rotational quantum states with non-fixed vibrational–rotational energies E_{vr} which are less than or equal to E_{vr}. This number can be obtained by summing separate contributions involving each of the possible vibrational levels, i.e. those with $E_v \leqslant E_{vr}$. The total number of vibrational–rotational states corresponding to a given vibrational level of energy E_v is the number $P(E_v)$ of vibrational states of this energy multiplied by the number of possible rotational states, which is the total number of rotational states with energy $E_r \leqslant E_{vr} - E_v$, denoted by

$$\sum_{E_r=0}^{E_{vr}-E_v} P(E_r) \equiv W_r(E_{vr} - E_v)$$

Thus the number of vibrational–rotational quantum states in which the vibrational energy is specifically E_v (and the total energy does not exceed E_{vr}) is $P(E_v) W_r(E_{vr} - E_v)$, and the total number of states counting all possible values of E_v is given by (5.2).

$$W(E_{vr}) \equiv \sum_{E_{vr}=0}^{E_{vr}} P(E_{vr}) = \sum_{E_v=0}^{E_{vr}} P(E_v) W_r(E_{vr} - E_v) \qquad (5.2)$$

As shown in section 4.11 [equation (4.44)], the *density* of vibrational–rotational states, $N(E_{vr})$, can now be obtained as the differential of (5.2) with respect to E_{vr}. It is simplest to consider again the contribution from a particular vibrational state of energy E_v, and this is obtained by differentiation of the term $P(E_v) W_r(E_{vr} - E_v)$. Since E_v and therefore $P(E_v)$ are constant for this contribution, and $dE_{vr} = d(E_{vr} - E_v)$, the contribution to the density of states at E_{vr} is therefore

$$P(E_v)\frac{\mathrm{d}}{\mathrm{d}E_{vr}}W_r(E_{vr}-E_v) = P(E_v)\frac{\mathrm{d}}{\mathrm{d}(E_{vr}-E_v)}W_r(E_{vr}-E_v)$$

$$= P(E_v)N_r(E_{vr}-E_v)$$

The total density of vibrational–rotational states at E_{vr}, counting all possible vibrational levels, is therefore given by (5.3).

$$N(E_{vr}) = \sum_{E_v=0}^{E_{vr}} P(E_v)N_r(E_{vr}-E_v) \tag{5.3}$$

The above derivations of (5.2) and (5.3) can readily be visualized in terms of Figure 5.1, which shows the vibrational and rotational energies

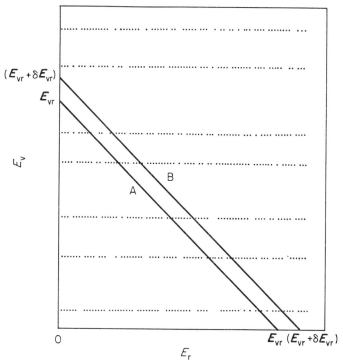

Figure 5.1 Distribution of vibrational–rotational quantum states, illustrating the derivation of (5.2) and (5.3)

of each quantum state by plotting the states as points in the (E_v, E_r) plane. Since the vibrational levels are coarsely quantized, while the rotational levels are closely spaced, the diagram takes the form of a series

5

of lines at constant E_v, each line comprising many closely spaced quantum states. If the vibrational levels are degenerate there are effectively $P(E_v)$ coincident lines at energy E_v, and each rotational point on the line at E_v occurs $P(E_v)$ times. The region of the plot corresponding to vibrational–rotational energies less than or equal to E_{vr} is the triangle to the left of the straight line A joining the intercepts $E_v = E_{vr}$ and $E_r = E_{vr}$ on the two axes; the equation of this line is $E_v + E_r = E_{vr}$. Thus the total number of quantum states with energy $E_{vr} \leqslant E_{vr}$ is the number of points in this triangle and is clearly given by (5.2). The density of states at energy E_{vr} is readily obtained by noting that the number of states with E_{vr} between E_{vr} and $E_{vr} + \delta E_{vr}$ is $N(E_{vr})\,\delta E_{vr}$, and is obtained from the diagram as the number of states between the lines A and B. On each horizontal line of states the number of rotational states in this band is

$$N_r(E_r)\,\delta E_{vr} = N_r(E_{vr} - E_v)\,\delta E_{vr}$$

and (5.3) follows by summation over all the relevant vibrational levels.

Finally, it is convenient to anticipate here a subsequent treatment of the vibrational states of an energized molecule A* in terms of a continuous distribution function $N_v(E_v) = (d/dE_v)\,W_v(E_v)$ in which case (5.3) can be replaced by an integration:

$$N(E_{vr}) = \int_{E_v=0}^{E_{vr}} N_v(E_v)\,N_r(E_{vr} - E_v)\,dE_v \tag{5.4}$$

5.2   CLASSICAL TREATMENT OF ROTATIONAL STATE DISTRIBUTIONS

Although it is proposed to treat any rotational degrees of freedom classically, it is convenient to approach this case as the limiting behaviour of quantized rotors at high energies, where certain approximations can be made to the quantum equations. On this basis we proceed to evaluate $W(E_r)$ or $\sum P(E_r)$, the sum of the numbers of rotational quantum states at all energies up to E_r, and obtain $N(E_r)$ by differentiation, this being justified by the general argument given in section 4.11.2. Only free rotations are considered here; the treatment of hindered rotations and the closely related problem of torsional oscillations have not received much attention and would seem to offer scope for future investigation.

5.2.1   Quantized rotations

It will be assumed that non-degenerate rotations (i.e. those with different moments of inertia) can be treated as being independent. Thus their energies are additive and all combinations of quantum states are possible; the validity of this assumption will be discussed in section 5.2.4. The permissible energy levels of an independent, unhindered, quantized, rigid rotor are given by (5.5) for a one-dimensional or singly degenerate rotor,[2,3] and by (5.6) for a two-dimensional or doubly degenerate rotor.[4,5]

$$\text{(singly degenerate)}\quad E_K = (h^2/8\pi^2 I)\,K^2 \quad (K = 0, 1, 2, ...) \qquad (5.5)$$

$$\text{(doubly degenerate)}\quad E_J = (h^2/8\pi^2 I)\,J(J+1) \quad (J = 0, 1, 2, ...) \quad (5.6)$$

Both terminologies are widely used; care must be taken not to confuse the degeneracy of a rotor with the degeneracies of its energy levels. In (5.5) and (5.6), h is Planck's constant, I the moment of inertia, and J and K are quantum numbers. For high values of the rotational quantum number the term $J(J+1)$ in (5.6) can be approximated by J^2, and the energy in the ith rotation of a molecule with several rotational degrees of freedom is therefore given approximately by (5.7), in which I_i and J_i are the appropriate moment of inertia and quantum number respectively.

$$E_i \approx (h^2/8\pi^2 I_i)\,J_i^2 \qquad (5.7)$$

For a one-dimensional rotor the number of states at any level except the ground state is two, corresponding to the two possible directions of rotation. For a two-dimensional rotor there are $(2J+1)$ states at the Jth energy level (see refs. 4, 5, and section 5.2.4). Thus if the degeneracy of the ith rotor is d_i, the number of quantum states at energy E_i is given by (5.8).

$$\left.\begin{array}{ll}(d_i = 1,\ K \neq 0) & P(E_i) = 2 \\ (d_i = 2) & P(E_i) = (2J_i+1)\end{array}\right\} \qquad (5.8)$$

At high energies $2J_i+1$ is approximately equal to $2J_i$ and the two cases can thus be combined in the single approximate expression (5.9). If there

$$P(E_i) \approx 2J_i^{d_i-1} \qquad (5.9)$$

are altogether p non-degenerate rotors then, for a given set of quantum numbers J_i, the total rotational energy E_r is given by (5.10), and the number of rotational states at this energy by (5.11). The total number

$$E_r = \sum_{i=1}^{p} E_i \approx \sum_{i=1}^{p}\left(\frac{h^2}{8\pi^2}\frac{J_i^2}{I_i}\right) \qquad (5.10)$$

$$P(E_r) = \prod_{i=1}^{p} P(E_i) \approx \prod_{i=1}^{p} 2J_i^{d_i-1} \qquad (5.11)$$

$$\left(\equiv 2J_1^{d_1-1}\cdot 2J_2^{d_2-1}\cdot\ ....\ 2J_p^{d_p-1}\right)$$

of states at energies up to and including E_r is therefore (5.12), in which the summation is taken over all combinations of $J_1, ..., J_p$ such that (5.13) is satisfied.

$$W(E_r) \equiv \sum_{E_r=0}^{E_r} P(E_r) = \sum_{J_p} \sum_{J_{p-1}} ... \sum_{J_1} \left(\prod_{i=1}^{p} 2J_i^{d_i-1} \right) \qquad (5.12)$$

$$\sum_{i=1}^{p} (h^2/8\pi^2 I_i) J_i^2 \leqslant E_r \qquad (5.13)$$

5.2.2 Classical rotations

For rotors which are so highly excited as to permit a classical treatment, the summation (5.12) is replaced by a multiple integration (5.14). The

$$W(E_r) \approx \int_{J_p=0}^{J'_p} \int_{J_{p-1}=0}^{J'_{p-1}} ... \int_{J_1=0}^{J'_1} \left(\prod_{i=1}^{p} 2J_i^{d_i-1} \right) dJ_1 \, dJ_2 ... dJ_p \qquad (5.14)$$

limits are again determined by (5.13), so that the lower limits are all zero and the upper limits J'_n are given by

$$(n = p) \quad \frac{h^2 J'^2_p}{8\pi^2 I_p} = E_r$$

$$(n \neq p) \quad \frac{h^2 J'^2_n}{8\pi^2 I_n} = E_r - \sum_{i=n+1}^{p} \frac{h^2 J_i^2}{8\pi^2 I_i}$$

It will be seen that these equations express the condition that the pth rotor cannot have energy E_p greater than the total available energy E_r, that the $(p-1)$th rotor cannot have more than the energy $(E_r - E_p)$ left after fixing E_p, and so on, the energy of the first rotor being limited to that which remains when all the other E_i are fixed.

The integration of (5.14) has not been published, at any rate not in the present context, and this is therefore carried out in some detail in Appendix 4. Three methods are given, differing in the amount of mathematical shorthand required; the third and longest requires little more than a knowledge of simple integral calculus. For present purposes, only the result is required and this is reproduced as (5.15). In this equation $\Gamma(x)$

$$W(E_r) \equiv \sum_{E_r=0}^{E_r} P(E_r) = \frac{Q_r}{\Gamma(1+\tfrac{1}{2}r)} \left(\frac{E_r}{kT} \right)^{\tfrac{1}{2}r} \qquad (5.15)$$

where

$$r = \sum_{i=1}^{p} d_i$$

is the gamma function, essential properties of which are discussed in

Appendix 7. The sum r of the degeneracies of the p rotors is simply the total number of rotational degrees of freedom involved, irrespective of whether they occur singly or in degenerate pairs. The quantity Q_r is the partition function for the rotational degrees of freedom under the same approximations (5.10) and (5.11) as were used in the derivation of (5.15); the derivation is given in Appendix 5 and the result is (5.16). It will be

$$Q_r = \left(\frac{8\pi^2 kT}{h^2}\right)^{\frac{1}{2}r} \prod_{i=1}^{p} \{I_i^{\frac{1}{2}d_i}\,\Gamma(\tfrac{1}{2}d_i)\} \equiv \prod_{i=1}^{p} \left\{\left(\frac{8\pi^2 I_i kT}{h^2}\right)^{\frac{1}{2}d_i}\Gamma(\tfrac{1}{2}d_i)\right\} \quad (5.16)$$

noted that the temperature dependence of Q_r is exactly cancelled by the term $(kT)^{-\frac{1}{2}r}$ in (5.15), so that $W(E_r)$ is temperature independent, as it obviously should be. Although the partition function is introduced into (5.15) mainly to simplify the expression, the result does suggest a useful approximate treatment of different rotational models (e.g. hindered rotations) by using (5.15) with the appropriate partition function expression inserted.[1]

In addition to the result (5.15) for the sum of states, an expression is also required for the density of states, and this is obtained by simple differentiation of (5.15) (cf. section 4.11.2). Since $\Gamma(1+\tfrac{1}{2}r) = \tfrac{1}{2}r\Gamma(\tfrac{1}{2}r)$ the result is (5.17).

$$N(E_r) = \frac{d}{dE_r} W(E_r) = \frac{Q_r}{(kT)^{\frac{1}{2}r}\,\Gamma(\tfrac{1}{2}r)} E_r^{\frac{1}{2}r-1} \quad (5.17)$$

5.2.3 Results for $N^*(E^*)$, $\sum P(E_{vr}^+)$, and k_{uni}

The required results for the density and sum of states used in the RRKM theory, using the classical treatment of the active rotations, are now obtained by simple substitution of (5.17) and (5.15) into (5.4) and (5.2) respectively, and restoring the appropriate nomenclature; the results are (5.18) and (5.19). Finally, the expression for k_{uni} assuming classical

$$N^*(E^*) = \frac{Q_r^*}{(kT)^{\frac{1}{2}r}\,\Gamma(\tfrac{1}{2}r)} \int_{E_v^*=0}^{E^*} (E^* - E_v^*)^{\frac{1}{2}r-1} N_v^*(E_v^*)\,dE_v^* \quad (5.18)$$

$$\sum_{E_{vr}^+=0}^{E^+} P(E_{vr}^+) = \frac{Q_r^+}{(kT)^{\frac{1}{2}r}\,\Gamma(1+\tfrac{1}{2}r)} \sum_{E_v^+=0}^{E^+} [(E^+ - E_v^+)^{\frac{1}{2}r} P(E_v^+)] \quad (5.19)$$

active rotations is obtained by substituting (5.18) and (5.19) into the general expression (4.17) for k_{uni}. The result is (5.20), in which $k_a(E_0+E^+)$ is given by (5.21) with (5.18) again substituted for $N^*(E_0+E^+)$. The equations (5.20) and (5.21) are identical with those of Marcus,[1] except

$$k_{uni} = \frac{L^{\ne} Q_1^+ Q_r^+ \exp(-E_0/kT)}{hQ_1 Q_2 (kT)^{\frac{1}{2}r} \Gamma(1+\frac{1}{2}r)}$$

$$\times \int_{E^+=0}^{\infty} \frac{\left\{ \sum_{E_v^+=0}^{E^+} (E^+ - E_v^+)^{\frac{1}{2}r} P(E_v^+) \right\} \exp(-E^+/kT)}{1 + k_a(E_0+E^+)/k_2[M]} \, dE^+$$

(5.20)

$$k_a(E_0+E^+) = \frac{L^{\ne} Q_1^+}{Q_1} \frac{Q_r^+}{(kT)^{\frac{1}{2}r} \Gamma(1+\frac{1}{2}r)} \frac{\sum_{E_v^+=0}^{E^+} (E^+ - E_v^+)^{\frac{1}{2}r} P(E_v^+)}{hN^*(E_0+E^+)}$$

(5.21)

for the treatment of reaction path degeneracy by the statistical factor $L^{\ne}$ and a minor point concerned with the possible existence of 'inactive' vibrations (see sections 4.2 and 4.12.1). They embody the crude treatment of adiabatic rotations, but the more sophisticated treatments of section 4.10 are easily substituted.

When there are active rotations (5.18) can be integrated and (5.19) summed using values of the sum and density of the vibrational states derived by methods to be discussed in sections 5.3–5.5. The summation and integration must normally be carried out by numerical methods, although an analytical integration of (5.18) may occasionally be possible. This method of combining the distributions of vibrational and rotational states provides an exact result for the classical rotational model if the exact count of vibrational states is used (see section 5.3). The numerical integration is awkward, however, and it is probably for this reason that approximate treatments of the combined vibrational–rotational problem have also been developed. These are commonly based on equations similar to (5.18) and (5.19), and will be dealt with at appropriate points in sections 5.4 and 5.5.

5.2.4 Validity of the classical independent rotor treatment

There are two main points which are worthy of consideration regarding the validity of the treatment outlined above. Firstly, some justification is needed for treating the various rotations independently of each other, and, secondly, the various mathematical approximations made in the derivation need to be examined.

The various points may conveniently be illustrated with reference to the overall rotations of a prolate symmetric top molecule, i.e. a molecule for which $I_A < I_B = I_C$. The stationary wave functions and energy levels of such a system are correctly obtained by solution of the Schrödinger equation for the molecule as a whole.[6] The energy levels are given by (5.22), in

$$
\left.
\begin{aligned}
E_{J,K,M} &= BhJ(J+1)+(A-B)hK^2 \\
J &= 0, 1, 2, \ldots, \quad K = 0, \pm 1, \pm 2, \ldots, \pm J \\
(M &= 0, \pm 1, \pm 2, \ldots, \pm J)
\end{aligned}
\right\} \tag{5.22}
$$

which $A = h/8\pi^2 I_A$, etc. and J, K, M are integral quantum numbers in the ranges shown. It will be noted that there are *three* quantum numbers, only two of which (J, K) appear in the expression for the energy. For a given value of J there are $2J+1$ acceptable values of M, so that each J, K combination comprises not one but $2J+1$ quantum states. The energy of a state $J, -K$ is the same as that of the state J, K, and the overall degeneracy of a given energy level E_r is therefore given by (5.23). The rotational

$$
\left.
\begin{aligned}
(K=0) \quad & P(E_r) = 2J+1 \\
(K \neq 0) \quad & P(E_r) = 2(2J+1)
\end{aligned}
\right\} \tag{5.23}
$$

energy levels for a given symmetric top can now be evaluated, together with their degeneracies, and plotted on a $\sum P(E_r)$ vs energy diagram similar to Figures 4.7 and 4.8. Such a plot is shown in Figure 5.2 (curve A) for the methyl chloride molecule; the data were calculated by a direct-count computer program similar to that described in section 5.3 for vibrational states.

We now wish to examine the effects of regarding the rotational degrees of freedom as independent or separable. The symmetric top molecule would thus be regarded as equivalent to a two-dimensional rotor with moment of inertia I_B plus an independent one-dimensional rotor with moment of inertia I_A. Their energies will be additive and their degeneracies multiplicative, with no restriction on the permissible quantum numbers for the two rotations. For a one-dimensional rotor the energy levels are (5.5) and the degeneracies are 1 $(K=0)$ or 2 $(K \neq 0)$. For a two-dimensional rotor the energy levels are (5.6) and the degeneracy of a given level is $(2J+1)$; this degeneracy arises from the existence of a second quantum number J' (usually called K) which takes the values $0, \pm 1, \pm 2, \ldots, \pm J$ but does not enter the expression for the energy (cf. M for a symmetric top). Thus the energy levels of the symmetric top with its rotations separated are given by (5.24), and the degeneracy of a given level is again given by (5.23). The

$$
\left.
\begin{aligned}
E_{J,J',K} &= BhJ(J+1)+AhK^2 \\
J &= 0, 1, 2, \ldots, \quad K = 0, 1, 2, \ldots \\
(J' &= 0, \pm 1, \pm 2, \ldots, \pm J)
\end{aligned}
\right\} \tag{5.24}
$$

resulting $\sum P(E_r)$ curve is plotted as curve B in Figure 5.2 for the methyl chloride molecule. It is clear that although the assumption of separability causes some slight redistribution of the energy levels there is no significant systematic disturbance of the $\sum P(E_r)$ curve. Current and Rabinovitch[7]

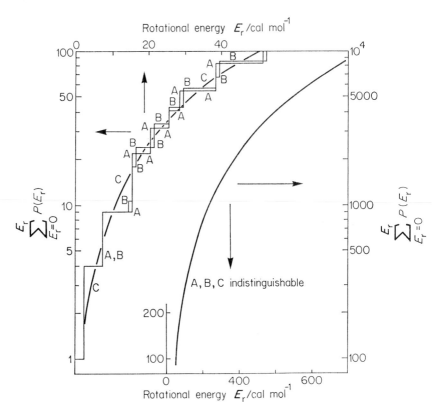

Figure 5.2 Distribution of rotational states for methyl chloride ($A = 1·50 \times 10^{11}$ Hz, $B = C = 1·33 \times 10^{10}$ Hz) using rigorous treatment (A), independent rotor approximation (B), and classical approximation [equation (5.15)] (smooth curve C)

reached similar conclusions for another symmetric top (having different moments of inertia) and also for the separation of an internal rotation from the overall rotation about the same axis. It thus appears that separability of rotations is a sufficiently accurate simplification for present purposes.

In order to obtain analytical results for the independent-rotor model, the approximations were introduced that $J(J+1) \approx J^2$ and $2J+1 \approx 2J$, and the summation (5.12) was replaced by the integration (5.14) (and similarly in derivation of the expression for Q_r). The accuracy of these approximations is illustrated for the present case by the smooth curve C in Figure 5.2. This curve, calculated from (5.25), is an excellent representation of the actual quantized behaviour even at very low rotational energies. Whitten

$$W(E_r) \equiv \sum_{E_r=0}^{E_r} P(E_r) = \left(\frac{8\pi^2 I_A}{h^2}\right)^{\frac{1}{2}} \left(\frac{8\pi^2 I_B}{h^2}\right) E_r^{\frac{3}{2}} \qquad (5.25)$$

and Rabinovitch[8, 9] similarly checked the corresponding equation for the vibrational–rotational case [equation (5.19)] against exact counts for some representative cases, and found only small discrepancies (generally less than 2% at energies above 300 cal mol^{-1}). Bearing in mind that the average rotational energy in three classical rotational degrees of freedom is $\frac{3}{2}kT = 900$ cal mol^{-1} at room temperature, it is clear that the classical approximation is entirely adequate for rate calculations.

There is of course the wider question of whether the assumptions of *rigid* rotors and *free* rotations are adequate. These are points to which little attention has been given so far, and whilst it is probable that the assumptions are adequate in many cases, there is perhaps room for investigation of these points.

5.3 DIRECT COUNT OF VIBRATIONAL STATES[7, 9–13]

We turn now to the more taxing problem of calculating the distribution of vibrational levels of the species of interest. Most workers have treated the molecules as sets of simple harmonic oscillators, and only occasional reference will be made to deviations from this ideal case (see also sections 7.2.5 and 7.3 concerning anharmonicity). In the present section the exact but laborious method of direct counting of states is described. It will emerge that the method, although indispensable at low energies, needs supplementing by less time-consuming approximate methods for use at higher energies, and these will be described in sections 5.4 and 5.5.

5.3.1 The basic method

The direct-count method is very simple in concept once the basic significance of $\sum P(E_v)$ has been grasped; see section 4.11. The method will be illustrated first for the very simple case of the SO_2 molecule, which has only $(3 \times 3) - 6 = 3$ vibrational modes, and for non-fixed energies up to the rather low level of 10 kcal mol^{-1}. The vibrational frequencies of the molecule are taken[14] as 1361, 1151 and 524 cm^{-1}, corresponding to energies per quantum of $N_A hc\bar{\nu}_i = 3{\cdot}891$, $3{\cdot}291$ and $1{\cdot}498$ kcal mol^{-1} respectively, and if the quantum numbers for the three vibrations are v_1, v_2 and v_3, the non-fixed vibrational energy is thus

$$3{\cdot}891v_1 + 3{\cdot}291v_2 + 1{\cdot}498v_3 \text{ kcal mol}^{-1}$$

It is convenient to describe the working in two stages, although these may be combined in practice. The first stage, illustrated in Table 5.1, is a systematic listing of all possible combinations of quantum numbers for which the total vibrational energy is below the required energy limit of 10 kcal mol^{-1}. Thus all oscillators are started in the ground state and the energy in the third vibrational mode is progressively increased, recording the total energy at each step, until a value greater than 10 kcal mol^{-1} is reached. The second mode is then excited by one quantum ($v_2 = 1$) and the process repeated, again starting with $v_3 = 0$. When $E_v > 10$ again, v_2 is raised to 2, and so on. When eventually any further increase in v_2 gives $E_v > 10$, even with $v_3 = 0$, v_1 is increased by one quantum and v_2 and v_3 are again started from zero. In this systematic way the energies of all the quantum states in the energy range of interest are recorded. A second table (Table 5.2) is now constructed, in which the given energy range is split up into as many small increments as required (20 steps of 0·5 kcal mol^{-1} in this illustration) and an entry is made in column 2 at the appropriate energy for each quantum state recorded in Table 5.1.† These entries are summed horizontally to give the total number of states within each incremental energy range (column 3) and progressive vertical summation of the different ranges gives the values for $\sum P(E_v)$ in column 4.

When some of the modes of vibration have the same frequency it is possible to save time by regarding these as one degenerate vibration having several quantum states at each energy level, cf. section 4.11. In general, the number of distinct ways of distributing j quanta in p oscillators is given by simple permutation theory as $(j+p-1)!/j!(p-1)!$ The first table must now be modified to show, in addition to the energy, the number of vibrational states for a given combination of quantum numbers, and the second column of the second table will contain numbers other than unity. An example is shown in Tables 5.3 and 5.4, which form part of a direct count for $COCl_2$ for energies up to 10 kcal mol^{-1}; even for this limited case the full tables would take up too much space to be worth

---

† We take care to define the energy ranges precisely as 0–0·4̇9, 0·5–0·9̇9, etc., and not loosely as 0–0·5, 0·5–1·0, etc. Failure to do so leads to considerable confusion in some applications, notably those using increments commensurable with the assumed frequencies. Note similarly the explanatory comment at the top of Table 5.3.

Table 5.1 Energies of lower vibrational quantum states of SO_2

Quantum numbers v_1 v_2 v_3	Total energy/ kcal mol^{-1}
0 0 0	0·00
0 0 1	1·498
0 0 2	2·996
0 0 3	4·49
0 0 4	5·99
0 0 5	7·49
0 0 6	8·99
0 0 7	> 10
0 1 0	3·29
0 1 1	4·79
0 1 2	6·29
0 1 3	7·79
0 1 4	9·28
0 1 5	> 10
0 2 0	6·58
0 2 1	8·08
0 2 2	9·58
0 2 3	> 10
0 3 0	9·87
0 3 1	> 10
0 4 0	> 10
1 0 0	3·89
1 0 1	5·39
1 0 2	6·89
1 0 3	8·39
1 0 4	9·88
1 0 5	> 10
1 1 0	7·18
1 1 1	8·68
1 1 2	> 10
1 2 0	> 10
2 0 0	7·78
2 0 1	9·28
2 0 2	> 10
2 1 0	> 10
3 0 0	> 10

Table 5.2 Numbers of quantum states of SO_2 in given energy ranges

Energy range/ kcal mol^{-1}	Entries from Table 5.1	States in range	$\sum P(E_v)$
0·00–0·4$\dot{9}$	1	1	1
0·50–0·9$\dot{9}$		0	1
1·00–1·4$\dot{9}$	1	1	2
1·50–1·9$\dot{9}$		0	2
2·00–2·4$\dot{9}$		0	2
2·50–2·9$\dot{9}$	1	1	3
3·00–3·4$\dot{9}$	1	1	4
3·50–3·9$\dot{9}$	1	1	5
4·00–4·4$\dot{9}$	1	1	6
4·50–4·9$\dot{9}$	1	1	7
5·00–5·4$\dot{9}$	1	1	8
5·50–5·9$\dot{9}$	1	1	9
6·00–6·4$\dot{9}$	1	1	10
6·50–6·9$\dot{9}$	1, 1	2	12
7·00–7·4$\dot{9}$	1, 1	2	14
7·50–7·9$\dot{9}$	1, 1	2	16
8·00–8·4$\dot{9}$	1, 1	2	18
8·50–8·9$\dot{9}$	1, 1	2	20
9·00–9·4$\dot{9}$	1, 1	2	22
9·50–9·9$\dot{9}$	1, 1, 1	3	25

Table 5.3 Energy and degeneracy of vibrational quantum states of $COCl_2$ [Frequencies 1810, 834, 571, 444 and 300 (2) cm^{-1}. Energies are rounded to the next lower multiple of 0·1 kcal mol^{-1}; i.e. the ranges taken are 0·00–0·09, 0·10–0·19, etc. Entries in bold type correspond to those in bold type in Table 5.4]

v_1	v_2	v_3	v_4	v_5	Energy/ kcal mol^{-1}	No. of states	v_1	v_2	v_3	v_4	v_5	Energy/ kcal mol^{-1}	No. of states
0	0	0	0	0	**0**	**1**	0	0	1	0	0	1·6	1
0	0	0	0	1	0·8	2	0	0	1	0	1	2·4	2
0	0	0	0	2	1·7	3	0	0	1	0	2	3·3	3
0	0	0	0	3	2·5	4	0	0	1	0	3	4·2	4
0	0	0	0	4	**3·4**	**5**	.	.	.	.	.	.	.
.	.	.	.	.	.	.	1	0	2	1	0	9·7	1
0	0	0	0	9	7·7	10	1	0	2	1	1	>10	—
0	0	0	0	10	8·5	11							
0	0	0	0	11	9·4	12	1	0	2	2	0	>10	—
0	0	0	0	12	>10	—							
							1	0	3	0	0	>10	—
0	0	0	1	0	1·2	1							
0	0	0	1	1	2·1	2	1	1	0	0	0	7·5	1
0	0	0	1	2	2·9	3	1	1	0	0	1	8·4	2
0	0	0	1	3	3·8	4	1	1	0	0	2	9·2	3
.	.	.	.	.	.	.	1	1	0	0	3	>10	—
0	0	0	1	8	8·1	9							
0	0	0	1	9	8·9	10	1	1	0	1	0	8·8	1
0	0	0	1	10	**9·8**	**11**	1	1	0	1	1	**9·6**	**2**
0	0	0	1	11	>10	—	1	1	0	1	2	>10	—
0	0	0	2	0	2·5	1	1	1	0	2	0	>10	—
0	0	0	2	1	**3·3**	**2**							
0	0	0	2	2	4·2	3	1	1	1	0	0	9·1	1
.	.	.	.	.	.	.	1	1	1	0	1	>10	—
0	0	0	5	3	8·9	4							
0	0	0	5	4	**9·7**	**5**	1	1	1	1	0	>10	—
0	0	0	5	5	>10	—							
							1	1	2	0	0	>10	—
0	0	0	6	0	7·6	1							
0	0	0	6	1	8·4	2	1	2	0	0	0	**9·9**	**1**
0	0	0	6	2	9·3	3	1	2	0	0	1	>10	—
0	0	0	6	3	>10	—							
							1	2	0	1	0	>10	—
0	0	0	7	0	8·8	1							
0	0	0	7	1	**9·7**	**2**	1	2	1	0	0	>10	—
0	0	0	7	2	>10	—							
							1	3	0	0	0	>10	—
0	0	0	8	0	>10	—							
							2	0	0	0	0	>10	—

Table 5.4 Quantum states of $COCl_2$ in given energy ranges (entries in bold type correspond to those in bold type in Table 5.3)

Energy range/ kcal mol^{-1}	Entries from Table 5.3	Total states in range	$\sum P(E_v)$
0·0–0·0̇9̇ **1**		1	1
0·1–0·1̇9̇		0	1
0·2–0·2̇9̇		0	1
· ·		· · ·	(22)
3·2–3·2̇9̇	1, 2	3	25
3·3–3·3̇9̇	**2, 3**	5	30
3·4–3·4̇9̇	**5**	5	35
3·5–3·5̇9̇		0	35
3·6–3·6̇9̇	1	1	36
· ·		· · ·	(897)
9·6–9·6̇9̇	3, 7, 4, 1, 8, 5, 1, 1, 2, **2**	34	931
9·7–9·7̇9̇	5, **2**, 9, 6, **1**, 2, 2, 3, 4, 3, 1	38	969
9·8–9·8̇9̇	**11**, 8, 3, 4, 1, 4, 1, 5, 5, 2	44	1013
9·9–9·9̇9̇	1, 5, 2, 5, 2, 6, 3, 7, **1**.	32	1045

Summary of complete count in 1 kcal mol^{-1} ranges

Energy/kcal mol^{-1}	1	2	3	4	5	6	7	8	9	10
$\sum P(E_v)$	3	8	22	43	88	164	274	441	701	1045

printing. The six modes of vibration of this molecule have frequencies† equivalent to 1810, 834, 571, 444 and 300 (2) cm^{-1}, the last being doubly degenerate and therefore giving rise to $(v_5 + 1)$ quantum states when there are v_5 quanta of this energy in the molecule. Thus when the five vibrational quantum numbers are $v_1, v_2, ..., v_5$, the total non-fixed vibrational energy is $5·17v_1 + 2·38v_2 + 1·63v_3 + 1·27v_4 + 0·86v_5$ kcal mol^{-1}, and the degeneracy is $1 \times 1 \times 1 \times 1 \times (v_5 + 1)$. The construction of Tables 5.3 and 5.4 should now be self-explanatory, and even the extension to anharmonic vibrations is not difficult.[12]

This sort of problem is ideally suited for working by digital computer, and indeed the work is prohibitively laborious if performed by hand for

† In this and several other examples we use the frequency assignments adopted by Whitten and Rabinovitch,[11] noting that these are typical data rather than best values. A number in parentheses following a frequency is the number of vibrational degrees of freedom having that frequency.

most molecules at the energies of interest. This is simply because of the enormous numbers of states which have to be counted; see, for example, Figures 4.7 and 4.8, and the data in Table 5.5 for the chloroform molecule. Computer programs written in various languages may be had on request from several authors,[16] and for those who prefer to write their own programs Appendix 3 gives essential details of the central state-counting section, which is the only likely source of programming difficulty.

Even with the aid of a fast computer, however, the direct count can become very time-consuming and therefore unduly expensive at high energies. This is illustrated in Table 5.5, which gives the times required for

Table 5.5 Direct count for $CHCl_3$ by Atlas I computer. [Frequencies:[15] 3033, 1205 (2), 760 (2), 667, 364, 260 (2) cm^{-1}]

Energy/kcal mol^{-1}	$\sum P(E_v)$	Computer time/s^a
10	$3 \cdot 041 \times 10^3$	0·1
15	$2 \cdot 527 \times 10^4$	0·3
20	$1 \cdot 348 \times 10^5$	0·9
30	$1 \cdot 787 \times 10^6$	5·3
40	$1 \cdot 302 \times 10^7$	21
50	$6 \cdot 560 \times 10^7$	63
60	$2 \cdot 570 \times 10^8$	152

^{a} State-counting time only, excluding compilation (ca 2 s), evaluation of degeneracies, etc., and input/output time.

an ICT Atlas I computer to count up to various energies for the $CHCl_3$ molecule. Atlas is a very fast computer (e.g. addition ca 1·6 μs) and the program used here is considered to be fairly efficient. The core of the program (Appendix 3) was written in machine code (this in itself reduces computing time by a factor of 3 because of the large amount of array-element accessing required), and the computing was done on a 'production run' basis with all run-time checks removed. Even so, the time required is becoming excessively large at 60 kcal mol^{-1} for this rather small molecule. Admittedly the chloroform molecule has some rather low-frequency vibrations and hence a relatively large number of states for a given energy. Methyl chloride, with the same number of vibrations, has the substantially higher frequencies[11] of 3047 (2), 2879, 1460 (2), 1355, 1020 (2) and 732 cm^{-1} (thus the zero-point energy, $\frac{1}{2}\sum h\nu$, is 8009 cm^{-1} compared with only 4262 cm^{-1} for chloroform). Methyl chloride has only $5 \cdot 96 \times 10^5$ vibrational states up to 60 kcal mol^{-1} and these can be counted in about 4 s. On the other hand, the energies of interest are likely to be substantially higher as well; for example, HCl elimination from methyl chloride is about

45 kcal mol^{-1} more endothermic than from chloroform.[17] Thus it will still be unjustifiably expensive to carry out a direct count for the energized molecules. The situation is obviously worse for larger molecules (see for example Figure 4.7), and it soon becomes clear that not all of the necessary calculations can reasonably be carried out by direct counting of the vibrational states.

5.3.2   Grouped-frequency models

For larger molecules considerable economy can be achieved (at the expense of some slight loss of realism) by artificially grouping the vibration frequencies into five or six groups of degenerate vibrations, chosen so that the geometric mean of the $3N-6$ frequencies remains the same. The calculation then follows the method already indicated for $COCl_2$ where two of the frequencies are naturally equal because of symmetry considerations. An example of the economy achieved for a given loss of accuracy is presented in Tables 5.6–5.8, which refer to the ethylene oxide molecule

Table 5.6   Frequency groupings for ethylene oxide[11]

Actual frequencies	7-freq. grouping	5-freq. grouping	3-freq. grouping	2-freq. grouping	1-freq. grouping
3080 3060 3008 (2)	3039 (4)	3039 (4)	3039 (4)	3039 (4)	
1492 1453	1472 (2)	1472 (2)			1402 (15)
1270	1270	1176 (5)	1254 (7)		
1168 (2) 1159 1122	1154 (4)			1058 (11)	
867	867	826 (3)			
807 (2)	807 (2)		785 (4)		
673	673	673			

$(CH_2)_2O$. Table 5.6 shows the frequency groupings examined by Whitten and Rabinovitch,[11] and Table 5.7 shows the effect of the grouping on the values of $\sum P(E_v)$ obtained at various energy levels. It will be seen that the more drastic groupings lead to considerable errors at the lower energies, but that the representations of the molecule as five or seven groups of vibrations give results very close to those calculated from the actual molecular vibration frequencies. Finally, Table 5.8 compares the computer

time required to count up to various energy levels for the different groupings. It appears that there is a considerable saving when the actual twelve
groups are collected into five or seven artificial groups, but that there is

Table 5.7 $\sum_{E_v=0}^{E_v} P(E_v)$ for ethylene oxide using frequency groupings of
Table 5.6[a]

| Energy/kcal mol^{-1} | 10·4 | 34·5 | 69·0 | 172·6 |
E_v/E_z[b]	0·300	1·00	2·00	5·00
1-freq. group	136	$0·490 \times 10^6$	$5·657 \times 10^8$	$2·975 \times 10^{13}$
2-freq. group	368	$1·053 \times 10^6$	$7·452 \times 10^8$	$3·395 \times 10^{13}$
3-freq. group	459	$1·110 \times 10^6$	$8·094 \times 10^8$	$3·493 \times 10^{13}$
5-freq. group	401	$1·162 \times 10^6$	$8·227 \times 10^8$	$3·520 \times 10^{18}$
7-freq. group	411	$1·158 \times 10^6$	$8·211 \times 10^8$	—
Actual freqs.	415	$1·158 \times 10^6$	—	—

[a] Present results show small unexplained differences from those in ref. 11.
[b] E_z = zero-point energy for actual frequencies (34·512 kcal mol^{-1}).

Table 5.8   Effect of grouping on computer time (Atlas I
computer; state counting time only)

| | Time/s to count up to E_v/E_z = | | | |
	0·300	1·00	2·00	5·00
1-freq. group	0·02	0·02	0·02	0·02
2-freq. group	0·02	0·02	0·03	0·07
3-freq. group	0·02	0·04	0·13	1·45
5-freq. group	0·03	0·36	5·5	297
7-freq. group	0·04	3·0	118	—
Actual freqs.	0·09	40·2	—	—

little to be gained by further grouping. Thus the best overall compromise
seems to be a regrouping of the frequencies into five to seven artificial
groups, and this sort of approach has been widely used in computer
calculations of $\sum P(E_v)$.

5.3.3   Commensurable-frequency models

For hand calculations it is advantageous to use an even more artificial
grouping in which the frequencies are all *commensurable*, i.e. are all
multiples of the lowest frequency. This procedure reduces all the arithmetic
to operations on integers, which is very convenient for hand calculations,

although it offers little advantage in a machine calculation. All the energy levels of the molecule are now multiples of the energy corresponding to the lowest frequency, and the number of states at a given energy level is easily worked out, virtually by inspection. For example, Wieder and Marcus[18] used a model of the activated complex for ethyl chloride decomposition in which the seventeen frequencies correspond to energies per quantum of $N_A h c \bar{\nu}_i = 8\cdot4\,(4),\ 4\cdot0\,(5),\ 3\cdot6\,(2),\ 2\cdot8\,(4)$ and $0\cdot4\,(2)$ kcal mol^{-1}, the figures in parentheses being the number of vibrations with the given frequency. These energies are all multiples of $0\cdot4$ kcal mol^{-1}, which is now taken as the unit of energy, so that the energies per quantum are 21 (4), 10 (5), 9 (2), 7 (4) and 1 (2) units. The evaluation of the number of states at a given energy level is illustrated in Table 5.9 by the level

Table 5.9   Illustration of hand-count for commensurable frequencies

No. of oscillators	4	5	2	4	2	No. of states for given combination of quantum numbers
Quantum energy	21	10	9	7	1	
	0	0	0	0	20	$1\times1\times1\times1\times21$ = 21
	0	0	0	1	13	$1\times1\times1\times4\times14$ = 56
	0	0	0	2	6	$1\times1\times1\times10\times7$ = 70
Possible combinations of	0	0	1	0	11	$1\times1\times2\times1\times12$ = 24
numbers of quanta to	0	0	1	1	4	$1\times1\times2\times4\times5$ = 40
give a total energy of 20	0	0	2	0	2	$1\times1\times3\times1\times3$ = 9
units	0	1	0	0	10	$1\times5\times1\times1\times11$ = 55
	0	1	0	1	3	$1\times5\times1\times4\times4$ = 80
	0	1	1	0	1	$1\times5\times2\times1\times2$ = 20
	0	2	0	0	0	$1\times15\times1\times1\times1$ = 15

Total states at $8\cdot0$ kcal mol^{-1} = 390

$8\cdot0$ kcal mol^{-1} = 20 units of $0\cdot4$ kcal mol^{-1}. The possible combinations of the number of quanta of the different frequencies are first written down by systematic inspection, then the number of states corresponding to each combination can be evaluated from the usual combinatorial expression given in section 5.3.1. The total degeneracy of the given energy level is then obtained by summing the contributions from all the possible combinations of quantum numbers. A more systematic application of the same basic idea obtains the degeneracies as the coefficients of the terms in a multinomial series in which each term represents a certain energy level.[19] This technique has also been extended to form the basis of procedures for handling the cases of non-commensurable frequencies[10] and anharmonic vibrations.[20]

5.3.4 Conclusions on the direct-count method

In summary, computer calculations are capable of giving an exact count of the vibrational states of small molecules and of simplified models of larger molecules at energies which are not too high. In practice, this often means that the methods described above are suitable for the determination of $\sum P(E_v^+)$ for thermally energized systems, since the non-fixed energies of the great majority of activated complexes are not very high (e.g. < 30 kcal mol^{-1}; see section 6.4 for an example). Even hand calculations may be feasible if the model involves commensurable frequencies, which may not involve too great a loss of realism. On the other hand, it is unlikely that the direct count will be a practicable way of determining $N^*(E^*)$ in most cases. The energies of interest are much higher here ($E^* = E^+ + E_0$) and the labour or expense of computing the vast numbers of states involved will generally be prohibitive. It therefore appears that there is a great need for less laborious methods for determining $N^*(E^*)$, and the development of these is dealt with in the following sections.

5.4 CLASSICAL TREATMENT OF VIBRATIONAL STATES, AND DERIVED SEMICLASSICAL APPROXIMATIONS

A natural approach to the derivation of simple expressions for the density and sum of vibrational states is to take the well-known expressions for classical harmonic oscillators (section 5.4.1) and introduce modifications designed to overcome the failings of these equations when applied to quantized vibrations. Such modifications lead to what might generally be called 'semiclassical' approximations, although the term is often used to refer to one specific expression of this type, developed in section 5.4.2. In section 5.4.3 the more satisfactory Whitten–Rabinovitch formulation is discussed, and its extension to vibrational–rotational systems appears in section 5.4.4.

5.4.1 Treatment of classical harmonic vibrations[21, 22]

The simplest approximation to the sum and density of vibrational states is obtained by treating the molecule as a set of classical harmonic oscillators, and as in the previous treatment of rotational states (section 5.2) the classical behaviour is most easily evaluated as a limiting case of the quantized behaviour.

For a set of s quantum oscillators of frequency v_i ($i = 1$ to s), each possible combination of the quantum numbers v_i gives rise to one quantum state only and the total number of states at all energies up to and including E_v is therefore given by

$$\sum_{E_v=0}^{E_v} P(E_v) = \sum_{v_s=0} \sum_{v_{s-1}=0} \cdots \sum_{v_1=0} 1 \tag{5.26}$$

where the summation is taken over all possible v_i such that the total vibrational energy is less than or equal to E_v:

$$\sum_{i=1}^{s} v_i h\nu_i \leqslant E_v$$

In the classical approximation this is replaced by a multiple integral:

$$\sum_{E_v=0}^{E_v} P(E_v) \approx \int_{v_s=0}^{v_s'} \int_{v_{s-1}=0}^{v_{s-1}'} \cdots \int_{v_1=0}^{v_1'} d v_1 \, d v_2 \ldots d v_s$$

in which the upper limits v_n' are specified in a similar way to those for the rotational case (section 5.2.2):

$$(n = s) \quad v_s' h\nu_s = E_v$$

$$(n \neq s) \quad v_n' h\nu_n = E_v - \sum_{i=n+1}^{s} v_i h\nu_i$$

The integral is similar to, but less complicated than, that for the rotational case, and is evaluated without further approximation in Appendix 6 with the result (5.27). Here, and subsequently, the product $\prod h\nu_i$ is over all the

$$W(E_v) \equiv \sum_{E_v=0}^{E_v} P(E_v) \approx \frac{E_v^s}{s! \prod h\nu_i} \tag{5.27}$$

vibrational degrees of freedom involved, i.e.

$$\prod h\nu_i \equiv \prod_{i=1}^{s} h\nu_i$$

The density of quantum states is obtained, as before, by differentiation:

$$N(E_v) = \frac{d}{dE_v} \sum P(E_v) = \frac{E_v^{s-1}}{(s-1)! \prod h\nu_i} \tag{5.28}$$

The validity of this classical approximation is indicated by some of the results for $\sum P(E_v)$ in Tables 5.10 and 5.11, which compare the classical results from (5.27) with the results of direct counts at various energies for a fairly small molecule ($CHCl_3$) and a larger molecule (cyclopropane). The other figures in these tables will be referred to later. It is clear from these results that the classical approximation is poor at low energies, although the discrepancies are less at the higher energies as would be expected from the Bohr Correspondence Principle, which implies that in the limit of high quantum numbers the classical and quantum theories become essentially equivalent.[23] The error at low energies is particularly clear in the case of a

Table 5.10 $\sum\limits_{E_v=0}^{E_v} P(E_v)$ for $CHCl_3$ by various methods (frequencies as in Table 5.5; $E_z = 12{\cdot}171$ kcal mol^{-1})

E_v/E_z	Classical (5.27)	Marcus–Rice (5.29)	Whitten–Rabinovitch (5.32)	Exact count
1·00	$0{\cdot}030 \times 10^3$	$15{\cdot}5 \times 10^3$	$8{\cdot}18 \times 10^3$	$8{\cdot}20 \times 10^3$
2·00	$0{\cdot}155 \times 10^5$	$5{\cdot}97 \times 10^5$	$4{\cdot}58 \times 10^5$	$4{\cdot}57 \times 10^5$
3·00	$0{\cdot}597 \times 10^6$	$7{\cdot}94 \times 10^6$	$6{\cdot}87 \times 10^6$	$6{\cdot}85 \times 10^6$
5·00	$0{\cdot}592 \times 10^8$	$3{\cdot}05 \times 10^8$	$2{\cdot}87 \times 10^8$	$2{\cdot}86 \times 10^8$
8·00	$0{\cdot}407 \times 10^{10}$	$1{\cdot}17 \times 10^{10}$	$1{\cdot}14 \times 10^{10}$	$1{\cdot}14 \times 10^{10}$

Table 5.11 $\sum\limits_{E_v=0}^{E_v} P(E_v)$ for cyclopropane by various methods [21 frequencies grouped[10] as 3221 (6), 1478 (3), 1118 (7), 878 (3), 749 (2) cm^{-1}; $E_z = 51{\cdot}061$ kcal mol^{-1}]

$E_v/\text{kcal mol}^{-1}$	Classical (5.27)	Marcus–Rice (5.29)	Whitten–Rabinovitch (5.32)	Exact count (ref. 10)
10	0·00	545×10^2	$7{\cdot}17 \times 10^2$	$8{\cdot}02 \times 10^2$
30	0·02	$20{\cdot}9 \times 10^6$	$2{\cdot}65 \times 10^6$	$2{\cdot}69 \times 10^6$
50	$8{\cdot}20 \times 10^2$	$21{\cdot}5 \times 10^8$	$6{\cdot}15 \times 10^8$	$6{\cdot}12 \times 10^8$
100	$1{\cdot}72 \times 10^9$	$9{\cdot}94 \times 10^{12}$	$5{\cdot}90 \times 10^{12}$	$5{\cdot}84 \times 10^{12}$
150	$8{\cdot}58 \times 10^{12}$	$4{\cdot}03 \times 10^{15}$	$3{\cdot}02 \times 10^{15}$	$3{\cdot}00 \times 10^{15}$
200	$3{\cdot}61 \times 10^{15}$	$4{\cdot}27 \times 10^{17}$	$3{\cdot}56 \times 10^{17}$	$3{\cdot}54 \times 10^{17}$

simple system containing s degenerate oscillators with the same frequency ν. At energy levels of 0 and $h\nu$ the classical expression predicts $\sum P(E_v) = 0$ and $1/s!$ respectively, whereas the correct values are 1 and $s+1$ respectively since there is one quantum state at $E_v = 0$ and there are s states at $E_v = h\nu$. It can be shown, in fact, that this approximation must always underestimate the number of states at a given energy level,[22] and although the error decreases with increasing energy it does not become small until impracticably high energies are reached. Better approximations are therefore required, and the development of these is discussed in the following sections.

It can easily be shown that insertion of the classical expressions (5.27) and (5.28) into the RRKM equations leads to Hinshelwood's expression for k_1/k_2 and to the widely used classical Kassel formula for $k_a(E)$. It follows that the accuracy of these results is limited by that of the implied classical treatment of the molecular vibrations.

5.4.2   Semiclassical (Marcus–Rice) approximation

Marcus and Rice[24] suggested that since a quantized oscillator with a non-fixed energy E_v has a total vibrational energy above the classical ground state of $E_v + E_z$ (where E_z is the zero-point energy, $\frac{1}{2}hv$), it should be compared with a classical oscillator at energy $(E_v + E_z)$ rather than E_v. Thus the classical expressions (5.27) and (5.28) were modified to give the 'semiclassical' or Marcus–Rice expressions (5.29) and (5.30). Tables 5.10 and 5.11 include some values of $\sum P(E_v)$ calculated from (5.29) for two

$$W(E_v) \equiv \sum_{E_v=0}^{E_v} P(E_v) = \frac{(E_v + E_z)^s}{s! \prod hv_i} \tag{5.29}$$

$$N(E_v) = \frac{(E_v + E_z)^{s-1}}{(s-1)! \prod hv_i} \tag{5.30}$$

where

$$E_z = \sum_{i=1}^{s} (\tfrac{1}{2}hv_i) \tag{5.31}$$

typical molecules, and the results confirm that this approach gives a more accurate approximation than the classical expression (5.27), although the number of states at any level is now always *over*-estimated. This semiclassical approach has occasionally been applied to the lower-frequency (and therefore more highly excited) vibrations of a molecule, while using an exact count for the highest-frequency (C—H) vibrations grouped at their geometric mean.[18] This approach is an improvement, but is still far from accurate.[25]

5.4.3   Whitten–Rabinovitch approximation

Although the Marcus–Rice expression is not in itself highly accurate, it leads directly to the development of a greatly improved semiclassical type of approach to be described in the present section. Rabinovitch and Diesen suggested[26] that only an appropriate fraction of the zero-point energy should be added to E_v; thus an empirical factor a was introduced into

(5.29) and (5.30), giving the expressions (5.32) and (5.33). The factor a

$$W(E_v) \equiv \sum_{E_v=0}^{E_v} P(E_v) = \frac{(E_v + aE_z)^s}{s! \prod h\nu_i} \tag{5.32}$$

$$N(E_v) = \frac{(E_v + aE_z)^{s-1}}{(s-1)! \prod h\nu_i} \tag{5.33}$$

has a value between 0 and 1, and by comparison of the right-hand side of (5.32) with the result of direct counts it is possible to determine the precise

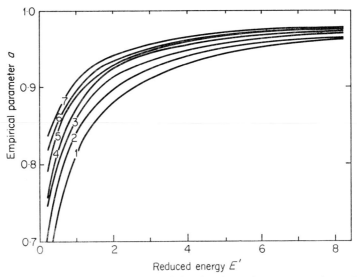

Figure 5.3 Values of a required for (5.32) to give correct values of $W(E_v)$ at various reduced energies for the molecules listed in Table 5.12

value of a required to give the correct result for a given molecule at any energy level. These values have been tabulated quite extensively,[11, 15, 26-28] and the importance of this approach depends on the consequent ability to predict suitable values of a for other cases.

Thus Figure 5.3[11] shows the typical variation of a with energy for a number of molecules.† The values always increase with increasing energy,

† The plots in Figure 5.3 are actually smoothed versions of a sawtooth type of variation. This behaviour arises because (5.32) predicts a smooth variation of $\sum P(E_v)$ with energy, whereas the actual variation is stepwise (e.g. Figure 4.7) and can therefore only be reproduced by having a stepwise variation of a with energy. Such effects are generally negligible for $E' > 1$, i.e. at energies where the density of states is high and $\sum P(E_v)$ increases fairly smoothly with increasing energy (see section 4.11).

approaching unity at high energies, and the values for different molecules lie in a fairly narrow band when plotted (as in Figure 5.3) against a reduced energy $E' = E_v/E_z$ rather than against E_v itself:

$$E' = E_v/E_z \qquad (5.34)$$

Furthermore, the position in this band of the curve for a particular molecule is found empirically to depend only on the distribution of the molecular vibration frequencies; the values of a are lower for a molecule whose vibrations have widely different frequencies than for one containing many vibrations of similar frequency. This dispersion of the frequencies can be measured by their standard deviation, and Whitten and Rabinovitch[11] define a closely related *modified frequency dispersion parameter* β by (5.35), in which $\langle v \rangle$ and $\langle v^2 \rangle$ are the mean frequency and

$$\beta = \frac{s-1}{s} \frac{\langle v^2 \rangle}{\langle v \rangle^2} \qquad (5.35)$$

mean-square frequency of the molecule, respectively. The value of the parameter β varies from about unity for molecules with largely similar frequencies, to about two for molecules with a wide spread of frequencies. For example, SF_6 with reciprocal wavelengths of 965 (3), 772, 642 (2), 617 (3), 525 (3) and 370 (3) cm^{-1} has $\beta = 1 \cdot 03$, while $Ni(CO)_4$ with $\bar{v}_i = 2057$ (3), 2040, 545 (3), 460 (2), 422 (3), 381, 300 (3) and 79 (5) cm^{-1} has $\beta = 2 \cdot 04$. Table 5.12 shows the values of β for the molecules under

Table 5.12   Modified frequency dispersion parameters β for the molecules in Figure 5.3

Curve in Figure 5.3	Molecule	β
1	$Ni(CO)_4$	2·04
2	$CHBr_3$, C_4H_2	1·76, 1·72
3	$(CN)_2$, $C_2H_2Cl_2$	1·47, 1·45
4	C_5H_5N, $\overline{(CH_2)_3CO}$	1·35, 1·34
5	c-C_3D_6, $(CH_2)_2O$	1·23, 1·22
6	CH_3F, B_2H_6	1·05, 1·15
7	SF_6	1·03

consideration in Figure 5.3, and illustrates the very good correlation of the relative positions of the various curves with the values of β for the molecules concerned.

It remains to discover a simple analytical relationship between a, β and E', and Whitten and Rabinovitch achieve this by writing

$$a = 1 - \beta . w(E') \tag{5.36}$$

where $w(E')$ is a unique function of E' shown in Figure 5.4 and described adequately by equations (5.37).

$$\left.\begin{array}{ll}(0\cdot1 < E' < 1\cdot0) & w = (5\cdot00E' + 2\cdot73E'^{0\cdot5} + 3\cdot51)^{-1} \\[2mm] (1\cdot0 < E' < 8\cdot0) & w = \exp(-2\cdot4191E'^{0\cdot25})\end{array}\right\} \tag{5.37}$$

This completes what we shall call the Whitten–Rabinovitch approximation for $\sum P(E_v)$, and the method of calculation can be summarized as

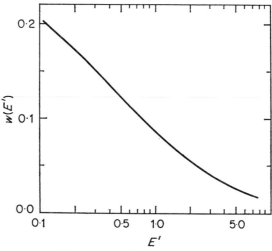

Figure 5.4   The function $w(E')$

follows. The molecular vibration frequencies are used to calculate β (5.35) and E_z (5.31), then for any non-fixed energy E_v of interest, the reduced energy $E' = E_v/E_z$ is used to obtain $w(E')$ (5.37), a (5.36), and hence $\sum P(E_v)$ (5.32). The density $N(E_v)$ may be obtained directly from (5.33) or, more accurately,[29] from the differential of (5.32), which is:

$$N(E_v) = \frac{(E_v + aE_z)^{s-1}}{(s-1)! \prod h\nu_i}\left[1 - \beta\left(\frac{dw}{dE'}\right)\right] \tag{5.38}$$

where dw/dE' is easily obtained as (5.39) by differentiation of (5.37).

$$\left.\begin{array}{ll}(0\cdot1 < E' < 1\cdot0) & dw/dE' = -(5\cdot00 + 1\cdot365E'^{-0\cdot5})\,w^2 \\[2mm] (1\cdot0 < E' < 8\cdot0) & dw/dE' = -(0\cdot60478E'^{-0\cdot75})\,w\end{array}\right\} \tag{5.39}$$

The Whitten–Rabinovitch approximation has been extensively tested by various authors, some typical results being shown in Tables 5.10 and

5.11 and Figure 5.5. The results are always in very good agreement with the direct count for non-fixed energies E_v greater than the zero-point energy E_z, and often for much lower energies. This is well illustrated by

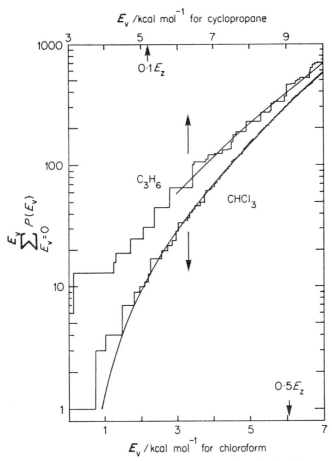

Figure 5.5 Comparison of Whitten–Rabinovitch $\sum P(E_v)$ [equation (5.32)] with exact counts for chloroform (lower curves) and a grouped-frequency model of cyclopropane[10] (upper curves). In each case the smooth curve represents (5.32)

Figure 5.6 (from Whitten and Rabinovitch[11]), the important feature of which is the negligible error at energies above $E' = 1$;† some other comparisons are noted in section 5.6. The errors at low energies can be

---

† The irregular variation at low energies arises from the stepwise variation of the exact count (compare Figure 5.5) and the detail of the plots is in part an artefact of the arbitrary energy intervals for which the exact counts were performed.

avoided to some extent by using a 'truncated model'. This means that vibrations for which the quantum energy $h\nu_i$ is greater than the available energy E_v, and which therefore cannot be excited, are removed from the

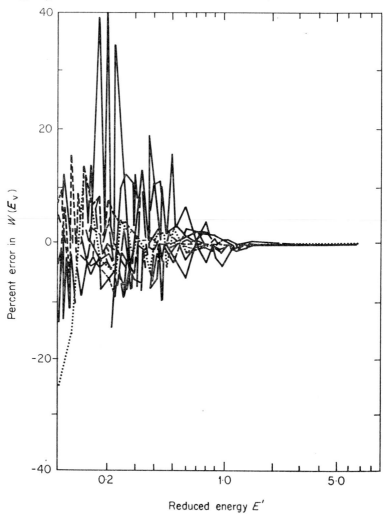

Figure 5.6 Error of the Whitten–Rabinovitch approximation at various reduced energies for the molecules listed in Table 5.12

model. Thus SO_2, with $s = 3$ and frequencies of 1361, 1151 and 524 cm^{-1}, would be treated as such for $E_v > 1361$ cm^{-1}, but at energies between 1361 and 1151 cm^{-1} it would be treated as two vibrations with frequencies of 1151 and 524 cm^{-1}, and between 1151 and 524 cm^{-1} as a single vibration of frequency 524 cm^{-1}. Below 524 cm^{-1}, of course, $\sum P(E_v) = 1$. Although this treatment does improve the Whitten–Rabinovitch approximation at low energies, it is scarcely worth while, since the errors are more easily avoided by carrying out a direct count up to an energy at which it agrees with the approximation (see section 5.6). The direct count up to this level should generally be accomplished without too much labour or expense.

Particular difficulties may be encountered with very large molecules ($s > 30$) at low energies, and Whitten[9, 30] has described a more accurate method of handling this situation. The high-frequency and medium-frequency vibrations are treated by direct count, but the low frequencies (e.g. < 700 cm^{-1}) are treated as a separate unit, and their combined degeneracy is evaluated by the Whitten–Rabinovitch method provided the energy in this part of the molecule is sufficiently high; otherwise the direct count is used for this part as well.

Other approaches of a semiclassical nature have also been suggested. In particular, Schlag and Sandsmark[10] obtained a useful approximation by manipulating the exact combinatorial expressions with suitable approximations; the result was a three-term expression in which the first term was the Marcus–Rice expression (5.29). Expressions derived by the inverse Laplace transformation technique (see section 5.5) can also often be approximated into a semiclassical form. Wahrhaftig and co-workers[22, 31] developed a method in which an exact count is used for any oscillators in their ground state and a semiclassical type of approach is used for excited oscillators, but the results are still inferior to those of a number of other approaches.[32]

The outstanding advantage of the Whitten–Rabinovitch approximation is its simplicity and ease of application. Its accuracy is good and compares favourably with those of other approaches, and the method is thus likely to find wide application for harmonic oscillator models (including those with rotational degrees of freedom—see next section).

5.4.4 Whitten–Rabinovitch treatment of vibrational–rotational systems

When there are active rotations to be taken into account, $N(E_{vr})$ and $\sum P(E_{vr})$ could be obtained by numerical integration of (5.4) and the continuous equivalent of (5.2). It is much simpler to use a single composite analytical expression, however, and Whitten and Rabinovitch have suggested (5.40), in which $a = 1 - \beta w(E_{vr}/E_z)$, and β is extended from

$$W(E_{vr}) \equiv \sum_{E_{vr}=0}^{E_{vr}} P(E_{vr}) = \frac{Q_r}{(kT)^{\frac{1}{2}r}\,\Gamma(1+s+\frac{1}{2}r)\prod hv_i}(E_{vr}+aE_z)^{s+\frac{1}{2}r} \quad (5.40)$$

(5.35), now becoming (5.41). The expression (5.40) reduces exactly to

$$\beta = \frac{s-1}{s} \frac{s+\frac{1}{2}r}{s} \frac{\langle v^2 \rangle}{\langle v \rangle^2} \qquad (5.41)$$

(5.32) if $r = 0$ and to (5.15) if $s = 0$ (and thus $E_z = 0$), but is otherwise an approximation to the true sum obtained from these two expressions. Whitten obtained (5.40) by replacing the lower limit in the integral

$$\int_{x = aE_z/(E_{vr}+aE_z)}^{1} x^{s-1}(1-x)^{\frac{1}{2}r} \, dx$$

by zero and replacing (5.35) by (5.41). It is also obtained by treating the r rotational degrees of freedom with partition function Q_r in the same way as $\frac{1}{2}r$ vibrations with the same average zero-point energy as the genuine vibrations of the molecule, and with the product $\prod h\nu_i$ set equal to $(kT)^{\frac{1}{2}r}/Q_r$. The change in the definition of β was apparently an arbitrary step found to give improved results in some test cases. The density of vibrational–rotational states is similarly obtained as (5.42) by differentiation of (5.40).

$$N(E_{vr}) = \frac{Q_r(E_{vr}+aE_z)^{s+\frac{1}{2}r-1}}{(kT)^{\frac{1}{2}r}\,\Gamma(s+\frac{1}{2}r)\prod h\nu_i}\left[1-\beta\left(\frac{dw}{dE'}\right)\right] \qquad (5.42)$$

The resulting equations (5.40) and (5.42) have been tested in a number of cases,[8, 29] and appear to be good approximations for a wide range of models (see section 5.6). Forst and Prášil[32] found a slightly modified version of (5.42) to be equally satisfactory.

5.5   INVERSE LAPLACE TRANSFORMATION OF THE PARTITION FUNCTION

The approach to be described in this section is more fundamentally based than the semiclassical type of treatment described above, and can lead, at least in principle, to precise expressions for the required density and sum of states. The treatment starts with the observation that the partition function Q for any system can be regarded as the Laplace transform[33] (5.43) of the energy-level density $N(E)$, using the transform parameter

$$Q(\beta) = \mathcal{L}\{N(E)\} = \int_0^\infty N(E)\exp(-\beta E)\,dE \qquad (5.43)$$

$\beta = 1/kT$.† The partition function is written $Q(\beta)$ to emphasize its

† The β in this section has no connection whatsoever with the Whitten–Rabinovitch parameter β.

functional dependence on β. It is written as an integral rather than a sum, so that the case of quantized rather than continuous energy levels must be dealt with if necessary by expressing $N(E)$ as a series of delta-functions (see section 4.11.2). The essential point of the present treatment now follows, that if the partition function can be written down as a function of β then it is possible, at least in principle, to obtain $N(E)$ by means of the inverse Laplace transformation (5.44). The number of states at all energies up to and including E is similarly given by (5.45). Most of the

$$N(E) = \mathscr{L}^{-1}\{Q(\beta)\} \qquad (5.44)$$

$$W(E) \equiv \sum_{E_n=0}^{E} P(E_n) = \mathscr{L}^{-1}\{Q(\beta)/\beta\} \qquad (5.45)$$

work in this area has in fact been directed in the first instance at the evaluation of $W(E)$, but either quantity follows easily from the other by virtue of (4.44). In useful cases the inverse transformations present a difficult problem which can rarely be solved exactly, but progress can be made in three ways which are described in the following sections.

5.5.1   Direct inversion of an approximate partition function

The technique was first applied to the present problem by Haarhoff,[34] and may be illustrated by his treatment of a collection of s quantized harmonic oscillators with frequencies ν_i. The partition function for this system is given by the well-known expression (5.46). This expression (even

$$Q(\beta) = \prod_{i=1}^{s} \left\{ \frac{\exp\left(\tfrac{1}{2}h\nu_i\right)}{1 - \exp\left(-h\nu_i\beta\right)} \right\} \qquad (5.46)$$

for a single oscillator) is not readily subjected to inverse transformation in its exponential form. It may, however, be expressed as a power series in β; it emerges that only alternate powers appear, and the appropriate series for $Q(\beta)/\beta$ is (5.47). In this series the a_n are constants involving the

$$\frac{Q(\beta)}{\beta} = \left(\prod_{i=1}^{s} h\nu_i\right)^{-1} \left(\frac{1}{\beta^{s+1}} - \frac{a_2}{\beta^{s-1}} + \frac{a_4}{\beta^{s-3}} - \frac{a_6}{\beta^{s-5}} + \ldots\right) \qquad (5.47)$$

frequencies ν_i together with numerical factors; for example

$$a_2 = (\langle \nu^2 \rangle / \langle \nu \rangle^2)\, E_z^2 / 6s$$

Haarhoff[34] gave detailed expressions for the first five terms in (5.47); their complexity increases rapidly, but the general term can be obtained from Thiele's formulation[35] in terms of the so-called 'Bernoulli numbers'.

The inverse transformation of $Q(\beta)/\beta$ is now considered term by term; it is a property of Laplace transformations that this is valid, i.e. that

$$\mathcal{L}^{-1}\{af(\beta)+bg(\beta)\} = a\mathcal{L}^{-1}\{f(\beta)\}+b\mathcal{L}^{-1}\{g(\beta)\}$$

where f and g are functions of β and a and b are constants. The infinite series in (5.47) consists of a limited number of terms having negative powers of β, after which all other terms have zero or positive powers of β. There are in fact $\frac{1}{2}s+1$ or $\frac{1}{2}(s+1)$ terms with negative powers of β according to whether s is even or odd, and the inverse transformation of these terms is readily accomplished by means of the standard equation (5.48).

$$\mathcal{L}^{-1}(\beta^{-m}) = E^{m-1}/(m-1)! \quad (m = \text{positive integer}) \qquad (5.48)$$

Since (5.46) is based on energies measured from the classical energy zero, the result (5.49) emerges in terms of energies on the same scale, i.e. E_v+E_z rather than E_v. The transformation (5.48) is not valid for $m \leqslant 0$,

$$W(E_v) = \mathcal{L}^{-1}\{Q(\beta)/\beta\} = \left(\prod h\nu_i\right)^{-1}\left[\frac{(E_v+E_z)^s}{s!} - \frac{a_2(E_v+E_z)^{s-2}}{(s-2)!} + \ldots\right]$$

$$= \frac{(E_v+E_z)^s}{s!\,\prod h\nu_i}\left[1 - \frac{a_2'}{(E_v+E_z)^2} + \frac{a_4'}{(E_v+E_z)^4} - \ldots\right]$$

$$[\tfrac{1}{2}s+1 \text{ or } \tfrac{1}{2}(s+1) \text{ terms}] \qquad (5.49)$$

but it is known that the inverse transformation of the remaining terms in (5.47) will produce a set of stepped functions of just the type required to reproduce the actual stepwise variation of $W(E_v)$ in a quantized system (see, for example, section 4.11). The terms in (5.49) represent a smooth-curve basis on which the stepped functions are superimposed, and Haarhoff's approximation assumes that this finite series provides a reasonable smooth-curve approximation to $W(E_v)$.

It is interesting to note that the first term in (5.49) is just the semiclassical expression (5.29), so that even this one term is a substantial improvement over the classical approximation. The second term in the brackets is given in detail by (5.50), which provides an interesting connection[35]

$$-\frac{a_2'}{(E_v+E_z)^2} = -\frac{s}{6}\frac{\langle \nu^2\rangle}{\langle \nu\rangle^2}\left(\frac{1}{1+E_v/E_z}\right)^2 \qquad (5.50)$$

with the deduction of Whitten and Rabinovitch[11] that the negative correction needed to the Marcus–Rice expression was a more-or-less universal function of the frequency dispersion parameter (5.35) and the reduced energy $E' = E_v/E_z$.

Haarhoff's expression (5.49)[34] has been found very satisfactory in practice by a number of workers, particularly when the full series as developed by Thiele[35] is used.[9, 30, 34, 36, 44] Haarhoff also derived the analytical equation (5.51) as an approximation to the series expression for

$$N(E_{\mathrm{v}}) = \left(\frac{2}{\pi s}\right)^{\frac{1}{2}} \frac{(1 - 1/12s)\,\lambda}{h\langle\nu\rangle(1+\eta)} \left[(1 + \tfrac{1}{2}\eta)(1 + 2/\eta)^{\frac{1}{2}\eta}\right]^{s}[1 - 1/(1+\eta)^{2}]^{\beta_{0}} \quad (5.51)$$

where

$$\eta = 1 + E_{\mathrm{v}}/E_{\mathrm{z}}, \quad \lambda = \prod_{i=1}^{s}(\langle\nu\rangle/\nu_{i})$$

and

$$\beta_{0} = \frac{(s-1)(s-2)}{6s} \frac{\langle\nu^{2}\rangle}{\langle\nu\rangle^{2}} - \frac{s}{6}$$

the density of states,[37] and extended the same treatment to vibrational–rotational systems, while Forst et al.[38] extended the series treatment to vibrational–rotational systems to give an excellent approximation illustrated in section 5.6.

5.5.2 Inversion by complex integration: Method of residues

The subject matter of this and the next section is mathematically difficult, and only the main features can be dealt with here. Details can be found in many textbooks; see, for example, ref. 33. An alternative method of carrying out the inverse Laplace transformation (5.44) or (5.45) is by means of the complex integration formula (5.52). In this equation β is a complex variable ($\beta = \beta' + i\beta''$) and the integration is a line integral along a path parallel to the imaginary axis (Figure 5.7) and lying to the right of

$$\mathscr{L}^{-1}\{f(\beta)\} = \frac{1}{2\pi i} \int_{\gamma - i\infty}^{\gamma + i\infty} f(\beta)\exp(\beta E)\,\mathrm{d}\beta = \frac{1}{2\pi i} \int_{\gamma - i\infty}^{\gamma + i\infty} \phi(\beta)\,\mathrm{d}\beta \quad (5.52)$$

all singularities of the integrand $\phi(\beta) = f(\beta)\exp(\beta E)$. *Singularities* are points at which $\phi(\beta)$ is not analytic, i.e. does not satisfy the conditions of being single-valued and having a definite differential coefficient at the point β. There are different types of singularity; the type encountered in the present context are *poles*, defined as follows. If, in the region of a point a, $\phi(\beta)$ can be expressed in the form $g(\beta)/(\beta - a)^{n}$ [where $g(a) \neq 0$], then $\phi(\beta)$ is said to have a pole of order n at $\beta = a$.

The integration in (5.52) can be attempted in some direct manner (see next section) or by means of the *method of residues*. A residue is a property of a pole, and is simply the coefficient of the term $1/(\beta - a)$ in the Taylor-series expansion of $g(\beta)/(\beta - a)^{n}$ about the point $\beta = a$. The method of residues applies the established theorem that the integral of a function $\phi(\beta)$ along a line such as that in (5.52) is equal to $2\pi i$ times the sum of the residues of $\phi(\beta)$ at all its poles. To obtain an expression for $W(E)$ we thus put $\phi(\beta) = Q(\beta)\exp(\beta E)/\beta$ [cf. (5.45)], and obtain the result (5.53); alternatively for $N(E)$ the residues of $Q(\beta)\exp(\beta E)$ are required [cf. (5.44)].

$$W(E) = \text{sum of residues of } Q(\beta)\exp(\beta E)/\beta \text{ at all poles of this function in the plane of complex } \beta. \tag{5.53}$$

The application of this approach is illustrated in a simple manner by Thiele's treatment[35] of a single harmonic oscillator. The partition function

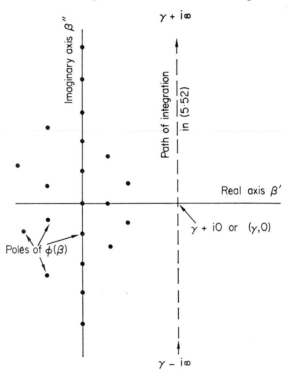

Figure 5.7 Illustration of the complex integration formula (5.52)

is (5.46) with $s = 1$, and $W(E_v)$ is therefore given by (5.54). On series expansion it is found that the poles are all on the imaginary axis, at

$$W(E_v) = \text{sum of residues at poles of } \frac{\exp(\tfrac{1}{2}h\nu\beta)\exp[\beta(E_v+E_z)]}{[1-\exp(-\beta h\nu)]\beta} \tag{5.54}$$

$\beta = (2\pi n i/h\nu)$ $(n = 0, \pm1, \pm2, ...)$, and substitution of these values gives (5.55) in which the first term is the residue at the origin ($\beta = 0$) and the other terms represent the other poles. The behaviour of this expression is illustrated in Figure 5.8; the first term gives a smooth approximation to

$$W(E_v) = \frac{(E_v + E_z)}{hv} + \sum_{n=1}^{\infty} \frac{(-1)^n}{\pi n} \sin\left[\frac{2\pi n(E_v + E_z)}{hv}\right] \tag{5.55}$$

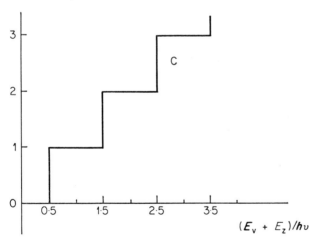

Figure 5.8   Illustration of equation (5.55);
A = variation of $(E_v + E_z)/hv$ term,
B = variation of $\sum [(-1)^n/\pi n] \sin [2\pi n(E_v + E_z)/hv]$,
C = sum of A and B

6

the true curve, while the other terms are the Fourier series giving a saw-tooth function which converts the smooth curve into the stepped variation appropriate to the quantized oscillator.

The assumption can now be made that for any system the residue of $Q(\beta) \exp(\beta E)/\beta$ at $\beta = 0$ will give a reasonable smooth-curve approximation for $W(E)$. Thiele[35] applied the technique to (5.46) and obtained a generalized form of the series (5.49) which Haarhoff derived by direct transformation. It will be seen in section 5.6 that the result gives a very good fit to exact-count data, even at quite low energies.

5.5.3 Evaluation of complex inversion integral by method of steepest descent[39]

The second method of obtaining an approximation to the inversion integral (5.52) is based on the theorem that a line integral of this type has the same value for *any* line joining two given points. Thus the path of integration between $\gamma - i\infty$ and $\gamma + i\infty$ may be chosen at will. Now the value of the integrand is small at the ends of the line but passes through larger values near the real axis, and it is this region of large values which contributes most to the integral. Along the real axis the integrand is infinite at the origin and at $\beta = \infty$, with finite values in between. There is thus a saddle at some point β^* on the real axis, through which the path of integration may be made to pass. If it crosses the saddle-point in the direction in which it falls away most steeply, a reasonable approximation to the integral can be made by expanding the integrand about the saddle-point (5.56) and integrating along this path of steepest descent, putting

$$\phi(\beta) = \phi(\beta^*) + (\beta - \beta^*)\,\phi'(\beta^*) + \tfrac{1}{2}(\beta - \beta^*)^2\,\phi''(\beta^*) + \dots \tag{5.56}$$

where $\phi(\beta)$ is the integrand of (5.52), $\phi'(\beta) = d\phi(\beta)/d\beta$, etc.

$\phi'(\beta) = 0$ (saddle-point) and ignoring terms beyond the second derivative.

This technique was originally applied to the present problem by Lin and Eyring[40] for the specific case of harmonic oscillators, and was extended to anharmonic oscillators and to the vibrational–rotational case by Tou and Lin.[41] There is, however, a certain artificiality in these treatments,[42] and a more powerful generalization of the method has recently been developed by Forst and Prášil[32, 46] and independently by Hoare and Ruijgrok.[42, 43] These workers showed that in the first-order approximation described above, the energy-level density $N(E)$ is given for any system by (5.57), in which β^* is the value of β which satisfies (5.58). The sum $W(E)$

$$N(E) \approx Q(\beta^*) \exp(\beta^* E)\,[2\pi(\partial^2 \ln Q/\partial\beta^2)_{\beta=\beta^*}]^{-\frac{1}{2}}$$

$$\text{or} \qquad N(E) \approx Q(\beta^*) \exp(\beta^* E)\,[2\pi C_v(\beta^*)/k\beta^{*2}]^{-\frac{1}{2}} \tag{5.57}$$

where $\qquad \left(\dfrac{\partial \ln Q}{\partial \beta}\right)_{\beta=\beta^*} = -E$ $\qquad$ (5.58)

and C_v is the specific heat at constant volume for the degrees of freedom under consideration

may be obtained accurately by numerical integration of $N(E)$, but in the same first-order approximation is given more simply by (5.59).

$$W(E) \approx N(E)/\beta^* \qquad (5.59)$$

The application of these equations is fairly straightforward. One has simply to write an expression for the partition function Q of the molecular system involved, and obtain the first and second differentials with respect to β ($= 1/kT$). For any energy of interest β^* is then determined from (5.58), and $N(E)$ and $W(E)$ follow from (5.57) and (5.59). An example is the case of s simple harmonic oscillators, for which the partition function is given by (5.46), and hence the differentials are given by (5.60) and (5.61). Thus $N(E_\mathrm{v})$ is given by (5.62), in which β^* is the value of β which satisfies (5.63) for the given energy E_v.

$$\frac{\partial \ln Q}{\partial \beta} = \sum_{i=1}^{s} \left[\tfrac{1}{2}h\nu_i - \frac{h\nu_i}{\exp(h\nu_i\beta)-1} \right] \qquad (5.60)$$

$$\frac{\partial^2 \ln Q}{\partial \beta^2} = \sum_{i=1}^{s} \frac{(h\nu_i)^2 \exp(h\nu_i\beta)}{[\exp(h\nu_i\beta)-1]^2} \qquad (5.61)$$

$$N(E_\mathrm{v}) = Q(\beta^*) \exp\left[\beta^*(E_\mathrm{v}+E_\mathrm{z})\right] \left\{ 2\pi \sum_{i=1}^{s} \frac{(h\nu_i)^2 \exp(h\nu_i\beta^*)}{[\exp(h\nu_i\beta^*)-1]^2} \right\}^{-\frac{1}{2}} \qquad (5.62)$$

$$\sum_{i=1}^{s} \frac{h\nu_i}{\exp(h\nu_i\beta^*)-1} = E_\mathrm{v} - \sum_{i=1}^{s} \tfrac{1}{2}h\nu_i \qquad (5.63)$$

The model can be indefinitely complicated, provided only that an expression for Q can be written and that the differentials are well-behaved single-valued functions. The only slight difficulty in use is that β^* is in general an implicit rather than analytical solution of (5.58) and must be found by an iterative root-finding procedure. A very precise calculation of β^* is required,[32, 42] but even so there is no real difficulty with machine computation. The method has been applied with impressive results to harmonic oscillators and vibrational–rotational systems[32, 42, 46] and to anharmonic oscillators,[42, 46] Hoare[43] has extended the treatment to higher-order approximations and found that second-order corrections are significant but small, while third- and higher-order corrections are negligible. Lin and co-workers have similarly found that second-order corrections give a significant improvement in their steepest-descent formulation.[47] It seems likely that this approach will find considerable application, particularly in the investigation of more sophisticated molecular models than have hitherto been used in reaction-rate calculations.

5.6   COMPARISONS AND CONCLUSIONS

Many comparisons have been made of the various approximations for $N(E_{vr})$ and $W(E_{vr})$ $[\equiv \sum P(E_{vr})]$, and the overall picture is summarized here by reference to some of the calculations of Tou[36] for harmonic oscillators and of Forst and Prášil[32] for harmonic oscillators coupled with free rotations. The latter authors particularly have given a useful review of the various equations for the sum and density of states of vibrational-rotational systems, and pointed out a number of errors and inconsistencies in earlier papers.

Figures 5.9–5.12 show the percentage deviations of various expressions from exact count results for two vibrational models and two vibrational-rotational models. In each case a small molecule (5–7 vibrations) and a medium-sized molecule (15–21 vibrations) is considered. Not all expressions are represented on each plot, and detailed examination of minor differences is in any case unprofitable. The overall conclusions are quite apparent, however, and have been clearly stated by Forst and Prášil.[32] Good results are obtained by the methods of Haarhoff, Whitten and Rabinovitch (modified or not), by inverse transformation as a series (Thiele, or Forst, Prášil, and St. Laurent), or by the generalized steepest-descent method (Forst and Prášil, Hoare and Ruijgrok). The expressions of Wahrhaftig and co-workers are less accurate. The Whitten–Rabinovitch expressions are the simplest to program and use the least computer time; there is little to choose between the original versions and the modified forms of Forst and Prášil. The inverse transformation series are difficult to program and a little more time-consuming. The generalized steepest-descent method is relatively simple to program but requires more computer time, mainly for the iterative calculation of β^* to the necessary accuracy.

The energy at which any given approximation becomes adequate can be found if necessary for any individual case by comparison with exact-count figures. It is clear from Figures 5.9–5.12 that the better expressions are correct within a few per cent at $E_{vr} = E_z$ for the small molecules and at considerably lower *reduced* energies for the larger molecules (e.g. $E_{vr} = \frac{1}{2}E_z$). In view of its simplicity in use we suggest that for most purposes the Whitten–Rabinovitch treatment is preferable, being checked

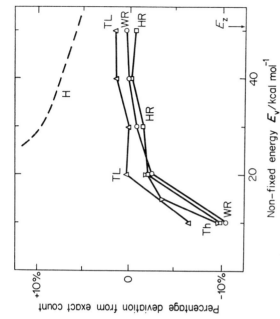

Figure 5.10 Percentage deviation from exact count[10] $W(E_v)$ for cyclopropane [3221 (6), 1478 (3), 1118 (7), 879 (3), 750 (2) cm^{-1}]: WR = Whitten–Rabinovitch expression;[45] Th = Thiele's series;[44] TL = Tou and Lin steepest descent;[41] HR = Hoare and Ruijgrok steepest descent;[42] H = Haarhoff's five-term series[36]

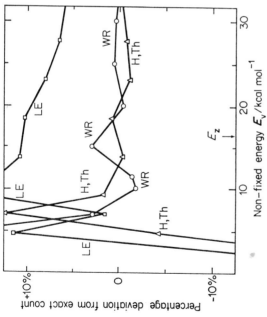

Figure 5.9 Percentage deviation from exact count[36] $W(E_v)$ for acetylene [3374, 3287, 1974, 729 (2), 612 (2) cm^{-1}]: H, Th = Haarhoff series and Thiele series (identical here);[36] LE = Lin and Eyring expression;[36] Vestal et al. is off-scale;[36] WR = Whitten–Rabinovitch expression[45]

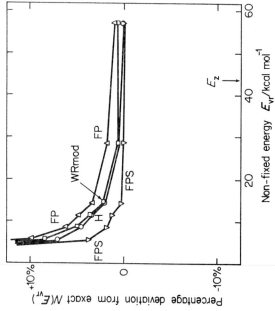

Figure 5.12 Percentage deviation from exact $N(E_{vr})$ for Forst and Prášil's Model B [3034 (6), 2000, 1415 (4), 983 (4) cm^{-1}] with four classical rotors; key as in Figure 5.11

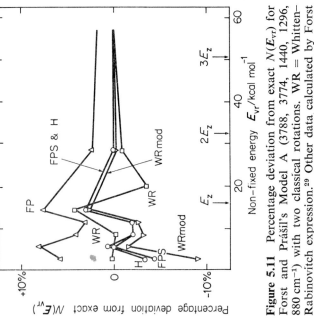

Figure 5.11 Percentage deviation from exact $N(E_{vr})$ for Forst and Prášil's Model A (3788, 3774, 1440, 1296, 880 cm^{-1}) with two classical rotations. WR = Whitten–Rabinovitch expression.[29] Other data calculated by Forst and Prášil:[32] H = Haarhoff algebraic expression; FPS = completed Haarhoff series; FP = generalized steepest descent; WRmod = slight modification of WR. For details see ref. 32

(and if necessary supplemented) by exact-count data if the energies of interest are very low. Very large molecules may need special treatment, however,[9, 30] and the generalized steepest-descent method seems to offer the greatest scope for development of models containing anharmonic vibrations and hindered or non-rigid rotations.

References

1. R. A. Marcus, *J. Chem. Phys.*, **20**, 359 (1952).
2. H. Eyring, D. Henderson, B. J. Stover and E. M. Eyring, *Statistical Mechanics and Dynamics*, Wiley, New York, 1964, p. 15.
3. T. M. Sugden and C. N. Kenney, *Microwave Spectroscopy of Gases*, Van Nostrand, London, 1965, p. 15.
4. Ref. 3, p. 18.
5. J. E. Wollrab, *Rotational Spectra and Molecular Structure*, Academic Press, New York, 1967, p. 13.
6. Ref. 3, p. 48; ref. 5, p. 15.
7. J. H. Current and B. S. Rabinovitch, *J. Chem. Phys.*, **38**, 783 (1963).
8. G. Z. Whitten and B. S. Rabinovitch, *J. Chem. Phys.*, **41**, 1883 (1964).
9. G. Z. Whitten, Ph.D. Thesis, University of Washington (1965).
10. E. W. Schlag and R. A. Sandsmark, *J. Chem. Phys.*, **37**, 168 (1962).
11. G. Z. Whitten and B. S. Rabinovitch, *J. Chem. Phys.*, **38**, 2466 (1963).
12. K. A. Wilde, *J. Chem. Phys.*, **41**, 448 (1964).
13. B. S. Rabinovitch and D. W. Setser, *Advan. Photochem.*, **3**, 1 (1964).
14. G. Herzberg, *Infrared and Raman Spectra of Polyatomic Molecules*, Van Nostrand, Princeton, 1945, p. 285.
15. B. S. Rabinovitch and J. H. Current, *J. Chem. Phys.*, **35**, 2250 (1961).
16. Refs. 11 and 44; the present authors can supply programs in Algol or Atlas Autocode.
17. For thermochemical data see S. W. Benson, *Thermochemical Kinetics*, Wiley, New York, 1968, pp. 195–204.
18. G. M. Wieder and R. A. Marcus, *J. Chem. Phys.*, **37**, 1835 (1962).
19. B. Steiner, C. F. Giese and M. G. Inghram, *J. Chem. Phys.*, **34**, 189 (1961); R. H. Fowler, *Statistical Mechanics*, Cambridge University Press, 2nd ed., 1936, § 2.3.
20. E. W. Schlag, R. A. Sandsmark and W. G. Valance, *J. Chem. Phys.*, **40**, 4461 (1964).
21. R. C. Tolman, *Foundations of Statistical Mechanics*, Oxford University Press, 1938, p. 492.
22. M. Vestal, A. L. Wahrhaftig and W. H. Johnston, *J. Chem. Phys.*, **37**, 1276 (1962); see comment in footnote 8 of ref. 32.
23. J. C. Slater, *Quantum Theory of Matter*, McGraw–Hill, New York, 1951.
24. R. A. Marcus and O. K. Rice, *J. Phys. and Colloid Chem.*, **55**, 894 (1951).
25. C. S. Elliott and H. M. Frey, *Trans. Faraday Soc.*, **62**, 895 (1966); H. M. Frey and B. M. Pope, *Trans. Faraday Soc.*, **65**, 441 (1968).
26. B. S. Rabinovitch and R. W. Diesen, *J. Chem. Phys.*, **30**, 735 (1959).
27. F. W. Schneider and B. S. Rabinovitch, *J. Amer. Chem. Soc.*, **84**, 4215 (1962).
28. R. W. Diesen, Ph.D. Thesis, University of Washington, 1958.

29. D. C. Tardy, B. S. Rabinovitch and G. Z. Whitten, *J. Chem. Phys.*, **48**, 1427 (1968).
30. M. J. Pearson, B. S. Rabinovitch and G. Z. Whitten, *J. Chem. Phys.*, **42**, 2470 (1965).
31. J. C. Tou and A. L. Wahrhaftig, *J. Phys. Chem.*, **72**, 3034 (1968); J. C. Tou, L. P. Hills and A. L. Wahrhaftig, *J. Chem. Phys.*, **45**, 2129 (1966).
32. W. Forst and Z. Prášil, *J. Chem. Phys.*, **51**, 3006 (1969).
33. J. C. Jaeger and G. H. Newstead, *An Introduction to the Laplace Transformation*, Methuen, London, 3rd ed., 1969; G. A. Korn and T. M. Korn, *Mathematical Handbook for Scientists and Engineers*, McGraw-Hill, New York, 1961.
34. P. C. Haarhoff, *Mol. Phys.*, **6**, 337 (1963).
35. E. Thiele, *J. Chem. Phys.*, **39**, 3258 (1963).
36. J. C. Tou, *J. Phys. Chem.*, **71**, 2721 (1967); the Haarhoff results in Table II are said to be erroneous.[32]
37. P. C. Haarhoff, *Mol. Phys.*, **7**, 101 (1963).
38. W. Forst, Z. Prášil and P. St. Laurent, *J. Chem. Phys.*, **46**, 3736 (1967).
39. Ref. 2, p. 161; R. H. Fowler and E. A. Guggenheim, *Statistical Thermodynamics*, Cambridge University Press, 1939, p. 34.
40. S. H. Lin and H. Eyring, *J. Chem. Phys.*, **39**, 1577 (1963); **43**, 2153 (1965).
41. J. C. Tou and S. H. Lin, *J. Chem. Phys.*, **49**, 4187 (1968); see comment in footnote 13 of ref. 32.
42. M. R. Hoare and Th. W. Ruijgrok, *J. Chem. Phys.*, **52**, 113 (1970).
43. M. R. Hoare, *J. Chem. Phys.*, **52**, 5695 (1970).
44. E. W. Schlag, R. A. Sandsmark and W. G. Valance, *J. Phys. Chem.*, **69**, 1431 (1965).
45. Calculated by present authors.
46. W. Forst and Z. Prášil, *J. Chem. Phys.*, **53**, 3065 (1970).
47. K. H. Lau and S. H. Lin, *J. Phys. Chem.*, **75**, 981 (1971); see also S. H. Lin and C. Y. Lin Ma, *Advan. Chem. Phys.*, **21**, 143 (1971).

6 Numerical Application of the RRKM Theory

The reader should now be almost in a position to carry out RRKM calculations, and a detailed numerical example of the working involved will be given in section 6.4. Before this, however, the necessary preliminary work will be discussed and the calculations will be described in a more general way. Thus section 6.1 describes the calculation of the high-pressure activation parameters $(\Delta S^{\ddagger}, A_\infty, E_\infty, E_0)$ resulting from any postulated model of the reaction. In section 6.2 the selection of models is discussed in detail, and in section 6.3 a suitable technique for carrying out the RRKM integration is described. Section 6.4 is a detailed illustration of the application of the RRKM theory, using the isomerization of 1,1-dichloro-cyclopropane as the example. Numerical details are given to illustrate the selection of a model for the activated complex and the actual calculation of k_{uni} as a function of pressure. Section 6.5 surveys the sensitivity of the calculated results to details of the computational procedure and (most importantly) to the properties of the model used. The overall picture obtained from the published results of such calculations will be discussed in Chapter 7.

6.1 CALCULATION OF ACTIVATION PARAMETERS FOR A POSTULATED MODEL OF THE REACTION

Before considering in detail the selection of a model for the reaction, it will be useful to consider how the postulated numerical properties of the reactant and complex can be used to calculate the resulting activation parameters for the reaction. The high-pressure A-factor is particularly significant, and in point of fact the model is almost invariably chosen so that the calculated A-factor agrees with the experimental value. The experimental high-pressure Arrhenius parameters are defined by (6.1) and (6.2). Since k_∞ is given in RRKM theory by the ART expression (6.3), it follows that the theoretical values of E_∞ and $\ln A_\infty$ are given by (6.4)–(6.7).

$$E_\infty = kT^2 \, \mathrm{d} \ln k_\infty / \mathrm{d}T \tag{6.1}$$

$$\ln A_\infty = \ln k_\infty + E_\infty / kT = \mathrm{d}(T \ln k_\infty)/\mathrm{d}T \tag{6.2}$$

$$k_\infty(\mathrm{RRKM}) = L^{\ddagger} \frac{kT}{h} \frac{Q_1^{+} Q_2^{+}}{Q_1 Q_2} \exp(-E_0/kT) \tag{6.3}$$

$$E_\infty(\mathrm{RRKM}) = E_0 + kT + kT^2 \, \mathrm{d}(\ln Q_1^{+} Q_2^{+}/Q_1 Q_2)/\mathrm{d}T \tag{6.4}$$

$$= E_0 + kT + \langle E^{+} \rangle - \langle E \rangle \tag{6.5}$$

$$\ln A_\infty(\mathrm{RRKM}) = \ln(L^{\ddagger} ekT/h) + \mathrm{d}(T \ln Q_1^{+} Q_2^{+}/Q_1 Q_2)/\mathrm{d}T \tag{6.6}$$

$$= \ln(L^{\ddagger} ekT/h) + \Delta S^{\ddagger}/R \tag{6.7}$$

Equation (6.5) is derived from (6.4) using the standard statistical-mechanical equation (6.8). The term kT in (6.5) may be identified with the

$$\langle E \rangle = kT^2 \, \mathrm{d} \ln Q/\mathrm{d}T \tag{6.8}$$

average translational energy of the complexes in the reaction coordinate, and $\langle E^{+} \rangle$ is the average internal energy in their other degrees of freedom, relative to the vibrational–rotational ground state of the complex. The quantity here denoted $\langle E \rangle$ is the average internal energy of *all A molecules* relative to their ground state. It has sometimes been denoted $\langle E^{*} \rangle$ and incorrectly called the average energy of the energized molecules, although the correct expression and value for $\langle E \rangle$ has been used. Equation (6.5) is therefore in accord with Tolman's classic theorem[1] that the high-pressure Arrhenius activation energy of a unimolecular reaction is the average internal energy of the species which are reacting minus the average internal energy of all the reactant molecules present. Equation (6.7) follows from (6.6) by virtue of the statistical-mechanical equation (6.9). In reverse, (6.7)

$$S = R \, \mathrm{d}(T \ln Q)/\mathrm{d}T = R \ln Q + RT \, \mathrm{d} \ln Q/\mathrm{d}T \tag{6.9}$$

is the usual equation for calculation of the entropy of activation from the experimental value of A_∞; convenient numerical versions of these equations are as follows:

$$\log_{10} A_\infty/\mathrm{s}^{-1} = 10 \cdot 753 + \log_{10}(L^{\ddagger} T/\mathrm{K}) + \Delta S^{\ddagger}/4 \cdot 576 \ \mathrm{cal} \, \mathrm{K}^{-1} \, \mathrm{mol}^{-1}$$

$$\Delta S^{\ddagger}/\mathrm{cal} \, \mathrm{K}^{-1} \, \mathrm{mol}^{-1} = 4 \cdot 576 \log_{10}[1 \cdot 766 \times 10^{-11}(A_\infty/\mathrm{s}^{-1})/(L^{\ddagger} T/\mathrm{K})]$$

Since the statistical factor is explicitly included in the present equations, the corresponding value of $\Delta S^{\ddagger}$ is the entropy of activation for a given specific reaction path; $\Delta S^{\ddagger} = S(\mathrm{A}^{+}) - S(\mathrm{A})$ with symmetry contributions[2] to $S(\mathrm{A}^{+})$ and $S(\mathrm{A})$ omitted. It must not be confused with the frequently quoted overall entropy of activation, which is greater by the contribution

$R \ln L^{+}$ from reaction-path degeneracy. In some applications it is useful to divide ΔS^{+} into vibrational and rotational contributions;

$$\Delta S_{v}^{+} = S_{v}(A^{+}) - S_{v}(A) \quad \text{and} \quad \Delta S_{r}^{+} = S_{r}(A^{+}) - S_{r}(A)$$

The expressions giving the entropy are usually taken to be those for quantum harmonic oscillators and classical rotations, (6.10)–(6.13), together with (6.9). The equation (6.12) for *independent* classical rotations

$$Q_{v} = \prod_{i=1}^{s} [1 - \exp(-h\nu_{i}/kT)]^{-1} \tag{6.10}$$

$$T \, d \ln Q_{v}/dT = \sum_{i=1}^{s} \{(h\nu_{i}/kT)[\exp(h\nu_{i}/kT) - 1]^{-1}\} = \langle E_{v} \rangle / kT \tag{6.11}$$

$$Q_{r} = (8\pi^{2} kT/h^{2})^{\frac{1}{2}r} \prod_{i=1}^{p} I_{i}^{\frac{1}{2}d_{i}} \Gamma(\tfrac{1}{2}d_{i}) \tag{6.12}$$

$$T \, d \ln Q_{r}/dT = \tfrac{1}{2}r = \langle E_{r} \rangle / kT \tag{6.13}$$

is not always strictly applicable and care is needed to be at least consistent; see section 5.2.4 and Appendix 2.

The appropriate critical energy E_{0} for a given model may be postulated empirically, but is more often calculated from the observed value of E_{∞} using (6.4) or (6.5), (6.11), and (6.13). If the model is chosen to reproduce the experimental A-factor, it will then automatically predict correctly the experimental values of k_{∞} and its temperature dependence at the temperature T used in the above calculations.

It is clear from the form of (6.4)–(6.7) that there are an infinite number of models which will reproduce a given set of Arrhenius parameters. Thus the Arrhenius parameters do not uniquely specify the model to be used, and it is usual to consider several different models for a given reaction. Factors influencing the choice of these models are discussed in the following sections.

6.2 SPECIFICATION OF A MODEL FOR THE REACTION

In order to carry out the calculation of k_{uni} as a function of pressure it is first necessary to set up models of the molecule A and the activated complex A^{+}. The necessary parameters for A are the details of its internal motion (i.e. its vibrations and internal rotations), its overall moments of inertia and its collision diameter. These data can always be obtained in principle from experimental measurements, and can often be found in the literature or can be estimated with reasonable certainty by empirical methods. The same parameters (except for collision diameter) must also be specified for the activated complex, but here there is no possibility of experimental determination. The information might in principle be obtained by a complete solution of the Schrödinger equation for the system, which would in fact give *all* the information required, including E_0. There is no prospect of such calculations being practicable in the foreseeable future, however, and more empirical methods have to be employed. Section 6.2.1 deals with the general classification of activated complexes as 'rigid' or 'loose', which is a prerequisite to any quantitative considerations. Section 6.2.2 describes the progress which has been made in semi-empirical estimation of the properties of complexes by comparison with related properties of normal molecules. These methods may provide useful assistance, although they are unlikely to be completely self-sufficient. The most widely used procedure depends on a purely empirical selection of properties for the complex in such a way as to reproduce correctly the experimental values of $\Delta S^{\neq}$ and A_∞, and the use of this approach is described in section 6.2.3.

6.2.1 Treatment of rotational degrees of freedom; 'rigid' and 'loose' complexes

The first step in the selection of a model is to decide how many rotational degrees of freedom will be included in the models of A and A^+, and which of these degrees of freedom will be taken to be active, i.e. to exchange their energy freely with the other degrees of freedom of the molecule. The overall rotations of reactant and complex are always included in principle, even if they are assumed to have no effect by putting $Q_1^+/Q_1 = 1$. A change in the number of internal rotational degrees of freedom on formation of the activated complex is commonly postulated to produce a model consistent with an unusually high or low A-factor.[3] This is possible since the density of rotational states is usually much higher than the density of vibrational states at the relevant energies, and the associated entropy is correspondingly higher (e.g. 10–12 compared with 0–2 cal K^{-1} mol^{-1} per degree of freedom). It should be realized, however, that changing a rotational degree of freedom from adiabatic to active (without changing the *number* of rotations) will have no effect on the *high-pressure* A-factor for the reaction.[4] The partition function for the rotation will be unchanged, and will merely enter (6.3) in Q_2 or Q_2^+ instead of in Q_1 or Q_1^+. Thus the magnitude of A_∞ will give guidance as to a possible change in the number of rotational degrees of freedom, but not as to the activity or otherwise of the rotations.

When choosing the number of rotations to be included it may be helpful to consider whether the complex is likely to be *rigid* or *loose*. A loose complex is one in which there is some degree of free internal rotation which was not present in the reactant molecule. An example is the complex often assumed for the decomposition of ethane into two methyl radicals; in the complex the incipient radicals have already a substantial degree of rotational freedom. Such complexes are usually postulated to explain high A-factors, and are generally associated with bond-fission processes of the type just exemplified. A rigid complex, on the other hand, has no internal rotations which were not present in the reactant molecule. An example is the commonly assumed four-centre activated complex for the decomposition of ethyl chloride. Wieder has suggested[5] that the nature of

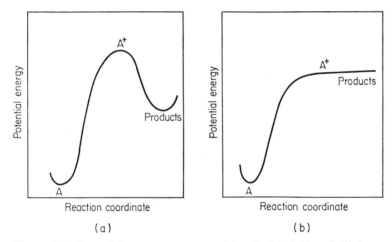

Figure 6.1 Potential energy curves resulting in (a) rigid and (b) loose activated complexes

the complex can be determined by examination of the potential energy curve for the reaction coordinate, and in particular by considering the formation of the complex from the reaction products rather than from the reactant molecule. Where there is a definite energy barrier, as in Figure 6.1(a), the formation of the complex from either reactant or products involves the partial transfer of an atom or group to a different point in the molecule, usually through a cyclic configuration, and it is difficult to see how a complex thus formed can have any new internal rotations. If, on the other hand, the potential energy profile is of the type shown in Figure 6.1(b), the activated complex can be formed from the freely rotating product fragments without any increase in potential energy, and it is likely that some of this free rotation will be retained as internal rotation in the activated complex.

In general, therefore, loose complexes will be plausible for reactions involving bond fission to give separated fragments, and the activated complexes for these reactions will have one or more internal rotational degrees of freedom. The reactant molecule may or may not have internal rotations, but it will have fewer than the complex. Rigid complexes will usually be assumed for reactions involving the simultaneous making and breaking of bonds. In some cases (e.g. six-centre rearrangements) the A-factor may be so low as to suggest that the reactant molecule has internal rotations which are lost in the activated complex.

Finally, the rotations must be classified as active or adiabatic. There is little theoretical guidance here, but the situation is fairly well defined empirically. All internal rotations appear to be active; this seems to be a reasonable postulate and has produced results in good agreement with experiment for a number of systems (see, e.g., sections 4.12.1 and 8.2). The overall rotations are usually treated as adiabatic, but there may sometimes be reasons for considering that one of the three will be strongly coupled with the internal degrees of freedom of the molecule and should therefore be taken to be active.[6]

6.2.2 Assignment of numerical properties to the activated complex

After a chemical structure has been postulated for the activated complex, and the treatment to be given to the rotational degrees of freedom has been settled, values must be assigned to the relevant vibration frequencies and moments of inertia of the complex. This could be done fairly accurately if the geometry of the complex was known (the determination of the moments of inertia of a molecule of known geometry is a standard procedure)[7] and if a complete set of force constants for the complex could be constructed. Although these aims cannot be met completely in practice, some progress can be made by consideration of analogous structures in normal molecules and free radicals, and by the use of semi-empirical correlations of bond lengths and force constants with bond order. When all else fails, a purely empirical approach can be used (see next section) and this is frequently used as a final adjustment in conjunction with the more theoretically based methods.

The transfer of data from analogous molecules is useful provided that truly analogous structures can be found. It is well illustrated by the use of the known vibration frequencies of the CH_3 and CF_3 radicals for the essentially separated CX_3 fragments in the complexes for C_2H_6 and C_2F_6 dissociation. This leaves only the C—C distance to be specified, and this can be determined by a special procedure mentioned below. In other cases less complete assistance may be obtained in a similar way. For the isomerization of methyl isocyanide, for example, the eight vibrations characteristic of the methyl group were assumed[8, 9] to have the same frequencies in the activated complex I as in the reactant molecule.

$$\text{I} \qquad\qquad \text{II}$$

Similarly the twelve CH_2 vibrations in the complex II were assumed[10] to have the same frequencies as in the ethyl chloride molecule. Vibrations of a complex which are obviously different from those of the reactant molecule may nevertheless find approximate analogies in different molecules. For the complex in reaction (6.14), for example, the B_1 ring deformation at

$$(6.14)$$

895 cm^{-1} was lowered to 650 cm^{-1} to correspond to a C—C=O bending vibration, and the ring-breathing vibration at 1266 cm^{-1} was raised to 1500 cm^{-1} to correspond to a C=O stretch.[4]

When as much progress as possible has been made in the above manner there are usually some bond lengths and frequencies as yet undetermined. One way of estimating these uses the empirical correlations of bond length and force constant with bond order which were developed by Badger,[11] Pauling[12] and others, and have been summarized by Johnston[13] and Pauling.[14] For present purposes it is required to predict bond lengths and force constants from reasonably assigned bond orders, these usually being allocated on the assumption that the total bond order in the system (and sometimes at a given atom) is constant. This is not strictly true, of course, but the correlations can be adjusted to use the intuitive 'bond number' n rather than the M.O. Theory bond order.† With this proviso, the lengths R and stretching force constants F of bonds between a pair of atoms i and j are given by (6.15) and (6.16). In these equations R_s and F_s are the

$$R = R_s - 0\cdot60 \log_{10} n \qquad (6.15)$$

$$F = F_s 10^{(R_s-R)/b_{ij}} = F_s n^{0\cdot60/b_{ij}} \qquad (6.16)$$

† For example, the bond number of the C—C bonds in benzene is 1·5, whereas[15] the bond order is 1·67, and the total bond order at each C atom is 4·33.

length and force constant for the corresponding single bond, and b_{ij} is a constant determined by the rows of the periodic table in which atoms i and j occur; $b_{ij} = 0.60 \pm 0.05$ for most cases of interest. There are unfortunately no similarly well-established correlations for bending or torsional force constants, and the estimation of these has remained rather arbitrary.[8, 10, 16, 17]

To the extent to which the vibration of a given bond is independent of the rest of the molecule, its vibration frequency will be given by (6.17). This

$$\nu/\nu_s = \sqrt{(F/F_s)} \quad \text{or} \quad \nu = \nu_s n^{0.30/b_{ij}} \tag{6.17}$$

simplification was used in the calculations on methyl isocyanide isomerization,[8, 9] although the dependence of F on n was assumed in this case to follow Badger's original rule, (6.18),[11] rather than (6.16); in fact these two equations give very similar values for F/F_s. Badger's rule and Pauling's

$$F = a'_{ij}(R - b'_{ij})^{-3} \tag{6.18}$$

equations are also said[17] to be the basis of the extensive tabulations by Benson and co-workers[17, 18] of frequencies for the stretching and bending vibrations involving bonds of $\frac{1}{2}$ and $1\frac{1}{2}$ order. These values have been used with remarkable success in predicting the high-pressure A-factors of unimolecular reactions,[17–19] and the activated complex models thus derived should be suitable for fall-off calculations. There are, in fact, substantial difficulties in setting up this type of model *a priori*,[20] but the approach of these workers should nevertheless find much application.

Although the simplification (6.17) can probably be used to derive vibration frequencies from the force constants without significant loss of accuracy, a more satisfying method involves the complete vibrational analysis of the complex or of relevant parts of the complex; in particular, the choice of reaction coordinate should emerge more naturally from this treatment. For example, in the calculations on ethyl chloride,[10] detailed analyses were carried out for the ring structure III in which the CH_2

$$(\underset{\vdots}{CH_2}){=\!\!=}(\underset{\vdots}{CH_2})$$
$$\text{H}\cdots\cdots\text{Cl}$$

III

groups were treated as the appropriate point masses. Various sets of bond numbers for the four bonds were assumed, e.g. 1.8, 0.8, 0.2, 0.2 for the C—C, C—Cl, Cl—H and H—C bonds respectively. The bond lengths were calculated from (6.15), apparently with 0.71 in place of the 0.6 although the latter is recommended,[21] especially for bond numbers less than unity. The ring was assumed to be planar, with the HCC and CCCl angles equal, thus completely defining the geometry. Force constants were calculated from Badger's rule (6.18), and the five in-plane vibration frequencies of the ring were then calculated by the Wilson **FG** matrix method.[22] The normal mode in which the C—C and H—Cl distances decreased and the C—H and C—Cl distances simultaneously increased always had a very low frequency (e.g. $100 \, cm^{-1}$) and was naturally chosen to represent the reaction coordinate. This left only the out-of-plane ring 'puckering' vibration frequency to be determined, and in the absence of any precedent this frequency was used as an adjustable parameter in the spirit of section 6.2.3.

It is worth noting that there are similar correlations of bond *energy* with bond order for diatomic molecules[23] and that these have been successfully used to estimate the activation energies of simple bimolecular abstraction reactions.[24] It seems likely that such techniques might be extended to estimate the activation energies of unimolecular reactions, but there has been little progress in this direction so far.

Finally, mention should be made of the special procedures which have been developed for determining the length of the critical bond in the loose activated complex for dissociation of a molecule into two fragments, e.g. $C_2H_6 \rightarrow 2CH_3$. The fragments are assumed to have the same geometry and vibration frequencies as the free radicals, and to have a large degree of rotational freedom. The reaction coordinate is the inter-fragment distance, and this is treated by considering the complex as a decomposing diatomic molecule in which the atoms have the masses of the actual fragments. The critical bond length R_c is that at which the force of attraction between the 'atoms' is just balanced by the centrifugal force tending to dissociate the molecule, i.e.

$$-6a/R_c^7 = J(J+1)\hbar^2/\mu R_c^3$$

where a is the constant in the attractive potential energy expression $V = -a/R^6$, J is the rotational quantum number and μ the reduced mass. Tschuikow-Roux considered[25] the distribution of rotational states in a manner similar to that of section 4.10 and showed that the average value of R_c for dissociating molecules was given by (6.19), in which an error function has justifiably been approximated to unity. The constant a was

$$\langle R_c \rangle \approx \frac{\pi^{\frac{1}{2}}}{\Gamma(\frac{2}{3})} \left(\frac{2a}{kT}\right)^{\frac{1}{6}} = 1 \cdot 309 \left(\frac{2a}{kT}\right)^{\frac{1}{6}} \tag{6.19}$$

evaluated for a series of complexes by summing the contributions from electrostatic (orientation), inductive and dispersion forces using estimated values of dipole moments and polarizabilities for the atoms and radicals involved. In all cases the dispersion forces were greatest, although the electrostatic contribution was not always negligible. The values so obtained for $\langle R_c \rangle$ were quite large (e.g. CH_3—CH_3 5·2 Å, CF_3—CF_3 6·3 Å, C_2H_5—H 4·6 Å, compared with 1·5, 1·6 and 1·1 Å respectively in the normal molecules), and this large extension can result in a substantial centrifugal effect on the rate of reaction (see section 4.10). Similar conclusions were reached in a closely related treatment by Forst,[26] and the subject has been reviewed by Waage and Rabinovitch.[27]

Bunker and Pattengill have suggested[28] that the 'critical molecular configuration' lies not at the top of the energy barrier, but at a slightly lower extension where the number of accessible quantum states per unit energy range is a minimum. This idea was applied to the decomposition of hypothetical classical triatomic molecules having various masses, vibration frequencies, bond angles and bond dissociation energies. The critical configuration occurred typically at a bond distance of 3·5 Å compared with 1·2 Å in the molecule and 4·3 Å at the top of the energy barrier. The use of the critical configuration gave significantly improved agreement between the RRKM formulation and the results of Monte-Carlo 'experiments' on the molecules, and it seems likely that this approach will be further developed in the future.

6.2.3   Empirical approach to specification of the activated complex

As has already been indicated, any attempt to deduce the properties of an activated complex from first principles almost inevitably leaves some features undetermined, and such properties are generally used as empirically adjustable parameters to make the model predict the correct value of A_∞. Many authors have made no attempt at all to deduce the properties of the complex in the above ways, but have adopted a wholly empirical approach to the problem, which will be described in this section. Such an attitude is in fact useful, since it will be seen later that the calculated fall-off behaviour is surprisingly insensitive to the details of the model, providing it leads to the correct value of A_∞. In this approach, a plausible-looking activated complex is constructed and the appropriate reaction-path degeneracy is selected. The partition function ratio Q_1^+/Q_1 for the adiabatic (overall) rotations is given some arbitrarily selected value, usually unity. This is quite reasonable for a reaction proceeding through a rigid activated complex because the complex will not be very different in size from the reactant molecule; since the moments of inertia enter as $(I_A^+ I_B^+ I_C^+/I_A I_B I_C)^{\frac{1}{2}}$ quite substantial changes in molecular size would be required before a serious effect on Q_1^+/Q_1 resulted. Similarly the simple treatment of the overall rotations by including the factor Q_1^+/Q_1 in k_{uni} is likely to be sufficiently accurate for such reactions (see section 4.10).

The entropy of activation (per single reaction path) is obtained from the experimental high-pressure A-factor using (6.7), and if necessary $R \ln (Q_1^+/Q_1)$ is subtracted to give the vibrational contribution ΔS_v^+ to the entropy of activation. The vibration frequencies of the molecule are generally known, and the vibrational entropy $S_v(A)$ of the reactant at the selected reaction temperature can thus be calculated from the harmonic oscillator equations (6.9)–(6.11). It is now required to construct an activated complex which has vibrational entropy $S_v(A^+) = S_v(A) + \Delta S_v^+$. One particular vibration of the molecule is chosen to represent the reaction coordinate and the corresponding frequency is removed, then the remaining frequencies are adjusted by trial and error in a largely empirical way so as to reproduce the required value of $S_v(A^+)$. At first sight this may appear involved in view of the complexity of (6.9)–(6.11). In fact a rapid convergence can be obtained by using in the first stage a coarse tabulation of harmonic oscillator entropies as a function of frequency and temperature (Table A2.2) and then a more precise tabulation as a function of the dimensionless quantity $h\nu/kT$ (Table A2.3). In the final stage one frequency or a number of frequencies can be adjusted to give a precise fit and this can be done by inverse use of Table A2.3 or by a simple computer program.

The detailed application of this approach to the isomerization of 1,1-dichlorocyclopropane is described in section 6.4.

6.3   THE RRKM INTEGRATION

Figure 6.2 contains a general outline of the integration procedure; a detailed computer program is not given since there are no unusual difficulties, and programs may be obtained on request from various reliable sources (see ref. 16 of Chapter 5). It is advantageous to have a number of possible treatments built into one program, and some of the alternatives are discussed in the present section.

The integration is carried out by a stepwise summation, and in view of the wide uncertainties involved in the model and its numerical parameters there is little to be gained by the use of a sophisticated integration procedure such as the Kutta–Merson method or even Simpson's rule. With machine integration it is easy in the present case to make the steps so small that a sufficiently accurate value for the integral is obtained from the simple summation (6.20) in which the summand for each step ΔE^+ has the value appropriate to the centre of the step. In (6.20), as in much of the work on this topic, energies have been expressed in $kcal\,mol^{-1}$, thus inviting classification with the 'crudest measurements';[29] other authors have used cm^{-1}. What is important is that the values used for the constants should

$$k_{\text{uni}} = \frac{L^+ Q_1^+}{h Q_1 Q_2} \exp(-E_0/RT) \Delta E^+ \sum_{i=1}^{i_{\max}} \left[\frac{\{\sum P(E_{\text{vr}}^+)\} \exp(-E^+/RT)}{1 + k_a(E^*)/\lambda Z p} \right]_i \quad (6.20)$$

where

$$k_a(E^*) = L^+ \frac{Q_1^+}{Q_1} \{\sum P(E_{\text{vr}}^+)\}/h N^*(E^*)$$

$$E^* = E_0 + E^+, \quad E^+ = (i + \tfrac{1}{2}) \Delta E^+$$

and k_{uni} will be in s^{-1} if the energies E_0, E^+ and ΔE^+ are expressed in kcal mol^{-1} and the constants h and R are given the values

$$h = 9 \cdot 5370 \times 10^{-14} \text{ kcal mol}^{-1} \text{s}$$

$$R = 0 \cdot 0019872 \text{ kcal mol}^{-1} \text{K}^{-1}$$

form a self-consistent set. The values quoted here are from McGlashan's compilation,[30] and the calories referred to are 'thermochemical calories';[31] thus $N_A hc = 0 \cdot 0028591$ cal(thermochem) mol^{-1} cm. Equation (6.20) corresponds to (4.17) with $k_2[M]$ replaced by the collisional deactivation rate-constant $\lambda Z p$. Adiabatic rotations are here treated by including the factor Q_1^+/Q_1; the extension to more refined treatments (see section 4.10) is obvious. In general a step-length ΔE^+ of $0 \cdot 1$–$0 \cdot 5$ kcal mol^{-1} seems to be small enough, and with machine computation it is simple enough to confirm this for a given case by checking the effect of varying the step length. The maximum energy to which the summation must be taken to obtain a sufficiently accurate result is also a matter for trial and error, and in any given case the question may be resolved by examining a print-out of the contributions to k_{uni} from the individual energy ranges. As a general guidance only, a maximum E^+ of $E_0 + s kT$, where s is the number of active degrees of freedom involved, will probably give a result within 1% in typical cases.

The reading in of the numerical data is reasonably straightforward and is conveniently delegated to 'subroutines' or 'procedures'. Some useful conversion factors and functions of universal constants are given in Appendix 2 (section A2.6). Since the computation of $\sum P(E_{\text{vr}}^+)$ tends to be much more time-consuming than the integration itself, it is usual to compute these values separately and to obtain the results in the form of tape or card output which can be read in as data for the integration program. It may be desirable to process the $\sum P(E_{\text{vr}}^+)$ data in a number of ways for use in the integration. If a series of energy steps and maximum energies is to be used it would be very uneconomical to compute separately a set of $\sum P(E_{\text{vr}}^+)$ values precisely as required for each choice. The obvious solution is to compute one set going to a high energy in small steps and to extract

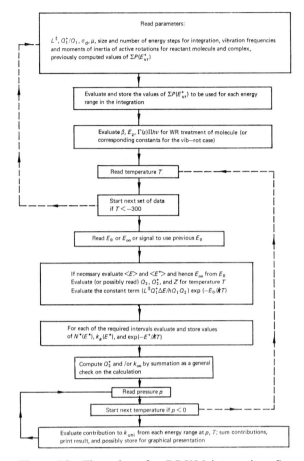

Figure 6.2 Flow chart for RRKM integration. See text for expansion and qualification

from this the data needed for each case to be treated. In one of the authors' programs a subroutine is used to pick out every jth value of $\sum P(E_{vr}^+)$ up to a required maximum energy, or to interpolate between values if the integration step is less than the intervals on the $\sum P(E_{vr}^+)$ tape. In addition the data can be treated as relevant to a commensurable model, in which case the $\sum P(E_{vr}^+)$ value for a step is taken as the value at the end of the step, or as a non-commensurable (essentially smooth-curve) function, in which case the value at the centre of the step is used.

The evaluation of the constants for the Whitten–Rabinovitch calculation of $N^*(E^*)$ is straightforward except for a tendency to produce numbers which are too large to be held in the computer. For example $\prod (\bar{\nu}/cm^{-1})$ in

(5.38) or (5.42) might well overflow in some cases, and the product $\Gamma(s) \prod \bar{\nu}$ can easily be of the order of 10^{100}. Such difficulties are easily avoided by suitable scaling; for example the product $\prod (N_A hc\bar{\nu})$ is quite manageable.

The critical energy E_0 is required and may be read in as such or calculated from E_∞ using (6.4), (6.11) and (6.13), with the given properties of the reactant and complex. If a series of integrations at different temperatures is to be carried out it may be convenient to calculate E_0 from E_∞ at one temperature then use the same value of E_0 at the other temperatures. E_0 is a true constant of course, and E_∞ varies with temperature according to (6.4). But the variation of E_∞ with temperature is experimentally inaccessible, and if the constant experimental value was used in (6.4) at different temperatures an apparent variation in E_0 would be created. It is thus useful to have these alternative sources of E_0 built into the program.

The partition functions Q_2 and Q_2^+ are obtained using (6.10) and (6.12). Previously calculated values could be read in, but since the frequencies and moments of inertia for at least the reactant are needed in the integration program [for $N^*(E^*)$] it is convenient to calculate Q_2 and Q_2^+ by subroutine in the same program. The collision number Z is given as a function of temperature by (6.21). Note that the factor $\frac{1}{2}$ which appears before this

$$Z = (\sigma_d{}^2 N_A/R)(8\pi N_A k/\mu)^{\frac{1}{2}}(1/T)^{\frac{1}{2}} \qquad (6.21)$$

where

Z will be in $\text{Torr}^{-1}\text{s}^{-1}$ (consistent with p in Torr and k_a in s^{-1}) when:

σ_d = collision diameter in cm

μ = reduced molar mass in g mol^{-1}

  $= (1/M_A + 1/M_B)^{-1}$ or $M_A/2$ if $M \equiv A$

T = temperature in K

$N_A = 6 \cdot 0225 \times 10^{23} \text{ mol}^{-1}$

$R = 6 \cdot 2362 \times 10^4 \text{ cm}^3 \text{ Torr K}^{-1} \text{mol}^{-1}$

$k = 1 \cdot 3805 \times 10^{-16} \text{ erg K}^{-1}$

equation when it refers to collisions between like molecules is omitted here, since even when $M \equiv A$ we are concerned with collisions between two distinguishable entities, A* and A. Since $M_{A*} = M_A$, however, μ becomes $M_A/2$ in this case. The selection of a self-consistent set of units for the quantities in (6.21) is curiously awkward, but the parameters prescribed here are in fact self-consistent and lead to the result as stated.

The last preparatory part is to evaluate and store the functions which will be needed to form the summand of (6.20); the same values will be used to evaluate (6.20) for each pressure required. The appropriate $\sum P(E_{vr}^+)$ has already been evaluated and stored. The energy E^+ for each range is taken as the value at the centre of the range. The density of states of the energized molecule, $N^*(E_0+E^+)$, is evaluated from (5.35)–(5.38) [or (5.41)–(5.42)], conveniently in a subroutine, $k_a(E^*)$ follows from (6.20), and it is also worth storing the values of $\exp(-E^+/RT)$.

Before actually carrying out the integration it is customary and worth while to perform a check on the setting up of the calculation by evaluating Q_2^+ and/or k_∞ by summation using the stored functions as above, and comparing the results with the accurately known values from the analytical expressions (6.10), (6.12) and (6.3). The approximate values are given by (6.22) and (6.23), and should closely reproduce the values from the exact

$$Q_2^+ \text{ (by summation)} = \frac{\Delta E^+}{RT} \sum_{i=1}^{i_{max}} \{\sum P(E_{vr}^+)\} \exp(-E^+/RT) \quad (6.22)$$

$$k_\infty \text{ (by summation)} = \frac{L^+ Q_1^+}{hQ_1 Q_2} \exp(-E_0/RT)\, \Delta E^+$$

$$\times \sum_{i=1}^{i_{max}} \{\sum P(E_{vr}^+)\} \exp(-E^+/RT) \quad (6.23)$$

expressions; it was shown in section 4.7 that Q_2^+ is given exactly by the integral to which the summation (6.22) is an approximation.

The integration can now proceed; for each pressure p the summand in (6.20) is evaluated for each energy range and the sum is formed. It is useful to have the option of printing out a selection of the contributions of the individual steps to the overall result, since this can give a further check that the integration is being taken to sufficiently high energy and can also give interesting information about the average energy of the molecules which are reacting at various pressures. In addition to printing out the results a graphical display is a useful adjunct and may perhaps include experimental data for direct comparison. An output of this type is reproduced in Figure 6.3.

6.4 APPLICATION OF RRKM THEORY TO THE ISOMERIZATION OF 1,1-DICHLOROCYCLOPROPANE

As an example of the application of RRKM theory, a set of calculations[32] for the isomerization of 1,1-dichlorocyclopropane to 2,3-dichloropropene will now be described in detail. The overall reaction (6.24) is homogeneous,

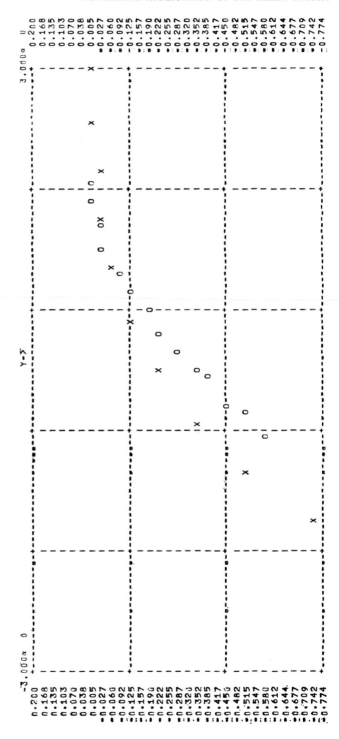

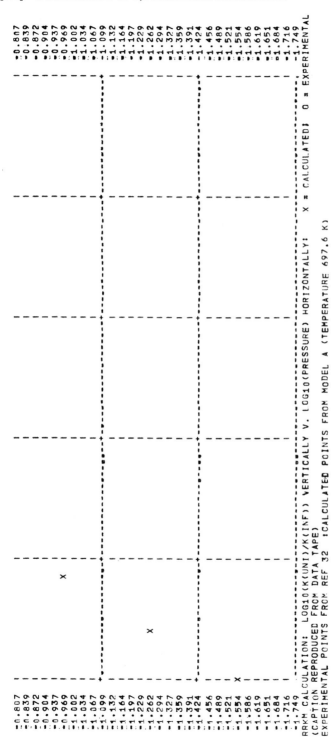

Figure 6.3   Computer-drawn graph of $\log_{10}(k_{uni}/k_\infty)$ vs $\log_{10} p$/Torr for 1,1-dichlorocyclopropane isomerization

first-order and unaffected by the presence of free-radical inhibitors, and has been concluded[33] to be a unimolecular reaction. The high-pressure Arrhenius equation (6.25) was established from studies at 20–120 Torr

$$\log_{10} k_\infty/\text{s}^{-1} = 15{\cdot}13 - 57{\cdot}81 \text{ kcal mol}^{-1}/2{\cdot}303 RT \qquad (6.25)$$

and 615–714 K, and experimental data for the fall-off at pressures down to 0·2 Torr were obtained in a separate series of studies[32] at 697·6 and 632·4 K.

The reaction is assumed to proceed through an activated complex of the type shown in (6.24), for which the statistical factor L^{+} is 4 (see section 4.9). The approach used for setting up models of the activated complex was the empirical method of section 6.2.3, and Q_1^{+}/Q_1 was accordingly taken to be unity. It was assumed that there were no active rotations, and the vibrational contribution to the entropy of activation, calculated from (6.7), was $\Delta S_v^{+} = 4{\cdot}266 \text{ cal K}^{-1} \text{ mol}^{-1}$ at the selected temperature of 697·6 K. The vibrational entropy of the reactant molecules at 697·6 K, calculated from (6.9)–(6.11) with the frequency assignment shown in Table 6.1, was

Table 6.1 Vibration frequencies and vibrational entropies of 1,1-dichloro-cyclopropane and a postulated activated complex

Vibrations of reactant molecule[a]		Corresponding vibrations of complex[b]	
Frequency/cm^{-1}	S_v (697·6 K)/ cal K^{-1} mol^{-1}	Frequency/cm^{-1} and (degeneracy)	S_v (697·6 K)/ cal K^{-1} mol^{-1}
3106, 3096 3048, 3022	0·104	3068 (4)	0·104
1454, 1409 1292, 1238	1·989	1346 (4)	1·977
1164, 1130 1037, 952	3·121	1068 (4)	3·106
874, 825, 772 500, 443	7·726	445 (5)	11·130
717	1·382	Reaction coordinate	
404, 300, 272	8·541	216·8 (3)	10·809
Total vibrational entropy/ cal K^{-1} mol^{-1}	[$S_v(A)$] 22·863		[$S_v(A^{+})$] 27·127

[a] The normal vibration frequencies are those quoted by Kay[34] and are not correct in detail,[35] although the errors will have little effect on the RRKM calculation and the data provide a perfectly valid illustration of the approach under discussion.

[b] Designated 'Complex B' in ref. 32 and section 6.5.

$S_v(A) = 22\cdot863$ cal K^{-1} mol^{-1}, and the required value of $S_v(A^+)$ was therefore $S_v(A) + \Delta S_v^+ = 27\cdot129$ cal K^{-1} mol^{-1}. In one of the complexes considered[32] the reaction coordinate was taken to be the CCl_2 anti-symmetric stretching vibration at 717 cm^{-1}. The remaining 20 frequencies were grouped as shown in Table 6.1, and a computer program was used to lower the frequencies of all but the C—H stretches in proportion to one another until $S_v(A^+)$ reached the required value with the frequencies shown in the third column. This model now reproduces the experimental value of A_∞, and with its derived value of E_0 (see section 6.1 and below) reproduces k_∞ and its temperature variation correctly. Other models for the same reaction will be considered in section 6.5.

The RRKM theory can now be applied to predict the fall-off behaviour for this model. Considerable detail of the calculation will be given, to illustrate a number of points which need attention and to provide a comprehensive set of data against which the reader's own computer programs may be checked. The basic parameters are shown in Table 6.2.

Table 6.2 Basic parameters

$L^+ = 4$ $Q_1^+/Q_1 = 1$ $\sigma_d = 6\cdot5 \times 10^{-8}$ cm $\mu = M_A/2 = 55\cdot5$ g mol^{-1}
Vibration frequencies of molecule and complex from Table 6.1
Calculated values: $\beta = 1\cdot4454$, $E_z = 38\cdot677$ kcal mol^{-1}
Integration up to $E_{max}^+ = 40\cdot0$ kcal mol^{-1} in steps of $\Delta E^+ = 0\cdot05$ kcal mol^{-1}

In this illustration there are 800 intervals for the integration (see comments in section 6.5), and the relevant functions for each energy range will be tabulated only for a few representative intervals. The $\sum P(E_v^+)$ tape was computed by direct count (section 5.3.1) and some extracts from it are given in Table 6.3 together with the values of $\sum P(E_v^+)$ actually used for each step of the integration; the latter are the linearly interpolated values at the centre of each range. Selection of the required temperature (697·6 K) then permits calculation of the parameters in Table 6.4. The calculation of $N^*(E^*)$ and $k_a(E^*)$ then follows; some of the results (again at the centre of each step) are shown in Table 6.5. Table 6.6 compares the values of Q_2^+ and k_∞ obtained by summation [(6.22) and (6.23)] with the accurate analytical values [(6.10) and (6.3)]; the very satisfactory agreement indicates that the integration is being taken to a high enough energy in sufficiently small steps and that the calculation generally is set up correctly.

The integration now proceeds; selected values of the contributions to k_{uni} from different energy ranges at various pressures are given in Table 6.7, together with the final results for k_{uni} and k_{uni}/k_∞. The way in which the contributions to k_{uni} vary with energy and pressure is in itself

Table 6.3   Values of $\sum\limits_{E_v^+=0}^{E^+} P(E_v^+)$

$E^+/\text{kcal mol}^{-1}$	0·00	0·05	0·10	0·15	0·20	...
$\sum P(E_v^+)$ (on tape)	1	1	1	1	1	...
$\sum P(E_v^+)$ (used for calc. in range shown)		1	1	1	1	...

$E^+/\text{kcal mol}^{-1}$	1·15	1·20	1·25	1·30	...
$\sum P(E_v^+)$ (on tape)	4	4	10	15	...
$\sum P(E_v^+)$ (used for calc. in range shown)		4	7	12·5	...

$E^+/\text{kcal mol}^{-1}$	19·80	19·85	19·90	...
$\sum P(E_v^+)$ (on tape)	$2 \cdot 4061 \times 10^7$	$2 \cdot 4297 \times 10^7$	$2 \cdot 4700 \times 10^7$	...
$\sum P(E_v^+)$ (used for calc. in range shown)		$2 \cdot 4179 \times 10^7$	$2 \cdot 4498 \times 10^7$	...

$E^+/\text{kcal mol}^{-1}$	39·85	39·90	39·95	40·00
$\sum P(E_v^+)$ (on tape)	$4 \cdot 4493 \times 10^{10}$	$4 \cdot 5163 \times 10^{10}$	$4 \cdot 5778 \times 10^{10}$	$4 \cdot 6429 \times 10^{10}$
$\sum P(E_v^+)$ (used for calc. in range shown)		$4 \cdot 4828 \times 10^{10}$	$4 \cdot 5471 \times 10^{10}$	$4 \cdot 6104 \times 10^{10}$

Table 6.4   Further parameters

Temperature 697·6 K   Use $E_\infty = 57 \cdot 81_0$ kcal mol^{-1}
$Q_2 = 115 \cdot 84$ $Q_2^+ = 568 \cdot 14$ $Z = 9 \cdot 4790 \times 10^6$ Torr^{-1} s^{-1}
$\langle E_v \rangle = 9 \cdot 361$ kcal mol^{-1} $\langle E_v^+ \rangle = 10 \cdot 131$ kcal mol^{-1}
whence $E_0 = 55 \cdot 65_4$ kcal mol^{-1}

Table 6.5   Values of $N^*(E^*)$ and $k_a(E^*)$

$E^+/\text{kcal mol}^{-1}$	0·025	5·025	10·025	...
$10^{-11} N^*(E^*)/(\text{kcal mol}^{-1})^{-1}$	1·3680	4·1832	12·003	...
$k_a(E^*)/\text{s}^{-1}$	$3 \cdot 0659 \times 10^2$	$1 \cdot 3169 \times 10^5$	$3 \cdot 0336 \times 10^6$	...

$E^+/\text{kcal mol}^{-1}$	20·025	30·025	39·975
$10^{-11} N^*(E^*)/(\text{kcal mol}^{-1})^{-1}$	83·944	489·58	2440·5
$k_a(E^*)/\text{s}^{-1}$	$1 \cdot 3249 \times 10^8$	$1 \cdot 4461 \times 10^9$	$7 \cdot 9232 \times 10^9$

Table 6.6 Q_2^+ and k_∞ from summation and from analytical expressions

	By summation	Analytical	Error
Q_2^+	568·19	568·14	
$10^3 \, k_\infty/\text{s}^{-1}$	1·0465	1·0464	0·010%

interesting and is discussed in section 4.8. The calculated variation of k_{uni} with pressure is compared with the experimental results in Figure 6.5; the comparison for this and other models of the reaction is discussed in detail in the following section.

Table 6.7 Contributions to the integral, and the final results

Energy range/ kcal mol^{-1}	10^6 (contribution to k_{uni}/s^{-1}) at pressures of			
	1000 Torr	10 Torr	0·1 Torr	0·001 Torr
0·00–0·05	0·0652	0·0652	0·0652	0·0632
2·00–2·05	0·6167	0·6166	0·6116	0·3387
4·00–4·05	1·6902	1·6892	1·5930	0·2379
6·00–6·05	2·7542	2·7468	2·1639	0·0974
8·00–8·05	3·4977	3·4647	1·7849	0·0361
10·00–10·05	4·1704	4·0423	0·9932	0·0130
12·00–12·05	4·0807	3·7526	0·4150	0·0046
16·00–16·05	2·2240	1·6010	0·0552	0·0006
20·00–20·05	0·9255	0·3914	0·0067	0·0000
24·00–24·05	0·3116	0·0625	0·0008	0·0000
30·00–30·05	0·0382	0·0027	0·0000	0·0000
39·95–40·00	0·0005	0·0000	0·0000	0·0000
10^3 (total k_{uni}/s^{-1})	1·0423	0·89393	0·31487	0·028394
k_{uni}/k_∞	0·99608	0·85426	0·30090	0·027134

6.5 SENSITIVITY OF THE RRKM CALCULATION TO COMPUTATIONAL DETAILS AND FEATURES OF THE MODEL

In this section comparisons will be made between the results of a number of RRKM calculations in order to define how (*a*) an accurate integration procedure can be selected for a given model and (*b*) a model can be set up which fits the experimental results for a given reaction. The isomerization of 1,1-dichlorocyclopropane again forms a suitable example for discussion.

6.5.1 Details of the computational procedure

With the simple integration procedure described in sections 6.3 and 6.4 the variable features are the step length ΔE^+, the maximum energy E_{max}^+ to which the integration is taken, and the related question of whether or not the summation is artificially cut off at such an energy that the resulting k_∞ agrees with the accurate value from (6.3). The results of a series of calculations to illustrate these points are summarized in Table 6.8. All are

Table 6.8 Effect of variation of details of the integration procedure. (All calculations are for complex B and associated parameters as in section 6.4)

Calcn.	E_{max}^+/kcal mol^{-1}	ΔE^+/kcal mol^{-1}	Cut off	Temp/K	$10^5 k_\infty$/s^{-1} (d)	k_{uni}/k_∞ at $\log_{10} p$/Torr =						
						−3	−2	−1	0	1	2	3
1	40	0·05	No	632·4	1·426	0·047	0·166	0·411	0·712	0·915	0·985	0·998
				697·6	104·7	0·027	0·107	0·301	0·598	0·854	0·970	0·996
2	40	0·05	Yes^a	632·4	1·426	0·047	0·166	0·411	0·712	0·915	0·985	0·998
				697·6	104·6	0·027	0·107	0·301	0·598	0·854	0·970	0·996
3	30	0·05	b	632·4	1·426	0·048	0·167	0·411	0·712	0·915	0·986	0·998
				697·6	104·4	0·027	0·107	0·302	0·599	0·856	0·971	0·996
4	20	0·05	b	632·4	1·390	0·049	0·171	0·422	0·729	0·931	0·991	0·999
				697·6	98·0	0·029	0·114	0·321	0·636	0·895	0·985	0·998
5	40	1·00	No	632·4	1·537	0·050	0·170	0·416	0·716	0·916	0·986	0·998
				697·6	111·4	0·028	0·109	0·305	0·601	0·856	0·970	0·996
6	40	1·00	Yes^c	632·4	1·435	0·053	0·182	0·445	0·761	0·950	0·994	0·999
				697·6	106·3	0·030	0·114	0·319	0·629	0·887	0·983	0·998

^{a} 32·35 kcal mol^{-1} at 632·4 K and 37·10 kcal mol^{-1} at 697·6 K.

^{b} Irrelevant since k_∞ (by summation) < k_∞ (accurate).

^{c} 21 kcal mol^{-1} at 632·4 K and 17 kcal mol^{-1} at 697·6 K.

^{d} By summation. Accurate values from (6.3): 1·426 at 632·4 K and 104·7 at 697·6 K.

based on the model described in section 6.4 ('activated complex B'), and the first calculation, labelled **1** in Table 6.8, is identical with that of section 6.4.

It will be seen immediately that none of the variations tested in Table 6.8 introduces a marked change in either k_∞ or the k_{uni}/k_∞ values. In calculation **2** the integration was cut off at the first step giving k_∞ greater than the accurate value. Since the error in k_∞ in calculation **1** was very small (0·01%) it is not surprising that the cut-off has no significant effect in this case. The upper limit of 40 kcal mol^{-1} for the integration is tested in calculations **3** and **4** by stopping at 30 and 20 kcal mol^{-1} respectively. Integration up to 30 kcal mol^{-1} gives results almost identical with calculation **1**; thus 40 kcal mol^{-1} is clearly more than adequate as an upper limit. Stopping at 20 kcal mol^{-1} does make a difference, although even this is not large. The value of k_∞ obtained by summation falls by about 7% at 697·6 K and $2\frac{1}{2}$% at 632·4 K, and the shape of the fall-off curve changes slightly. In Table 6.8 the values of k_{uni}/k_∞ are calculated using the k_∞ values obtained by summation; this is self-consistent and tends to offset the errors introduced by a poor choice of integration parameters. The actual k_{uni} values have a maximum error, as above, at high pressures where the average energy of the reacting molecules is highest and where a low value of E_{max}^+ therefore has most effect (cf. Table 6.7). At 10^{-3} Torr pressure the error in k_{uni} is only 0·6–0·8%.

Integration in coarse steps of 1 kcal mol^{-1} (calculation **5**) gives k_∞ values which are too high by 6–8%, but k_{uni}/k_∞ values which have hardly changed. If the integration is cut off to minimize the error in k_∞ (calculation **6**) the k_{uni}/k_∞ values also change, the maximum error being about 13%. This is not surprising in view of the low energy cut-off required (ca 20 kcal mol^{-1}; see the remarks above on calculation **4**). Even these gross deviations from calculation **1** scarcely produce an experimentally significant change in the results, however, and it is clear that steps of 0·1 kcal mol^{-1} or less are perfectly adequate for the present calculation.

In conclusion, the integration is surprisingly insensitive to variations in the computational detail, and there is no practical difficulty in carrying out the integration so as to obtain an accurate result.

6.5.2 Variation of the model

It was noted in section 6.4 that the calculations based on activated complex B did not give a particularly good representation of the experimental results. This is a typical state of affairs, and the question is then how the model can be modified to give better agreement. We deal first with some variations that are *not* likely to have much effect; the relevant results

Table 6.9 Variations in model which have little effect on the results. (Integrations to 40×0.05 kcal mol^{-1}. For details of models see Table 6.10; all use $E_\infty = 57.81$ kcal mol^{-1} and give within 0.7% the same values of k_∞.)

Calcn.	Complex	Features explored	Temp/K	k_{uni}/k_∞ at $\log_{10} p/\text{Torr} =$						
				-3	-2	-1	0	1	2	3
7	A		632.4	0.050	0.170	0.415	0.713	0.915	0.985	0.998
			697.6	0.028	0.109	0.303	0.598	0.853	0.969	0.996
1	B	Different frequency assignments	632.4	0.047	0.166	0.411	0.712	0.915	0.985	0.998
			697.6	0.027	0.107	0.301	0.598	0.854	0.970	0.996
8	C		632.4	0.057	0.183	0.427	0.717	0.914	0.985	0.998
			697.6	0.032	0.116	0.310	0.598	0.850	0.967	0.995
9	D		632.4	0.063	0.193	0.436	0.721	0.914	0.984	0.998
			697.6	0.035	0.121	0.314	0.598	0.847	0.965	0.995
10	E	$L^{\neq} = 1$	632.4	0.049	0.169	0.413	0.713	0.915	0.985	0.998
			697.6	0.028	0.108	0.302	0.598	0.854	0.969	0.996
11	F[a]	Commensurable model	632.4	0.052	0.175	0.419	0.714	0.914	0.985	0.998
			697.6	0.030	0.111	0.305	0.597	0.851	0.968	0.996
12	G	Active rotation	632.4	0.041	0.155	0.399	0.708	0.916	0.986	0.998
			697.6	0.024	0.100	0.295	0.597	0.858	0.971	0.996

[a] Integration to 40.0274×0.5718 kcal mol^{-1}; $Q_1^+/Q_1 = 1.083$ to produce correct k_∞.

are summarized in Table 6.9 and details of the models are given in Table 6.10.

The obvious feature which can be varied is the assignment of vibration frequencies to the activated complex, especially when the empirical approach of section 6.2.3 is used for determining these parameters. Calculations **7** and **8** are based on two activated complexes (A and C)[32] selected in a similar way to model B, i.e. to give the same value of A_∞, but differing in the details of their vibration frequencies (see Table 6.10). The fall-off data are virtually indistinguishable from those for calculation **1**. A more extreme frequency pattern, with no particular physical significance, is tested in calculation **9**. The activated complex (model D) has all its frequencies except the C—H stretches in the narrow range 500–700 cm^{-1}. Again, however, the results are very similar to those of calculation **1**.

The assignment of statistical factors is sometimes difficult, and the effect of a change in this parameter is investigated in calculation **10**, which uses $L^{\ddagger} = 1$ and a slightly looser activated complex (model E) to restore agreement with the experimental A_∞. Again the fall-off curve is scarcely altered. Since the equations used contain $L^{\ddagger}$ as the product $L^{\ddagger}Q_1^+/Q_1$, calculation **10** could also represent the effect of a marked change (in this case a decrease) in the molecular moments of inertia on formation of the complex. Since Q_1^+/Q_1 is given by $(I_A^+ I_B^+ I_C^+/I_A I_B I_C)^{\frac{1}{2}}$ it appears that even a several-fold increase in all three moments of inertia would not affect the fall-off curves significantly, provided the A_∞ value is correctly predicted. The approximation used here exaggerates the effect on the fall-off curve (see section 4.10), and the treatment based on the assumption that $Q_1^+/Q_1 \approx 1$ is thus likely to be quite satisfactory for reactions having rigid complexes.

Calculation **11** explores possible errors due to the use of a commensurable model of the activated complex (model F). In this type of model the natural stepwise variation of the $\sum P(E_{vr}^+)$ curve is greatly exaggerated; notwithstanding, the fall-off curve is much the same. Models of this type sometimes have the unit of energy considerably larger than the 0·5 kcal mol^{-1} of model F. In this case a low-energy cut-off is sometimes used to give the correct value of k_∞, and this may well introduce errors of the type discussed above. A better approach would be to use smaller integration steps; there is not in fact any inconsistency in using steps much smaller than the energy unit implicit in the model.

Calculation **12** introduces a major change in the type of model, by assuming that the activated complex (model G) has an active internal rotation. The moment of inertia for this rotation is simply a 'typical value', since the calculation is presented as a general illustration of the effects of active rotations rather than a useful advance for the particular reaction under discussion. The vibration frequencies were selected in the

7

Table 6.10 Details of models for Tables 6.9 and 6.11

Complex	Vibration frequencies/cm^{-1}	$L^{\ddagger}\,Q_1^+/Q_1$	E_0/kcal mol^{-1}	Q_2^+	$\Delta S^{\ddagger}_{697 \cdot 6}$/cal K^{-1} mol^{-1} (a)
A	3068 (4), 1383 (3), 1068 (4), 628·6 (2), 437·7 (4), 237·3 (3)	4	55·49	503·0	7·015
B	3068 (4), 1346 (4), 1068 (4), 445 (5), 216·8 (3)	4	55·65	568·1	7·019
C	3068 (4), 1008 (4), 800 (4), 636·1 (2), 442·9 (4), 248·3 (2)	4	55·04	365·7	7·018
D	3068 (4), 700 (4), 639·4 (4), 580 (4), 500 (4)	4	54·54	252·9	7·014
E	3068 (4), 1346 (4), 1068 (4), 385 (5), 171 (3)	1 4^b	55·17 57·00^b	1604·3 1604·3	7·014 9·769
F	3000 (4), 1200 (3), 1000 (2), 800 (5), 400 (4), 200 (2)	4·332^c	55·41	439·9	7·021
G	3068 (4), 1346 (4), 1068 (4), 550 (5), 299·5 (2) + active internal rotation, $\sigma = 1$, $I = 10^{-38}$ g cm^2	1	56·76	5031·9	7·016

^{a} Overall entropy of activation, $\Delta S^{\ddagger}_{\mathrm{tr}} + R \ln L^{\ddagger}$ ($R \ln 4 = 2\cdot755$ cal K^{-1} mol^{-1}).
^{b} Uses $E_{\infty} = 59\cdot64$ kcal mol^{-1}, cf. 57·81 kcal mol^{-1} for the other calculations.
^{c} $L^{\ddagger} = 4$ and $Q_1^+/Q_1 = 1\cdot083$ to produce the correct A_{∞}; the latter factor will have no visible effect on the fall-off curves (cf. calculation 10).

usual way so as to produce the correct value of A_∞ when the entropy of the active rotation was included. The results of the calculation are virtually as before, the only difference being a very slight shift upwards of the fall-off curve at low temperature and low pressure. The insensitivity to this change in the model can be traced to the fact that the sum of states up to any given

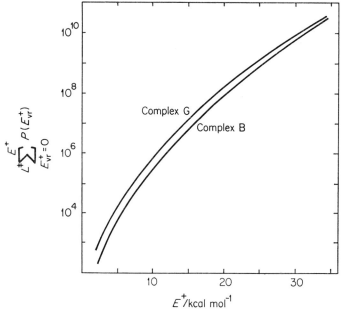

Figure 6.4 Comparison of $\sum P(E_{vr}^+)$ as a function of E^+ for complex B (basic calculation) and complex G (which has an active rotation) (see Table 6.10). The $\sum P(E_{vr}^+)$ values are multiplied by L^+ (4 and 1 respectively) to provide a valid comparison

energy is not greatly affected (Figure 6.4). This follows from the construction of the two models to give the same k_∞; the remaining slight difference between the curves is compensated for when the different values of E_0 are used in (6.20). The net effect on k_{uni} is small, even at low pressures, and the shift in the fall-off curve is considerably less than the likely experimental error. It is often loosely said that complexes with active rotations lead to fall-off at relatively high pressures, but it is clear from the above calculations that this is not necessarily the case. The real cause lies in the unusually high $\sum P(E_{vr}^+)$ values which follow when such a model is used to reproduce a high experimental A-factor, and the early fall-off is associated with the high A-factor rather than with the active rotations; this point will be further illustrated below.

The above calculations thus confirm the conclusion, which has been reached by numerous other authors, that the results of the RRKM calculation are strikingly insensitive to the details of the model used for the activated complex. The same fall-off curve will be predicted by virtually all models considered *provided they give the same values of* $\Delta S^{\ddagger}$ *and hence* A_{∞}; changes in the vibration frequencies, statistical factor, moments of inertia of the adiabatic rotations, or the number of active rotations will not lead to any significant shift of the fall-off curves.

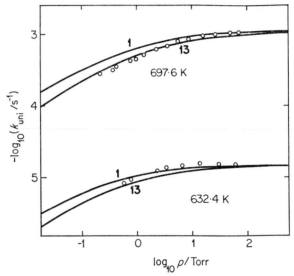

Figure 6.5 Effect of variation in λ; calculations **1** and **13** use $\lambda = 1\cdot0$ and $0\cdot4$ respectively (see Tables 6.10 and 6.11)

One might wonder at this stage if there are any features at all to which the results *are* sensitive; there are, in fact, two features of the model and one additional factor. The first significant variable is the de-energization rate-constant λZ, where Z is given by (6.21). This term appears in the expression for k_{uni} only as the product λZp, so that a change in λZ simply causes an inversely proportional change in the pressure required to produce a given value of k_{uni} or k_{uni}/k_{∞}. Thus the plot of $\log k_{uni}/k_{\infty}$ vs $\log p$ is shifted horizontally by a distance equal to the change in $\log \lambda Z$, without any change in shape. Uncertainties in λZ can only arise from the values of λ and the collision cross-section σ_d^2, and the latter can usually be guessed with reasonable certainty even if it is not known experimentally for the molecule in question. Table 6.11 (calculation **13**) and Figure 6.5 show the

Table 6.11 Variations of model which have a significant effect on the results (Integrations to 40×0.05 kcal mol^{-1}; parameters as in previous calculations except where otherwise stated)

Calcn.	Complex	Features of calcn.	Temp/K	$10^5 \, k_\infty/\text{s}^{-1}$	k_{uni}/k_∞ at $\log_{10} p/\text{Torr} =$						
					-3	-2	-1	0	1	2	3
1	B	Standard	632·4	1·426	0·047	0·166	0·411	0·712	0·915	0·985	0·998
			697·6	104·7	0·027	0·107	0·301	0·598	0·854	0·970	0·996
13	B	$\lambda = 0.4$	632·4	1·426	0·026	0·105	0·299	0·595	0·852	0·969	0·996
			697·6	104·7	0·014	0·064	0·208	0·474	0·767	0·939	0·991
14	E with $L^{\ddagger} = 4$	Different Arrhenius parameters[a]	632·4	1·328	0·026	0·103	0·292	0·584	0·843	0·965	0·995
			697·6	111·6	0·014	0·062	0·201	0·462	0·755	0·933	0·989

[a] $\log_{10} A_\infty/\text{s}^{-1} = 15.73$, $E_\infty = 59.64$ kcal mol^{-1}.

effect of assuming $\lambda = 0\cdot4$ but otherwise using the basic calculation as
in **1**. It is seen that a significant shift of the curves is produced by this
modest expedient, and the curve for $697\cdot6$ K is now a good fit to the
experimental results. Unfortunately the curve for $632\cdot4$ K is now a worse
fit, but this feature of the particular reaction should not mask the fact
that λZ is an important variable parameter, and many authors have introduced λ values between $0\cdot1$ and 1 to obtain agreement with their results.
Strictly, a failure of the strong collision assumption requires a more sophisticated treatment than the introduction of a constant λ (see Chapter 10
and section 4.12.2), but it emerges from these treatments that the simple
correction is adequate for many purposes.

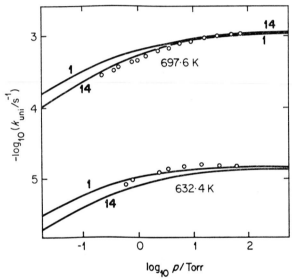

Figure 6.6 Effect of variations in A_∞ and E_∞;
$\log_{10} A_\infty/\mathrm{s}^{-1}$ and $E_\infty/\mathrm{kcal\,mol}^{-1}$ are $15\cdot13$ and $57\cdot81$
respectively for calculation **1**, and $15\cdot73$ and $59\cdot64$
respectively for calculation **14** (see Tables $6\cdot10$ and
$6\cdot11$)

Secondly, and perhaps very significantly, the fall-off curves are quite
sensitive to the values taken for the experimental A_∞ and E_∞ which the
model is adjusted to fit. This is illustrated by calculation **14** (Table 6.11
and Figure 6.6), in which A_∞ has been increased by a factor of four and
E_∞ has correspondingly been raised by $1\cdot83$ kcal mol^{-1}. These changes give
the same value of k_∞ at the experimental mean temperature, and values
which are not drastically changed at the two temperatures used for the

fall-off experiments. The effect of these changes is to produce a considerable displacement of the fall-off curve to higher pressures. The change in the k_∞ values produces simultaneous small vertical shifts, but these are less significant. The observed displacement along the $\log p$ axis is in good agreement with the value $\Delta \log_{10} p = 0.44$ predicted by (6.26), which is an approximate equation derived by Wieder and Marcus.[36]

$$\left.\begin{aligned}&\Delta \log p = (1-\theta)\log A_\infty/A'_\infty\\[6pt]&\text{where}\\[6pt]&\theta = (s+\tfrac{1}{2}r-1)\,kT/(E_0+E_z+\langle E^+\rangle)\\[6pt]&\text{with the symbols defined as before}\end{aligned}\right\} \qquad (6.26)$$

The present calculation, like **13**, produces a good fit to the experimental results at 697·6 K, but is less satisfactory at 632·4 K. It is not suggested that the adjustments made to A_∞ and E_∞ are fully justified, but they are certainly not large in view of the extent of disagreement commonly found between different investigators' results for the same reaction (e.g. Table 7.35). One particular source of error in A_∞ and E_∞ arises from the frequently unappreciated difficulty of obtaining the true high-pressure values.[37] The present calculations indicate that the isomerization of 1,1-dichlorocyclopropane is significantly into the fall-off region in the pressure range 20–120 Torr used for the 'high-pressure' work, and the Arrhenius activation energy may be less than E_∞ by more than 1 kcal mol^{-1} in parts of this range. It seems that many other workers may have erred in this way, and that incorrect values of A_∞ and E_∞ may be a common source of difficulty in fitting the RRKM theory to experimental observations.

A further factor which may lead to disagreement of the simple RRKM theory with experimental results for thermal reactions has been discussed by Lin and Laidler.[38] This is the fact that the product molecule in a unimolecular isomerization reaction is automatically energized for the back reaction to reform the reactant, until it is de-energized by collision (see also the introduction to Chapter 8). Thus the more complete reaction scheme (6.27) may be required for cases where the back reaction cannot be

$$A \underset{Zp}{\overset{k_1[\mathrm{M}]}{\rightleftarrows}} A^* \underset{k_a'(E)}{\overset{k_a(E)}{\rightleftarrows}} B^* \xrightarrow{Zp} B \qquad (6.27)$$

neglected. Steady-state treatment of this scheme leads to (6.28). RRKM calculations show that for endothermic or thermoneutral isomerizations

$$`k_{\mathrm{uni}}' = \frac{\mathrm{d}[B]/\mathrm{d}t}{[A]} = \int_{E=E_0}^{\infty} \frac{k_a(E)\,\mathrm{d}k_1/k_2}{1+[k_a(E)+k_a'(E)]/Zp} \qquad (6.28)$$

the correction can be important. For example, the simple RRKM treatment[36] of the isomerization of *cis-* to *trans-*but-2-ene required a λ factor of about 0·02 to fit the experimental data of Rabinovitch and Michel.[39] Allowance for the back reaction[38] shifted the fall-off curve to higher pressures by a factor of 8 in the pressure,† and the simultaneous use of more recent Arrhenius parameters gave a calculated curve in reasonable agreement with the experimental results assuming $\lambda = 1$. Similarly, the correction for cyclobutene moved the fall-off curve to higher pressures by a factor of 4, again producing good agreement with experiment. It thus appears that this correction can be quite important in many cases, although it is inapplicable to decomposition reactions and will not be significant for appreciably exothermic reactions (e.g.[38] cyclopropane isomerization with $\Delta E^0 = -8 \text{ kcal mol}^{-1}$).

In conclusion, it seems that there are no great difficulties in carrying out the RRKM integration satisfactorily, and that the selection of a model for the activated complex is a very non-critical operation. The features of the model can be varied widely without any significant effect on the results, provided the model reproduces correctly the required high-pressure Arrhenius parameters. The only changes which are likely to affect the results significantly are relaxation of the strong-collision assumption, alteration of the numerical values of the Arrhenius parameters which the model is made to fit, and allowance for the back-reaction of the energized products in appropriate cases.

References

1. R. C. Tolman, *Statistical Mechanics with Applications to Physics and Chemistry*, Chemical Catalog Co. Inc., New York, 1927, Chap. 21.
2. S. W. Benson, *Thermochemical Kinetics*, Wiley, New York, 1968, p. 37.
3. S. W. Benson, *Advan. Photochem.*, **2**, 1 (1964).
4. D. W. Setser, *J. Phys. Chem.*, **70**, 826 (1966).
5. G. M. Wieder, Ph.D. Thesis, Polytechnic Institute of Brooklyn, 1961.
6. See, for example, W. Forst and P. St. Laurent, *Can. J. Chem.*, **45**, 3169 (1967); K. M. Maloney and B. S. Rabinovitch, *J. Phys. Chem.*, **73**, 1652 (1969).
7. G. J. Janz, *Thermochemical Properties of Organic Compounds*, Academic Press, New York, revised edn., 1967, p. 21.
8. F. W. Schneider and B. S. Rabinovitch, *J. Amer. Chem. Soc.*, **84**, 4215 (1962).
9. F. J. Fletcher, B. S. Rabinovitch, K. W. Watkins and D. J. Locker, *J. Phys. Chem.*, **70**, 2823 (1966).
10. J. C. Hassler and D. W. Setser, *J. Chem. Phys.*, **45**, 3246 (1966).
11. R. M. Badger, *J. Chem. Phys.*, **2**, 128 (1934); **3**, 710 (1935); *Phys. Rev.*, **48**, 284 (1935); parameters updated by D. R. Herschbach and V. W. Laurie, *J. Chem. Phys.*, **35**, 458 (1961).

† Since $k_a{}'(E) \approx k_a(E)$ in this case, however, it is not obvious how the factor of 8 arises.

12. L. Pauling, *J. Amer. Chem. Soc.*, **69**, 542 (1947).

13. H. S. Johnston, (*a*) *Gas-phase Reaction Rate Theory*, Ronald, New York, 1966, Chap. 4; (*b*) *Advan. Chem. Phys.*, **3**, 131 (1961).

14. L. Pauling, *The Nature of the Chemical Bond*, Cornell University Press, Ithaca, 3rd edn., 1960, Chap. 7.

15. A. Streitwieser, *Molecular Orbital Theory for Organic Chemists*, Wiley, New York, 1962; L. Salem, *The Molecular Orbital Theory of Conjugated Systems*, Benjamin, New York, 1966.

16. Ref. 13(*a*), p. 82.

17. H. E. O'Neal and S. W. Benson, *J. Phys. Chem.*, **71**, 2903 (1967).

18. S. W. Benson, *Thermochemical Kinetics*, Wiley, New York, 1968.

19. H. E. O'Neal and S. W. Benson, *J. Phys. Chem.*, **72**, 1866 (1968); S. W. Benson and H. E. O'Neal, *Kinetic Data on Gas-phase Unimolecular Reactions*, Nat. Bur. Standards, NSRDS–NBS 21, 1970.

20. R. M. Noyes, *J. Amer. Chem. Soc.*, **91**, 3110 (1969).

21. Ref. 14, p. 255.

22. E. B. Wilson, J. C. Decius and P. C. Cross, *Molecular Vibrations*, McGraw–Hill, New York, 1955.

23. Ref. 13(*a*), p. 77 et seq.

24. See, for example, E. Jakubowski, H. S. Sandhu, H. E. Gunning and O. P. Strausz, *J. Chem. Phys.*, **52**, 4242 (1970); S. W. Mayer, *J. Phys. Chem.*, **73**, 3941 (1969); S. W. Mayer and L. Schieler, *J. Phys. Chem.*, **72**, 2628 (1968); and references cited therein.

25. E. Tschuikow-Roux, *J. Phys. Chem.*, **72**, 1009 (1968); see also comments in ref. 27.

26. W. Forst, *J. Chem. Phys.*, **48**, 3665 (1968).

27. E. V. Waage and B. S. Rabinovitch, *Chem. Rev.*, **70**, 377 (1970).

28. D. L. Bunker and M. Pattengill, *J. Chem. Phys.*, **48**, 772 (1968).

29. M. L. McGlashan, *R.I.C. Monographs for Teachers No. 15*, 1968, p. 13.

30. Ref. 29, Appendix 3.

31. Ref. 29, p. 23.

32. K. A. Holbrook, J. S. Palmer, K. A. W. Parry and P. J. Robinson, *Trans. Faraday Soc.*, **66**, 869 (1970).

33. K. A. W. Parry and P. J. Robinson, *J. Chem. Soc.* (*B*), 49 (1969).

34. M. I. Kay, Ph.D. Thesis, Rensselaer Polytechnic Institute, 1962.

35. J. M. Freeman and P. J. Robinson, *Can. J. Chem.*, **49**, 2533 (1971).

36. G. M. Wieder and R. A. Marcus, *J. Chem. Phys.*, **37**, 1835 (1962).

37. I. Oref and B. S. Rabinovitch, *J. Phys. Chem.*, **72**, 4488 (1968).

38. M. C. Lin and K. J. Laidler, *Trans. Faraday Soc.*, **64**, 94 (1968); see also R. A. Marcus, *J. Chem. Phys.*, **43**, 2658 (1965).

39. B. S. Rabinovitch and K. W. Michel, *J. Amer. Chem. Soc.*, **81**, 5065 (1959).

7 Experimental Data for Thermal Unimolecular Reactions

In order to compare the various theoretical treatments of unimolecular reactions it is necessary to be able to refer to experimental data which are reliable. As shown in Chapter 1, these are not always easily established and many of the early examples of so-called unimolecular reactions have since been discredited. The present chapter is an attempt to survey the various types of unimolecular reaction which have received much attention during the past ten to twenty years. The list of reactions given is fairly comprehensive and includes in particular those reactions which are most suited to the application of unimolecular reaction rate theories. The literature up to late 1970 is covered.

Since it is only molecules of moderate complexity (e.g. five to twelve atoms per molecule, see Figure 7.1) which have readily observable transition pressures, it is evident that most unimolecular reactions are studied in their high-pressure first-order regions. Such experimental studies are described in section 7.1. The differences between the various theoretical treatments are less marked at high pressures, and the main interest in this type of study lies in establishing the reaction mechanisms. Section 7.2 deals more specifically with molecules whose unimolecular reactions have been extensively studied in the fall-off region, and the applications of theoretical treatments, particularly RRKM calculations, are described. There are many unimolecular reactions which produce free radicals, and large free radicals can often decompose unimolecularly to give smaller radicals or atoms. It is convenient to discuss these types of reaction separately in sections 7.3 and 7.4 respectively.

7.1 UNIMOLECULAR REACTIONS AT HIGH PRESSURES

The major groups of unimolecular reactions which have been investigated to date are the pyrolyses of small-ring compounds and other hydrocarbons (reviewed by Frey[1] and Frey and Walsh[2]) and the pyrolyses of alkyl halides and esters (reviewed by Maccoll[3, 4]). It is well established that homogeneous unimolecular reactions can occur for many molecules in these groups, and the details of these investigations are important in establishing the unimolecular mechanisms. In the following sections the data are reviewed for various groups classified primarily according to the structure of the reactants.

7.1.1 Cyclopropane and its derivatives (Tables 7.1–7.4)

The isomerization of cyclopropane to propene has played an important part in the development of unimolecular reaction rate theories and it has already been discussed in this connection in earlier sections. The isomerization (7.1) appears to be free of complications and the rate-constant at

$$\begin{array}{c} CH_2 \\ / \quad \backslash \\ CH_2-CH_2 \end{array} \longrightarrow CH_3-CH=CH_2 \tag{7.1}$$

420–535 °C is given by (7.2);[5] in this and subsequent Arrhenius equations

$$\log_{10} k_\infty/s^{-1} = 15\cdot45 - 65\cdot6/\theta \tag{7.2}$$

θ is given by (7.3). The interest in the reaction at high pressures concerns

$$\theta = 4\cdot576 \times 10^{-3} T/K \tag{7.3}$$

the nature of the transition state for this reaction, and whether indeed it is a simple one-step reaction or whether it proceeds by two steps involving the trimethylene biradical as an intermediate. Attempts to identify the intermediate have proved unsuccessful[6] and it has been claimed[7] that the lifetime is too short (about $10^{-10\cdot5}$ s at 445 °C) for identification by normal methods.† Alkyl-substituted cyclopropanes undergo structural isomerizations with activation energies which tend to be lower than that observed for the isomerization of cyclopropane to propene (see Table 7.1); this would be expected on the basis of a biradical intermediate which would be stabilized by methyl substitution.

In addition to structural isomerization, dialkyl- and dideuterio-substituted cyclopropanes undergo a faster geometrical isomerization,

$$\begin{array}{c} \triangle \;\rightleftarrows\; \triangle^D \\ | \quad | \qquad | \\ D \;\; D \qquad D \end{array} \tag{7.4}$$

e.g. (7.4). The similarity of the Arrhenius parameters for the structural and geometrical isomerizations (see Tables 7.1 and 7.2) and the similar

† For discussion of possible biradical structures see Y. Jean and L. Salem, *Chem. Commun.*, 382 (1971), and references cited therein.

low-pressure behaviour of the rate-constants (see section 7.2.1) suggest that the two reactions may proceed through a common biradical intermediate.

Ingenious attempts have been made to clarify the mechanism of the geometrical isomerization by studying optically active 1,2-disubstituted cyclopropanes, which can also undergo racemization, (7.5). If the CRS groups invert individually, geometrical isomerization will predominate,

$$(7.5)$$

whereas a concerted inversion of the two CRS groups will lead to predominant racemization of the initial enantiomer. A biradical mechanism could lead to either result depending on the assumptions made about the stereochemistry of the ring-opening and ring-closure and the relative rates of rotation of the groups in the biradical. The results for the three compounds studied[8, 9] do not seem at first sight to be consistent with a common mechanism, but can in fact be reconciled with a common scheme involving trimethylene biradical intermediates.[9] The different observed behaviours are explained on the basis that recyclization of the biradicals and isomerization (by rotation about the single bonds) are delicately balanced processes whose relative rates are sensitive to the nature of the substituents present. A recent paper concerning the thermal isomerization of cis-2,3-dideuteriovinylcyclopropane[10] supports the biradical mechanism for this compound.

O'Neal and Benson[11, 12] have carried out an extensive set of ART calculations based on biradical mechanisms for the reactions of cyclopropane and cyclobutane derivatives. The initial ring-opening (ro) step was assumed to be at equilibrium in the basic scheme:

The high-pressure rate-constant for formation of product was calculated as $k_\infty = K_{ro} k_{Hmig}$, so that

$$A_\infty = \exp(\Delta S_{ro}^0/R)\{(ekT/h)\exp(\Delta S_{Hmig}^{\pm}/R)\}$$

The entropy changes were estimated semiempirically by assigning vibration frequencies and rotational data for the species involved, and hence the A-factors were predicted. Activation energies were similarly derived from a small number of energy parameters chosen empirically to fit the data for particular compounds. Various types of reaction were considered, including the structural and geometrical isomerization of simple cyclopropanes, the ring-expansion of vinyl cyclopropanes (see below), and ring-opening of cyclobutanes, and the calculated A-factors were in very good general agreement with the experimental values. It should be pointed out, however, that the success of such calculations does not in itself provide any real evidence that the assumed biradical mechanism is correct. If the biradical intermediate was deleted from the above scheme and the same activated complex was reached directly in a single step from the reactant, the predicted A-factor would be

$$A_\infty = (ekT/h)\exp\{[S(\text{complex}) - S(\text{reactant})]/R\}$$

which is identical with the previous expression since

$$S(\text{complex}) - S(\text{reactant}) = \Delta S_{\text{ro}}^0 + \Delta S_{\text{Hmig}}^{\ddagger}$$

Among the halogen-substituted cyclopropanes, the Arrhenius parameters for the isomerization of monofluorocyclopropane are comparable with those for the alkyl-substituted cyclopropanes and, in view of the strength of the carbon–fluorine bond, it seems reasonable to assume that an H-atom migration mechanism is also applicable in this case. The chloro- and bromo-derivatives isomerize at substantially greater rates, chlorocyclopropane giving only 3-chloropropene and bromocyclopropane giving 3-bromopropene which isomerizes further to 1-bromopropene. In view of the much lower activation energies for these isomerizations compared with that of cyclopropane itself, it has been suggested that migration of the halogen atom rather than a hydrogen atom occurs.[13] Chlorine atom migration has also been proposed for the isomerization of 1,1-dichlorocyclopropane which occurs according to the equation (7.6).

$$\begin{array}{c} CH_2 \\ | \quad \diagdown CCl_2 \\ CH_2 \diagup \end{array} \longrightarrow CH_2Cl \cdot CCl = CH_2 \qquad (7.6)$$

The structure of the observed product does not support a biradical mechanism, since it could not be formed from the most stable biradical intermediate $\overset{\cdot}{C}Cl_2CH_2\overset{\cdot}{C}H_2$.[14] The thermal isomerization of cis-1,1-dichloro-2,3-dimethylcyclopropane is a thousand times faster than that of 1,1-dichlorocyclopropane, and produces stereospecifically trans-3,4-dichloropent-2-ene with no geometrical isomerization of the reactant.[15]

It has been concluded that the results support a concerted mechanism rather than a biradical mechanism for this compound.

The ring-expansion of vinylcyclopropanes to cyclopentenes [e.g. (7.7)[16]] has been extensively studied (see Table 7.3) and is believed to proceed through a biradical intermediate which can ring-close to give either the 3- or 5-membered ring. The lower activation energies for these reactions as compared with ring opening of the simple cyclopropanes (Table 7.1) can be interpreted quantitatively in terms of resonance stabilization of the biradical intermediate;[11, 12] it is unlikely to be coincidental that the lowering of the activation energy is about equal to the resonance energy of the allyl radical.

$$\text{(7.7)}$$

Other types of reaction which have been observed for substituted cyclopropanes (see Table 7.4) are (a) the rearrangement of methylenecyclopropanes [e.g. (7.8)[17]], (b) the conversion of cis-1-methyl-2-vinylcyclopropane to cis-hexa-1,4-diene by 1,5-hydrogen shift with ring-rupture,[18]

$$\text{(7.8)}$$

(c) the pyrolysis of cyclopropylamine to N-propylidenecyclopropylamine[19] with a unimolecular rate-determining step as in (7.9), and (d) the reactions

$$\triangleright\text{—NH}_2 \longrightarrow \overset{CH_3}{\underset{CH_2}{|}}\text{—CH=NH} \xrightarrow{\triangleright\text{—NH}_2} \overset{CH_3}{\underset{CH_2}{|}}\text{—CH=N—}\triangleleft + NH_3 \quad \text{(7.9)}$$

of various fully fluorinated compounds such as perfluorocyclopropane[20] which decomposes according to (7.10), and perfluoroallylcyclopropane[21] which similarly loses difluorocarbene to give perfluoropenta-1,4-diene.

$$(CF_2)_3 \rightarrow C_2F_4 + CF_2 \quad \text{(7.10)}$$

The idea that chlorine substitution might stabilize free-radical intermediates led to a study of chlorine-substituted vinyl cyclopropanes.[22] The isomerization of 1,1-dichloro-2-methyl-2-vinylcyclopropane yields the expected 4,4-dichloro-1-methylcyclopentene and the rate of isomerization is much greater than that of the corresponding non-chlorinated compound, in support of a biradical mechanism (7.11).

$$\text{(7.11)}$$

Table 7.1 High-pressure Arrhenius parameters for ring-opening reactions of simple cyclopropane derivatives

Reactant	Products	$\log_{10} A_\infty/$ s^{-1}	$E_\infty/$ kcal mol^{-1}	Ref.
cyclopropane	propene	15·5	65·6	a
cis-cyclopropane-d_2	propene-d_2	15·1	65·4	81
methylcyclopropane	but-1-ene	14·4	62·8	
	cis-but-2-ene	14·2	62·7	
	trans-but-2-ene	14·6	65·2	b
	isobutene	14·3	65·1	
	overall	14·8	63·2	
1,2-dideuterio-3-methylcyclopropane	dideuteriobutenes[k]	14·4	62·3	87
ethylcyclopropane	pentenes	14·4	61·6	86
1,1-dimethylcyclopropane	methylbutenes	15·1	62·6	c
cis-1,2-dimethylcyclo-propane	cis-pent-2-ene	13·9	61·4	
	trans-pent-2-ene	14·0	61·2	
	2-methylbut-1-ene	13·9	61·9	d
	2-methylbut-2-ene (see also Table 7.2)	14·1	62·3	
trans-1,2-dimethyl-cyclopropane	cis-pent-2-ene	14·4	63·6	
	trans-pent-2-ene	14·3	62·9	
	2-methylbut-1-ene	13·9	61·9	d
	2-methylbut-2-ene (see also Table 7.2)	14·1	62·3	
1,1,2,2-tetramethyl-cyclopropane	2,4-dimethylpent-2-ene	15·5[e]	64·7[e]	e
1,1-diethylcyclo-propane	3-ethylpent-1-ene	15·0	63·8	
	3-ethylpent-2-ene	14·8	63·4	
	2-ethylbuta-1,3-diene + CH$_4$	15·4	65·9	f
	overall	15·7	64·9	
trifluoromethyl-cyclopropane	4,4,4-trifluorobut-1-ene	14·4	67·3	
	cis-1,1,1-trifluorobut-2-ene	14·2	65·2	
	trans-1,1,1-trifluorobut-2-ene	14·0	65·0	g
	trifluoroisobutene	13·9	69·5	
	overall	14·6	65·6	
2,2,2-trifluoroethyl-cyclopropane	5,5,5-trifluoropent-1-ene	13·9	63·4	
	cis-5,5,5-trifluoropent-2-ene	13·8	63·3	
	trans-5,5,5-trifluoro-pent-2-ene	13·7	63·1	g
	overall	14·4	63·6	
fluorocyclopropane	fluoropropenes	14·6	61·0	h
chlorocyclopropane	3-chloropropene	14·8	56·2	13

(*cont.*)

Table 7.1   *(cont.)*

Reactant	Products	$\log_{10} A_\infty/$ s^{-1}	$E_\infty/$ kcal mol^{-1}	Ref.
bromocyclopropane	bromopropenes	13·5	47·3	13
1,1-dichlorocyclopropane	2,3-dichloropropene	15·1	57·8	14
cis-1,1-dichloro-2,3- dimethylcyclopropane	*trans*-3,4-dichloropent- 2-ene	13·7	44·6	15
(2-methylpropenyl)- cyclopropane	5-methylhexa-1,4-diene	14·6	56·7	*i*
	cis-2-methylhexa-2,4-diene	13·3	53·0	
	trans-2-methylhexa-2,4- diene	13·3	52·1	
	overall (see also Table 7.3)	14·3	53·8	
cyclopropene	propyne	12·1[j]	35·2[j]	*j*
1-methylcyclopropene	but-2-yne (+ca 10% butadienes)	11·4[j]	34·7[j]	*j*

[a] Ref. 5; see also ref. 78; E. S. Corner and R. N. Pease, *J. Amer. Chem. Soc.*, **67**, 2067 (1945); D. W. Johnson, Ph.D. Thesis, University of Oklahoma, 1969 [cf. *Diss. Abs.* (*B*), **30**, 1658 (1969–70)]; and F. D. Tabbutt, *U.S. Clearinghouse Fed. Sci. Tech. Inform.*, *Report AD* 660591 (1967). [b] Parameters calculated by Placzek and Rabinovitch[g] from data of Setser and Rabinovitch,[87] probably with correction to k_∞. Similar results were calculated[g] from the data of Chesick.[85] [c] M. C. Flowers and H. M. Frey, *J. Chem. Soc.*, 3953 (1959); see also ref. 88. [d] M. C. Flowers and H. M. Frey, *Proc. Roy. Soc.* (*A*), **260**, 424 (1961). [e] H. M. Frey and D. C. Marshall, *J. Chem. Soc.*, 3052 (1962), recalculated by H. M. Frey and R. Walsh, ref. 2; C. Blumstein, D. Henfling, C. M. Sharts and H. E. O'Neal, *Internat. J. Chem. Kinetics*, **2**, 1 (1970), have discussed complicating decomposition reactions and suggested $\log_{10} k/\mathrm{s}^{-1} = 14\cdot8\text{–}62\cdot3/\theta$ for the isomerization. [f] H. M. Frey and D. C. Marshall, *J. Chem. Soc.*, 191 (1965). [g] D. W. Placzek and B. S. Rabinovitch, *J. Phys. Chem.*, **69**, 2141 (1965). [h] F. Casas, J. A. Kerr and A. F. Trotman-Dickenson, *J. Chem. Soc.*, 3655 (1964). [i] C. S. Elliott and H. M. Frey, *J. Chem. Soc.*, 345 (1965). [j] R. Srinivasan, *J. Amer. Chem. Soc.*, **91**, 6250 (1969); results at 60 Torr. [k] The product distribution is very similar to that from methylcyclopropane; individual Arrhenius parameters have been given.[87]

Table 7.2   High-pressure Arrhenius parameters for the geometrical isomerization of cyclopropane derivatives

Reactant	Products	$\log_{10} A_\infty/$ s^{-1}	$E_\infty/$ kcal mol^{-1}	Ref.
trans-1,2-dideuterio-cyclopropane	cis-compound	16·1[81]	65·1	81
cis-1,2-dimethyl-cyclopropane	trans-compound	15·3	59·4	a
cis-1-ethyl-2-methyl-cyclopropane	trans-compound	15·1	58·9	b
cis-1,2,3-trimethyl-cyclopropane	trans-compound	15·8	61·0	c
cis,trans-1,2-dideuterio-3-methylcyclopropane	cis,cis- and trans,trans-compounds	15·4[d]	60·5[d]	87
trans-1-methyl-2-vinyl-cyclopropane	cis-compound[e]	14·7	48·6	e
methyl chrysanthemate[f]	trans-compound	12·5	43·1	f

[a] M. C. Flowers and H. M. Frey, *Proc. Roy. Soc.* (*A*), **257**, 122 (1960). [b] C. S. Elliott and H. M. Frey, *J. Chem. Soc.*, 900 (1964). [c] H. M. Frey and D. C. Marshall, *J. Chem. Soc.*, 5717 (1963). [d] The measured rate-constants include the rate-constants for interconversion of the two products.[87] [e] See also Table 7.3. *Trans → cis* isomerization determines the rate of production of *cis*-hexa-1,4-diene (cf. reaction of *cis*-compound, Table 7.4).[18] [f] T. Hanafusa, M. Ohnishi, M. Mishima and Y. Yukawa, *Chem. Ind.* (*Lond.*), 1050 (1970); the compound is *cis*-1-carbomethoxy-2,2-dimethyl-3-(2-methylpropenyl)-cyclopropane.

Table 7.3   High-pressure Arrhenius parameters for ring expansion of vinylcyclopropanes

Reactant	Products	$\log_{10} A_\infty/$ s^{-1}	$E_\infty/$ kcal mol^{-1}	Ref.
vinylcyclopropane	cyclopentene [reaction (7.7)]	13·5	49·6	16
perfluorovinylcyclo-propane	perfluorocyclopentene	13·9	34·6	21
trans-but-1-enyl-cyclopropane	3-ethylcyclopentene	13·8	50·0	a
isopropenylcyclo-propane	1-methylcyclopentene	13·9	50·9	b
trans-buta-1,3-dienyl-cyclopropane	3-vinylcyclopentene	13·4	44·5	c
		14·0	51·3	d
		14·3	51·1	d
1-methyl-1-vinyl-cyclopropane	1-methylcyclopentene	14·1	49·4	e
1-methyl-1-isopropenyl-cyclopropane	1,2-dimethylcyclo-pentene	14·1	50·5	f
trans-1-methyl-2-vinyl-cyclopropane	4-methylcyclopentene[g]	13·7	48·6	g
(2-methylpropenyl)-cyclopropane	{ 3,3-dimethylcyclopentene { See also Table 7.1	14·0	54·6	h

[a] R. J. Ellis and H. M. Frey, J. Chem. Soc., 4188 (1964). [b] H. M. Frey and D. C. Marshall, J. Chem. Soc., 3981 (1962). [c] H. M. Frey and A. Krantz, J. Chem. Soc. (A), 1159 (1969). [d] G. R. Branton and H. M. Frey, J. Chem. Soc. (A), 1342 (1966). [e] R. J. Ellis and H. M. Frey, J. Chem. Soc., 959 (1964). [f] R. J. Ellis and H. M. Frey, J. Chem. Soc., 4289 (1965). [g] Ref. 18; see Table 7.2 for an alternative reaction. [h] C. S. Elliott and H. M. Frey, J. Chem. Soc., 345 (1965).

Table 7.4 High-pressure Arrhenius parameters for other reactions of cyclopropane derivatives

Reactant	Products	$\log_{10} A_\infty/$ s^{-1}	$E_\infty/$ kcal mol^{-1}	Ref.
(a) *Ring opening by 1,5 hydrogen migration in 1-alkyl-2-vinylcyclopropanes*				
cis-1-methyl-2-vinyl-cyclopropane	*cis*-hexa-1,4-diene	11·0	31·2	18
1,1-dimethyl-2-vinyl-cyclopropane	2-methyl-*cis*-hexa-1,4-diene	11·4	33·5	*a*
1,1-diethyl-2-vinyl-cyclopropane	3-ethylhepta-*trans*-2,*cis*-5-diene	11·3	33·7	*b*
	cis,*cis*-diene	11·3	33·6	
(b) *Other reactions*				
perfluorocyclopropane	$C_2F_4 + \ddot{C}F_2$	13·3	38·6	20
perfluoroallylcyclo-propane	perfluoropenta-1,4-diene $+ \ddot{C}F_2$	14·8	42·7	21
ethylidenecyclopropane	1-methyl-2-methylene-cyclopropane	14·0	40·7	*c*
[reaction (7.8)]				
reverse reaction, (−7.8)		13·9	40·1	
cyclopropylamine	propylideneamine, (7.9)	15·1	57·8	19

a H. M. Frey and R. K. Solly, *Internat. J. Chem. Kinetics*, **1**, 473 (1969). *b* H. M. Frey and R. K. Solly, *J. Chem. Soc.* (*B*), 996 (1970). *c* Arrhenius parameters for forward and reverse reactions calculated from data of ref. 17. See also J. J. Gajewski, *J. Amer. Chem. Soc.*, **90**, 7178 (1968), and W. von E. Doering and H. D. Roth, *Tetrahedron*, **26**, 2825 (1970), for discussions of related reactions.

7.1.2 Cyclobutane and its derivatives (Table 7.5)

The pyrolysis of cyclobutane is a homogeneous unimolecular reaction yielding two molecules of ethylene. The high-pressure rate-constant is given by (7.12) over the temperature range 420–468°C. The major interest in the high-pressure region again concerns the nature of the activated

$$\log_{10} k_\infty/s^{-1} = 15\cdot60 - 62\cdot5/\theta \qquad (7.12)$$

complex. A concerted mechanism involving the simultaneous breaking of two opposite ring bonds has been considered, but is difficult to reconcile with the high *A*-factor. More recently the theoretically based Woodward–Hoffman rules[23] have shown that a simple concerted process is forbidden on grounds of conservation of orbital symmetry. Biradical mechanisms are supported by the fact that suitably substituted derivatives of cyclobutane (e.g. 1,2-dimethylcyclobutane) can undergo *cis–trans* isomerization as well as pyrolysis, although this reaction is relatively less important than in the case of cyclopropane derivatives. Similarly, the decomposition of the cis- and trans-compounds (1) and (2) proceeded partly to give but-2-enes

and cyclopentene, and the butenes were formed neither with retention nor inversion of configuration.[24] The forbidden simple concerted process would proceed with retention of configuration, while an allowed high-energy concerted process would give 100% inversion. Pyrolysis of the cis-compound (3) gave predominantly trans-CHD:CHD, however,[25] and

it may well be that the presence of methyl groups in (1) and (2) actually affects the course of the reaction, as in the case of optically active cyclopropanes (section 7.1.1). O'Neal and Benson[11, 12] have successfully predicted the Arrhenius parameters for the reactions of cyclobutane derivatives on the basis of biradical mechanisms, in a similar manner to the calculations for cyclopropane derivatives discussed in section 7.1.1.

Vinylcyclobutane does not appear to have been studied, but isopropenylcyclobutane undergoes a ring-expansion reaction similar to the cyclopropane analogue, producing 1-methylcyclohexene.[26] In addition, an alternative reaction path leads to the formation of ethylene and isoprene. The similarity of the Arrhenius parameters observed for the two processes suggests that both proceed through the same resonance-stabilized biradical as shown in (7.13). The allylic stabilization results in an activation energy

$$(7.13)$$

(ca 50 kcal mol^{-1}) which is substantially lower than that for the normal cyclobutane decomposition (63 kcal mol^{-1}). The isomerizations of 1,2-dimethylenecyclobutane-d_4 (7.14) and related compounds are also thought to involve a resonance-stabilized tetramethylene radical.[27]

Theoretical treatments of tetramethylene biradical structures have been given.[28]

$$(7.14)$$

The other reactions noted for cyclopropane derivatives, viz. the ethylidenecyclopropane rearrangement and the *cis*-1-methyl-2-vinylcyclo-propane isomerization, do not appear to have been observed for similar derivatives of cyclobutane. The decomposition of methylenecyclobutane is a unimolecular reaction producing allene and ethylene.[29, 30]

Table 7.5 High-pressure Arrhenius parameters for the reactions of cyclo-butane and its derivatives

Reactant	Products	$\log_{10} A_\infty/$ s^{-1}	$E_\infty/$ kcal mol^{-1}	Ref.
cyclobutane	ethylene	15·6	62·5	a
cyclobutane-d_8	ethylene-d_4	16·0	63·9	b
methylcyclobutane	ethylene + propene	15·4	61·2	c
ethylcyclobutane	ethylene + but-1-ene	15·6	62·0	d
propylcyclobutane	ethylene + pent-1-ene	15·5	61·6	e
isopropylcyclobutane	ethylene + 3-methyl-but-1-ene	15·6	62·6	f
t-butylcyclobutane	ethylene + 3,3-dimethyl-but-1-ene	15·9	63·9	g
cis-1,2-dimethyl-cyclobutane	*trans*-compound	14·8	60·1	h
	propene	15·5	60·4	i, j
	ethylene + but-2-ene	15·6	63·0	i, j
trans-1,2-dimethyl-cyclobutane	*cis*-compound	14·6	61·3	h
	propene	15·4	61·6	h, j
	ethylene + but-2-ene	15·5	63·4	h, j
1,1,3,3-tetramethyl-cyclobutane	isobutene	16·3	65·2	k
perfluorocyclobutane	perfluoroethylene	16·0	74·1	l
chlorocyclobutane	ethylene + vinyl chloride	14·8	60·1	m
	HCl + buta-1,3-diene	13·6	55·2	
cyclobutanol	ethylene + acetaldehyde	15·1	60·1	n
▱=O	ethylene + CH$_2$CO	14·6	52·0	o
	cyclopropane + CO	14·4	58·0	p
▱—CHO	ethylene + CH$_2$:CH·CHO	14·4	53·3	q
▱—CO·Me	ethylene + CH$_2$:CH·CO·Me	14·5	54·5	r

(cont.)

Table 7.5 (*cont.*)

Reactant	Products	$\log_{10} A_\infty/$ s^{-1}	$E_\infty/$ kcal mol^{-1}	Ref.
(cyclobutane)—CO·Et	ethylene $+ CH_2:CH \cdot CO \cdot Et$	14·5	54·2	s
(cyclobutane)—CO$_2$Me	ethylene $+ CH_2:CH \cdot CO_2Me$	14·9	57·3	t
(cyclobutane)=CH$_2$	ethylene $+ CH_2:C:CH_2$	15·7	63·3	29, 30
(methylenecyclobutane)	{ (cyclohexene)	14·5	51·0	} 26
	ethylene $+ CH_2:CH \cdot CMe:CH_2$	14·6	51·0	

^{a} Ref. 91; see also F. D. Tabbutt, *U.S. Clearinghouse Fed. Sci. Tech. Inform., Report A D* 660591 (1967). ^{b} Isotopic ratios of J. Langrish and H. O. Pritchard, *J. Phys. Chem.*, **62**, 761 (1958), combined with absolute measurements of ref. *a*. ^{c} M. N. Das and W. D. Walters, *Zeit. Phys. Chem.* (*Frankfurt*), **15**, 22 (1958). ^{d} R. E. Wellman and W. D. Walters, *J. Amer. Chem. Soc.*, **79**, 1542 (1957); see also J. Aspden, N. A. Khawaja, J. Reardon and D. J. Wilson, *J. Amer. Chem. Soc.*, **91**, 7580 (1969). ^{e} S. M. Kellner and W. D. Walters, *J. Phys. Chem.*, **65**, 466 (1961). ^{f} M. Zupan and W. D. Walters, *J. Phys. Chem.*, **67**, 1845 (1963). ^{g} A. T. Cocks and H. M. Frey, *J. Chem. Soc.* (*A*), 2566 (1970). ^{h} H. R. Gerberich and W. D. Walters, *J. Amer. Chem. Soc.*, **83**, 4884 (1961). ^{i} H. R. Gerberich and W. D. Walters, *J. Amer. Chem. Soc.*, **83**, 3935 (1961). ^{j} For confirmatory data on overall disappearance of reactant see P. J. Conn, Ph.D. Thesis, University of Oregon, 1966 [cf. *Diss. Abs.* (*B*), **27**, 2311 (1966–7)]. ^{k} A. T. Cocks and H. M. Frey, *J. Chem. Soc.* (*A*), 1671 (1969); see also T. A. Babcock, *J. Amer. Chem. Soc.*, **91**, 7622 (1969). ^{l} B. Atkinson and A. B. Trenwith, *J. Chem. Phys.*, **20**, 754 (1952); J. M. Simmie, W. J. Quiring and E. Tschuikow-Roux, *J. Phys. Chem.*, **73**, 3830 (1969). ^{m} A. T. Cocks and H. M. Frey, *J. Amer. Chem. Soc.*, **91**, 7583 (1969). ^{n} P. M. Stacy, Ph.D. Thesis, University of Rochester, 1968 [cf. *Diss. Abs.* (*B*), **29**, 1632 (1968–9)]. ^{o} M. N. Das, F. Kern, T. D. Coyle and W. D. Walters, *J. Amer. Chem. Soc.*, **76**, 6271 (1954). ^{p} A. T. Blades, *Can. J. Chem.*, **47**, 615 (1969). ^{q} B. C. Roquitte and W. D. Walters, *J. Amer. Chem. Soc.*, **84**, 4049 (1962). ^{r} L. G. Daignault and W. D. Walters, *J. Amer. Chem. Soc.*, **80**, 541 (1958). ^{s} B. C. Roquitte and W. D. Walters, *J. Phys. Chem.*, **68**, 1606 (1964). ^{t} M. Zupan and W. D. Walters, *J. Amer. Chem. Soc.*, **86**, 173 (1964).

7.1.3 Cyclobutene and its derivatives (Table 7.6)

Cyclobutene isomerizes to buta-1,3-diene by a homogeneous uni-molecular reaction. The rate-constant over the temperature range 131–175°C is given[31] by (7.15). Substituted cyclobutenes undergo analogous reactions, and Table 7.6 shows that the A factors are normal, while the

$$\log_{10} k_\infty/s^{-1} = 13·40 - 32·9/\theta \qquad (7.15)$$

activation energies are very low compared with those for the pyrolyses of the cyclobutanes. Frey[1] has proposed an activated complex involving a simultaneous twisting of the cyclobutene ring and stretching of the carbon–carbon bond opposite the double bond (7.16). Benson, on the

$$\text{(7.16)}$$

other hand, has suggested[32] that an ion-pair or semi-ion-pair intermediate might be involved. Substitution of an alkyl group on carbon atoms C_1 or C_2 is found to increase the activation energy, whereas substitution on C_3 or C_4 decreases it by roughly the same amount. Cyclobutenes substituted on both C_3 and C_4 are found to isomerize exclusively by a conrotatory mechanism as predicted by the Woodward–Hoffman rules.[23] This is illustrated in (7.17) for the *cis*-3,4-disubstituted compounds which yield only the *cis-trans*-diene product. The reverse process of formation of a cyclobutene from a 1,3-diene is now known to occur measurably in some systems.[33]

$$\text{(7.17)}$$

Table 7.6 High-pressure Arrhenius parameters for the isomerization of cyclobutene and its derivatives

Reactant	Product	$\log_{10} A_\infty/$ s^{-1}	$E_\infty/$ kcal mol^{-1}	Ref.
cyclobutene	buta-1,3-diene	13·3	32·7	99
		13·4	32·9	31
cyclobutene-d_6	buta-1,3-diene-d_6	13·7	33·8	103
1-methylcyclobutene	2-methylbuta-1,3-diene	13·8	35·1	a
3-methylcyclobutene	penta-1,*trans*-3-diene	13·5	31·6	b
1-ethylcyclobutene	2-ethylbuta-1,3-diene	13·8	34·8	c
1-propylcyclobutene	2-propylbuta-1,3-diene	13·6	34·5	d
1-isopropylcyclobutene	2-isopropylbuta-1,3-diene	13·6	34·7	d
1-allylcyclobutene	3-methylenehexa-1,5-diene	13·5	34·2	d
1-cyclopropylcyclobutene	2-cyclopropylbuta-1,3-diene	13·5	34·1	d

(*cont.*)

Table 7.6 (*cont.*)

Reactant	Product	$\log_{10} A_\infty/$ s^{-1}	$E_\infty/$ kcal mol^{-1}	Ref.
1,2-dimethylcyclobutene	2,3-dimethylbuta-1,3-diene	13·8	36·0	e
1,3-dimethylcyclobutene	2-methylpenta-1,*trans*-3-diene	13·7	33·0	f
1,4-dimethylcyclobutene	3-methylpenta-1,*trans*-3-diene	13·5	33·4	f
3,3-dimethylcyclobutene	4-methylpenta-1,3-diene	13·9	36·1	g
3-ethyl-3-methylcyclo-butene	*cis*-4-methylhexa-1,3-diene	13·5	35·2	h
	trans-4-methylhexa-1,3-diene	13·5	35·9	
3,3-diethylcyclobutene	4-ethylhexa-1,3-diene	13·5	34·7	h
1,3,3-trimethylcyclobutene	2,4-dimethylpenta-1,3-diene	13·9	37·0	g
cis-3,4-dimethyl-cyclobutene	*cis,trans*-hexa-2,4-diene	13·9	34·0	i
1,2,3,*cis*-4-tetramethyl-cyclobutene	3,4-dimethylhexa-*cis*-2,*trans*-4-diene	14·2	37·4	j
1,2,3,*trans*-4-tetramethyl-cyclobutene	3,4-dimethylhexa-*trans*-2,*trans*-4-diene	13·9	33·6	j
perfluorocyclobutene	perfluorobuta-1,3-diene	14·1[k]	47·1[k]	k
perfluoro(1,2-dimethyl-cyclobutene)	perfluoro(2,3-dimethyl-buta-1,3-diene)	13·6[l]	46·0[l]	l

[a] H. M. Frey, *Trans. Faraday Soc.*, **58**, 957 (1962). [b] H. M. Frey, *Trans. Faraday Soc.*, **60**, 83 (1964). [c] H. M. Frey and R. F. Skinner, *Trans. Faraday Soc.*, **61**, 1918 (1965). [d] D. Dickens, H. M. Frey and R. F. Skinner, *Trans. Faraday Soc.*, **65**, 453 (1969). [e] H. M. Frey, *Trans. Faraday Soc.*, **59**, 1619 (1963). [f] H. M. Frey, *Trans. Faraday Soc.*, **61**, 861 (1965). [g] H. M. Frey, B. M. Pope and R. F. Skinner, *Trans. Faraday Soc.*, **63**, 1166 (1967). [h] H. M. Frey and R. K. Solly, *Trans. Faraday Soc.*, **65**, 448 (1969). [i] R. Srinivasan, *J. Amer. Chem. Soc.*, **91**, 7557 (1969). [j] G. R. Branton, H. M. Frey and R. F. Skinner, *Trans. Faraday Soc.*, **62**, 1546 (1966). [k] E. W. Schlag and W. B. Peatman, *J. Amer. Chem. Soc.*, **86**, 1676 (1964); $\log_{10} k_{reverse}/s^{-1} = 12\cdot6-36\cdot6/\theta$. [l] J. P. Chesick, *J. Amer. Chem. Soc.*, **88**, 4800 (1966); $\log_{10} k_{reverse}/s^{-1} = 12\cdot0-35\cdot4/\theta$. See also B. Atkinson and P. B. Stockwell, *J. Chem. Soc. (B)*, 984 (1966).

7.1.4 Polycyclic systems (Tables 7.7–7.11)

The high-pressure Arrhenius parameters for a number of unimolecular reactions involving polycyclic compounds are listed in Tables 7.7–7.11. Such compounds possess even more strain energy than the simple cyclo-propane and cyclobutane derivatives, and the release of this strain on formation of an activated complex allows the molecular path to be favoured energetically compared with the radical-chain mechanisms commonly observed in straight-chain hydrocarbon decompositions. O'Neal and Benson have discussed reactions of polycyclic molecules in terms of an ART approach similar to that used for simple ring compounds.[142]

In the tables an attempt has been made to classify the reactions as basically reactions of the cyclopropane, cyclobutane or cyclobutene ring (Tables 7.7, 7.8 and 7.9 respectively). For example, the isomerization of bicyclo[3.2.0]hept-6-ene to cis,cis-cyclohepta-1,3-diene, (7.18), both in the

$$\text{[image: structure]} \longrightarrow \text{[image: structure]} \qquad (7.18)$$

gas phase and in dimethylphthalate solution, is basically a reaction of the cyclobutene system.[34] The reactant is a cis-3,4-disubstituted cyclobutene and would normally be expected to give a cis,trans-diene as product. The cis,cis-stereochemistry of the observed product, together with energetic considerations, shows that the most likely mechanism in this case is the 'forbidden' disrotatory process, and this explains why the activation energy is about 10 kcal mol^{-1} higher than those for the isomerizations of the simple cyclobutenes (46 kcal mol^{-1}, cf. 33–37 kcal mol^{-1}, see Table 7.6).

The situation is not always so simple, however, and some reactions which fall into more than one of the above categories are listed in Table 7.11, along with some reactions which are of different types altogether. An example is the isomerization of bicyclobutane to butadiene, (7.19), which is homogeneous and unimolecular in 'aged' vessels.[35] The A-factor

$$\text{[image: structure]} \longrightarrow \text{[image: structure]} \qquad (7.19)$$

($10^{14 \cdot 0}$ s^{-1}) is smaller than would be expected for complete breaking of a 3-membered ring (cf. cyclopropane, $A = 10^{15 \cdot 5}$ s^{-1}) and the activation energy (40·6 kcal mol^{-1}) is larger than one would expect if all of the difference in strain energy between reactant and product were released in the activated complex. It was therefore suggested[35] that the reaction involves the simultaneous breaking of bonds C_1C_4 and C_2C_3 with the formation of double bonds C_1C_2 and C_3C_4 to create cis-butadiene in a strained configuration.

A recently opened field of study concerns the valence-bond isomerization of benzene derivatives. Dewar benzenes and one prismane have so far been studied; the results are listed in Table 7.10.

Table 7.7 High-pressure Arrhenius parameters for reactions of polycyclic molecules involving predominantly reactions of a cyclopropane ring

Reactant	Products	$\log_{10} A_\infty/$ s^{-1}	$E_\infty/$ kcal mol^{-1}	Ref.
		13·3	57·4	a
		13·9	61·2	
		13·3[b]	38·6[b]	b
		15·9	57·6	c
		14·8[d]	51·5[d]	d
		12·9	33·8	e
		14·0	36·6	f
		14·0	37·8	f
		13·4	33·7	f
		14·8	37·2	f

[a] H. M. Frey and R. C. Smith, *Trans. Faraday Soc.*, **58**, 697 (1962). [b] W. Grimme, *Chem. Ber.*, **98**, 756 (1965); $k = k_f + k_r$. [c] M. C. Flowers and H. M. Frey, *J. Chem. Soc.*, 5550 (1961); for methyl derivative see M. C. Flowers and A. R. Gibbons, *J. Chem. Soc. (B)*, 612 (1971), and for discussion of related matters see also J. J. Gajewski, *J. Amer. Chem. Soc.*, **92**, 3688 (1970). [d] G. J. Carb, *Tetrahedron*, **25**, 1459 (1969); $k = k_f + k_r$. [e] W. R. Roth and K. Enderer, *Annalen*, **733**, 44 (1970). [f] M. J. Jorgenson, T. J. Clark and J. Corn, *J. Amer. Chem. Soc.*, **90**, 7020 (1968).

Table 7.8 High-pressure Arrhenius parameters for reactions of polycyclic molecules involving predominantly reactions of a cyclobutane ring

Reactant	Products	$\log_{10} A_\infty/$ s^{-1}	$E_\infty/$ kcal mol^{-1}	Ref.
		13·9	38·7	a
		14·4	39·2	
		13·4	36·0	b
		11·3	31·0	c
		15·4	64·0	d
	$C_2H_4 + $ ... $[\rightarrow$ ... $+ H_2]$	14·8	60·7	
		15·7	55·0	e
		15·0	56·0	c
		15·5	55·9	f
		15·2^g	49·0^g	g
		16·2^g	53·0^g	g

(*cont.*)

Table 7.8 (*cont.*)

Reactant	Products	$\log_{10} A_\infty/$ s^{-1}	$E_\infty/$ kcal mol^{-1}	Ref.
		12·8	33·5	*h*
		17·2^i	57·4^i	*i*

^{a} J. P. Chesick, *J. Amer. Chem. Soc.*, **84**, 3250 (1962). ^{b} C. Steel, R. Zand, P. Hurwitz and S. G. Cohen, *J. Amer. Chem. Soc.*, **86**, 679 (1964). ^{c} R. Srinivasan, *Internat. J. Chem. Kinetics*, **1**, 133 (1969). ^{d} R. J. Ellis and H. M. Frey, *J. Chem. Soc.*, 4184 (1964). ^{e} R. Srinivasan and A. A. Levi, *J. Amer. Chem. Soc.*, **85**, 3363 (1963). ^{f} R. Srinivasan and A. A. Levi, *J. Amer. Chem. Soc.*, **86**, 3756 (1964). ^{g} R. Srinivasan, *J. Amer. Chem. Soc.*, **90**, 2752 (1968); results at 3–5 Torr pressure. ^{h} H. M. Frey, *J. Chem. Soc.*, 365 (1964). ^{i} Ref. *c*; the unimolecular reaction shown is dominated by an alternative surface-dependent reaction giving 1,3-dimethylcyclopentene.

Table 7.9   High-pressure Arrhenius parameters for reactions of polycyclic molecules involving predominantly reactions of a cyclobutene ring

Reactant	Product	$\log_{10} A_\infty/$ s^{-1}	$E_\infty/$ kcal mol^{-1}	Ref.
		13·3	31·6	a
		14·3	45·5	34
		14·1	43·2	b
		14·3	46·7	c
		14·6	44·2	c
		14·2	44·2	c
		14·7	40·5	c

[a] A. T. Cocks and H. M. Frey, *J. Chem. Soc. (B)*, 952 (1970). [b] G. R. Branton, H. M. Frey and R. F. Skinner, *Trans. Faraday Soc.*, **62**, 1546 (1966). [c] H. M. Frey, J. Metcalfe and J. M. Brown, *J. Chem. Soc. (B)*, 1586 (1970).

Table 7.10 High-pressure Arrhenius parameters for reactions of valence-bond isomers of benzene derivatives

Reactant	Product	$\log_{10} A_\infty/$ s^{-1}	$E_\infty/$ kcal mol^{-1}	Ref.
[structure] $-F_6$	[structure] $-F_6$	13·2 13·7	27·6 28·5	a b
[structure] X = H Me CF$_3$ (see footnote c)	[structure]	12·5 14·2 13·8	25·5 30·6 28·2	c
[structure] X = H Me CF$_3$ OMe (see footnote c)		12·4 14·4 13·4 11·7	25·6 29·8 29·6 25·9	
[structure]	[structure]	15·1 15·0, 15·2	37·2 37·2, 37·7	d e
[structure]	[structure]	14·2 14·5 16·6	31·4 33·8 35	d e f
[structure]	[structure]	13·9	33·2	e

[a] E. Ratajczak and A. F. Trotman-Dickenson, *J. Chem. Soc. (A)*, 509 (1968). [b] I. Haller, *J. Phys. Chem.*, **72**, 2882 (1968). [c] P. Cadman, E. Ratajczak and A. F. Trotman-Dickenson, *J. Chem. Soc. (A)*, 2109 (1970); identification of the *reactants* was not conclusive. [d] W. Adam and J. C. Chang, *Internat. J. Chem. Kinetics*, **1**, 487 (1969). [e] J. F. M. Oth, *Rec. Trav. Chim. Pays-Bas*, **87**, 1185 (1968); *Angew. Chem. internat. Edit.*, **7**, 646 (1968). [f] H. Hogeveen and H. C. Volger, *Chem. Commun.*, 1133 (1967); the A-factor (calculated from the quoted $\Delta S^{\ddagger}$) seems unduly high.

Table 7.11 High-pressure Arrhenius parameters for other reactions of polycyclic molecules

Reactant	Products	$\log_{10} A_\infty/$ s^{-1}	$E_\infty/$ kcal mol^{-1}	Ref.
		14·0	40·6	35
		14·5	41·4	a
		14·5	43·3	b
		14·6	46·6	c
		14·1	45·6	d
		14·4	52·3	d
		14·2	26·9	e
		14·0	35·2	f, g
		14·4	35·3	g
		13·9	58·5	h
		13·3	35·6	i
		15·1	46·7	i
		13·4	14·6	j

[a] R. Srinivasan, A. A. Levi and I. Haller, *J. Phys. Chem.*, **69**, 1775 (1965). [b] J. P. Chesick, *J. Phys. Chem.*, **68**, 2033 (1964). [c] M. L. Halberstadt and J. P. Chesick, *J. Amer. Chem. Soc.*, **84**, 2688 (1962). [d] C. Steel, R. Zand, P. Hurwitz and S. G. Cohen, *J. Amer. Chem. Soc.*, **86** 679 (1964). [e] D. M. Golden and J. I. Brauman, *Trans. Faraday Soc.*, **65**, 464 (1969); see also indications of a non-biradical mechanism for the 2-methyl derivative in J. E. Baldwin and A. H. Andrist, *Chem. Commun.*, 1561 (1970). [f] H. M. Frey, R. G. Hopkins, H. E. O'Neal and F. T. Bond, *Chem. Commun.*, 1069 (1969). [g] H. M. Frey and R. G. Hopkins, *J. Chem. Soc.* (*B*), 1410 (1970). [h] L. W. Gay, Ph.D. Thesis, University of Oregon [cf. *Diss. Abs.* (*B*), **29**, 960 (1968–9)]. [i] L. M. Dané, J. W. de Haan and H. Kloosterziel, *Tetrahedron Lett.*, 2755 (1970). [j] H. J. Reich, E. Ciganek and J. D. Roberts, *J. Amer. Chem. Soc.*, **92**, 5166 (1970).

7.1.5   Olefins and polyolefins (Tables 7.12–7.17)

Olefins and polyolefins undergo a variety of unimolecular reactions, most of which are isomerizations typified by the simple *cis–trans* geometrical isomerizations and molecular rearrangements such as cyclizations and the Cope rearrangement. The high-pressure Arrhenius parameters for a number of such reactions are given in Tables 7.12–7.17.

Cis–trans isomerizations (Table 7.12) would appear to be possible examples of genuine unimolecular mechanisms, but unfortunately the experimental data are not always reliable, owing to surface effects and free-radical decomposition reactions. Cundall[36] and Lin and Laidler[37] have reviewed this type of reaction, and much of the earlier work has been discounted on the above grounds. The simplest *cis–trans* olefin isomerization is that of *trans*-1,2-dideuterioethylene.[38] The high-pressure A-factor of $10^{13 \cdot 0}$ s^{-1} is 'normal' and the high-pressure activation energy of 65 kcal mol^{-1} is close to the value which can be calculated from the potential-energy curves for the ground electronic states of the two isomers. The data for other *cis–trans* isomerizations are in the main quite similar (Table 7.12), and it seems likely that the low Arrhenius parameters originally observed in some cases, and attributed to non-adiabatic reaction mechanisms, were due to surface effects or free-radical reactions.

A number of conjugated polyenes undergo cyclization reactions (Table 7.13), an example of such a reaction being (7.20). The fact that

$$\text{(structures)} \longrightarrow \text{(structure)} \longrightarrow \text{(structure)} \qquad (7.20)$$

hexa-1,*cis*-3,5-triene cyclizes to cyclohexa-1,3-diene, whereas hexa-1,*trans*-3,5-triene under similar conditions does not cyclize, is taken as evidence that the reaction proceeds through a cyclic activated complex rather than a linear biradical intermediate.[39]

A cyclic complex is also involved in the rearrangement of 1,3-dienes by 1,5-hydrogen shift (Table 7.14), an example being (7.21).[40] The Arrhenius

$$\begin{array}{ccc}
\text{H}-\text{CH}_2 & \text{H}\cdots\text{CH}_2 & \text{H} \quad \text{CH}_2 \\
\text{CH}_2 \quad \text{CH} & \text{CH}_2 \quad \text{CH} & \text{CH}_2 \quad \text{CH} \\
\text{C}-\text{CH} & \text{C}\cdots\text{CH} & \text{C}=\text{CH} \\
\text{CH}_3 & \text{CH}_3 & \text{CH}_3
\end{array} \qquad (7.21)$$

parameters are very similar to those for the conversion of *cis*-1-methyl-2-vinylcyclopropane to *cis*-hexa-1,4-diene, which is thought to occur by a similar mechanism[18] (see section 7.1.1).

Many 1,5-dienes undergo unimolecular isomerization by the Cope rearrangement (Table 7.15). This reaction can be regarded as the 1,3-shift of an allyl group via a 6-centred activated complex illustrated in (7.22) for

$$\text{(7.22)}$$

3-methylhexa-1,5-diene.[41] The loss of entropy associated with the formation of this cyclic complex is assumed to be responsible for the low A-factors (ca 10^{10} s^{-1}) observed for this type of reaction. Some analogous reactions of allyl esters and vinyl allyl ethers are listed in Tables 7.26 and 7.30.

Diels–Alder adducts are cyclic olefins, and the reverse decompositions of a number of these compounds have been found to exhibit unimolecular characteristics (Table 7.16). Finally, some miscellaneous reactions of cyclic olefins are listed in Table 7.17. These include the elimination of molecular hydrogen, which also occurs in some heterocyclic systems described in the next section.

Table 7.12 High-pressure Arrhenius parameters for *cis–trans* isomerization of olefins

Reactant	Product	$\log_{10} A_\infty /$ s^{-1}	$E_\infty /$ kcal mol^{-1}	Ref.
trans-CHD:CHD	*cis*-CHD:CHD	13·0	65·0	38
cis-MeCH:CHD	*trans*-MeCH:CHD	13·2	61·3	*a*
cis-but-2-ene	*trans*-but-2-ene	13·8	62·8	*b*
		13·4	61·6	*c*
cis-CHF:CHF	*trans*-CHF:CHF	13·2	62·8	*c*
perfluoro-*trans*-but-2-ene	*cis*-compound	13·5	56·4	*d*
cis-CH$_3$CH:CHCN	*trans*-compound	11·0	51·3	*e*
cis-CH$_3$CH:CHCO$_2$Me	*trans*-compound	13·2	57·8	*f*
hepta-1,*trans*-3,-*trans*-5-triene	hepta-1,*cis*-3,*trans*-5-triene	12·3	42·4	*g*
cis-CHCl:CHCl	*trans*-CHCl:CHCl	12·8	56·0	*h*
trans-CHCl:CHCl	*cis*-CHCl:CHCl	12·7	55·3	*h*
2,3-dimethylpenta-1,*cis*-3-diene	*trans*-compound	12·3[i]	45·0[i]	*i*

a M. C. Flowers and N. Jonathan, *J. Chem. Phys.*, **50**, 2805 (1968); **52**, 1623 (1970). *b* B. S. Rabinovitch and K. W. Michel, *J. Amer. Chem. Soc.*, **81**, 5065 (1959). *c* P. M. Jeffers and W. Shaub, *J. Amer. Chem. Soc.*, **91**, 7706 (1969). *d* E. W. Schlag and E. W. Kaiser, *J. Amer. Chem. Soc.*, **87**, 1171 (1965). *e* J. N. Butler and R. D. McAlpine, *Can. J. Chem.*, **41**, 2487 (1963). *f* J. N. Butler and G. J. Small, *Can. J. Chem.*, **41**, 2492 (1963). *g* K. W. Egger and T. Ll. James, *Trans. Faraday Soc.*, **66**, 410 (1970). *h* L. D. Hawton and G. P. Semeluk, *Can. J. Chem.*, **44**, 2143 (1966); see also B. S. Rabinovitch and M. J. Hulatt, *J. Chem. Phys.*, **27**, 592 (1957). *i* Rate parameters refer to overall conversion proceeding through reversible formation of 1,2,3-trimethylcyclobutene.[33]

8

Table 7.13 High-pressure Arrhenius parameters for cyclization reactions of polyolefins

Reactant	Product	$\log_{10} A_\infty/$ s^{-1}	$E_\infty/$ kcal mol^{-1}	Ref.
		11·9	29·9	39
		11·5	29·4	a
		11·2	'33'	a
		10·5	31·8	b
		11·0	17·0	c
		10·7	22·4	d

^{a} B. S. Schatz, Ph.D. Thesis, Oregon State University, 1967 [cf. *Diss. Abs. (B)*, **27**, 3870 (1967)]. ^{b} K. W. Egger, *Helv. Chim. Acta*, **51**, 422 (1968). ^{c} T. D. Goldfarb and L. Lindquist, *J. Amer. Chem. Soc.*, **89**, 4588 (1967). ^{d} R. Huisgen, A. Dahmen and H. Huber, *J. Amer. Chem. Soc.*, **89**, 7130 (1967).

Table 7.14 High-pressure Arrhenius parameters for 1,5-hydrogen transfer in 1,3-dienes

Reactant	Product	$\log_{10} A_\infty/$ s^{-1}	$E_\infty/$ kcal mol^{-1}	Ref.
(CD₂, H diene)	(CD₂H diene)	11·9	36·3	a
(D, CD₂ diene)	(CH₂D, CD₂ diene)	11·9	37·7	a
(H diene)	(diene)	10·8	32·5	b
(H diene)	(diene)	11·2	32·8	}40
(H diene)	(diene)	11·7	36·2	
(H diene)	(diene)	10·8	31·6	}c
(H diene)	(diene)	11·0	34·5	
(H diene)	(diene)	10·3	24·6	d
(H cyclic diene)	(cyclic diene)	12·6	33·3	e
(H diene)	(diene)	11·1	33·2	f

[a] W. R. Roth and J. König, *Ann.*, **699**, 24 (1966). [b] H. M. Frey and B. M. Pope, *J. Chem. Soc. (A)*, 1701 (1966). [c] H. M. Frey and R. K. Solly, *J. Chem. Soc. (A)*, 733 (1969). [d] R. Srinivasan, quoted by L. Skattebøl, *Tetrahedron*, **25**, 4933 (1969). [e] K. W. Egger, *J. Amer. Chem. Soc.*, **89**, 3688 (1967). [f] H. E. O'Neal and H. M. Frey, *Internat. J. Chem. Kinetics*, **2**, 343 (1970).

Table 7.15 High-pressure Arrhenius parameters for Cope rearrangement of 1,5-dienes

Reactant	Product	$\log_{10} A_\infty/$ s^{-1}	$E_\infty/$ kcal mol^{-1}	Ref.
CD_2 CD_2 1,5-hexadiene (3,3,4,4-d_4)	CD_2-CD_2	11·1	35·5	a
(3-methyl-1,5-hexadiene)		10·6	34·2	
		10·5	35·7	
				41
		10·4	35·3	
		10·7	36·7	
		11·1	34·6	
				b
		10·4	36·9	
(cyclohexenyl dinitrile)		10·8	26·2	c
		10·9	25·8	c
		10·4	28·6	c

Table 7.15 (*cont.*)

Reactant	Product	$\log_{10} A_\infty/$ s^{-1}	$E_\infty/$ kcal mol^{-1}	Ref.
		10·0	28·5	d
		12·5	31·7	e
		10·7	22·4	f

[a] V. Toscano and W. von E. Doering, quoted by W. von E. Doering and J. Gilbert, *Tetrahedron, Supplement no.* **7**, 397 (1966). [b] H. M. Frey and R. K. Solly, *Trans. Faraday Soc.*, **65**, 1372 (1969). [c] E. G. Foster, A. C. Cope and F. Daniels, *J. Amer. Chem. Soc.*, **69**, 1893 (1947). [d] H. M. Frey and D. M. Lister, *J. Chem. Soc.* (*A*), 26 (1967). [e] H. M. Frey and A. M. Lamont, *J. Chem. Soc.* (*A*), 1592 (1969). [f] G. S. Hammond and C. D. DeBoer, *J. Amer. Chem. Soc.*, **86**, 899 (1964).

Table 7.16 High-pressure Arrhenius parameters for reverse Diels–Alder reactions

Reactant	Products	$\log_{10} A_\infty/$ s^{-1}	$E_\infty/$ kcal mol^{-1}	Ref.
		12·9 15·3	57·5 66·9	a b
		15·1	66·6	b
		15·2 15·7	62·0 61·8	b c
		13·8	42·8	d
		14·1	46·0	e
		13·0	33·7	f
		14·6[g]	50·2[g]	g

[a] L. Küchler, *Trans. Faraday Soc.*, **35**, 874 (1939). [b] W. Tsang, *Internat. J. Chem. Kinetics*, **2**, 311 (1970); *J. Chem. Phys.*, **42**, 1805 (1965). [c] N. E. Duncan and G. J. Janz, *J. Chem. Phys.*, **20**, 1644 (1952). [d] W. C. Herndon, W. B. Cooper and M. J. Chambers, *J. Phys. Chem.*, **68**, 2016 (1964). [e] W. C. Herndon and J. M. Manion, *J. Org. Chem.*, **33**, 4504 (1968). [f] J. B. Harkness, G. B. Kistiakowsky and W. H. Mears, *J. Chem. Phys.*, **5**, 682 (1937). [g] W. C. Herndon and L. L. Lowry, *J. Amer. Chem. Soc.*, **86**, 1922 (1964); see Table 7.17 for alternative reactions of the same molecule.

Table 7.17 High-pressure Arrhenius parameters for some miscellaneous reactions of cyclic olefins

Reactant	Products	$\log_{10} A_\infty/$ s^{-1}	$E_\infty/$ kcal mol^{-1}	Ref.
	+ H_2	12·0	42·5	a
	+ H_2	12·5	43·0	b
cis-	+ H_2	12·1	40·0	c
	+ H_2	13·0	58·8	d
	+ H_2	12·8	49·9	e
		14·2	52·5	
		12·2	'36'	f
		11·7	'34'	f
		10·9	19·9	g
$-d_5$	$-d_5$	11·9	22·4	g
	4 isomers	13·7	45·1	h

(cont.)

Table 7.17 (cont.)

Reactant	Products	$\log_{10} A_\infty/$ s^{-1}	$E_\infty/$ kcal mol^{-1}	Ref.
		14·2	53·1	i
		14·7	50·6	
		13·5	51·1	j
$-d_6$	$-d_6$	13·5^k	51·4^k	k
Me (equilibrated isomers)	benzene derivatives^l	14·0^l	51·7^l	l

[a] L. W. Gay, Ph.D. Thesis, University of Oregon [cf. Diss. Abs. (B), 29, 960 (1968–9)].
[b] H. M. Frey and D. H. Lister, J. Chem. Soc. (A), 1800 (1967). [c] H. M. Frey, A. Krantz and I. D. R. Stevens, J. Chem. Soc. (A), 1734 (1969). [d] D. W. Vanas and W. D. Walters, J. Amer. Chem. Soc., 70, 4053 (1948); see also C. J. Grant and R. Walsh, Chem. Commun., 667 (1969). [e] H. M. Frey and J. Metcalfe, J. Chem. Soc. (A), 2529 (1970). [f] B. S. Schatz, Ph.D. Thesis, Oregon State University, 1967 [cf. Diss. Abs. (B), 27, 3870 (1967)]. [g] S. McLean, C. J. Webster and R. J. D. Rutherford, Can. J. Chem., 47, 1555 (1969). [h] W. C. Herndon and J. M. Manion, J. Org. Chem., 33, 4504 (1968). [i] W. C. Herndon and L. L. Lowry, J. Amer. Chem. Soc., 86, 1922 (1964); see Table 7.16 for reverse Diels–Alder reaction of this molecule. [j] K. N. Klump and J. P. Chesick, J. Amer. Chem. Soc., 85, 130 (1963). [k] R. Atkinson and B. A. Thrush, Proc. Roy. Soc. (A), 316, 131 (1970); based on Arrhenius parameters of ref. j for the parent molecule. [l] Arrhenius parameters are quoted here only for overall conversion to xylenes, etc.; for details of individual reactions see K. W. Egger, J. Amer. Chem. Soc., 90, 6 (1968).

7.1.6 Heterocyclic compounds (Table 7.18)

There is not much reliable information available concerning the unimolecular reactions of small-ring heterocyclic compounds. The pyrolysis of ethylene oxide yields a large variety of products, and it has been suggested that unimolecular rearrangement produces acetaldehyde molecules which are sufficiently excited to undergo subsequent free-radical reactions which are not fully suppressed by the addition of propene.[42, 43]

The complications of concurrent chain and surface modes of decomposition make this an unsuitable compound from the point of view of unimolecular reaction study.

Tetrafluoroethylene oxide,[44] on the other hand, appears to undergo a unimolecular reaction to produce carbonyl fluoride and difluorocarbene:

$$\underset{O}{\overset{CF_2-CF_2}{\diagdown\diagup}} \longrightarrow COF_2 + \ddot{C}F_2$$

The relative unreactivity of the carbene leads to its disappearance mainly by the reactions

$$2\,\ddot{C}F_2 \longrightarrow C_2F_4$$

$$\ddot{C}F_2 + C_2F_4 \longrightarrow cyclo{-}C_3F_6$$

Some other unimolecular reactions of heterocyclic compounds are listed in Table 7.18.

Table 7.18 High-pressure Arrhenius parameters for reactions of some heterocyclic compounds

Reactant	Products	$\log_{10} A_\infty/$ s^{-1}	$E_\infty/$ kcal mol^{-1}	Ref.
3-membered rings				
$(CH_2)_2O$	CH_3CHO, CH_2O, CH_2CO	$14\cdot1^a$	$56\cdot9^a$	*a*
$(CF_2)_2O$	$COF_2 + \ddot{C}F_2$	$13\cdot7$	$31\cdot6$	44
$(Me_2C)_2O$	$Me_3C\cdot CO\cdot Me$	$13\cdot8$	$56\cdot7$	
	propene + acetone	$14\cdot8$	$59\cdot2$	*b*
	$Me_2C(OH)\cdot CMe{:}CH_2$	$10\cdot9$	$47\cdot5$	
$Me_2C\underset{N}{\overset{N}{\diagup\!\!\!\diagdown}}\|$	$N_2 + [Me_2\ddot{C}] \rightarrow MeCH{:}CH_2$	$14\cdot0$	$33\cdot2$	*c*
$\overline{Et_2\dot{C}\cdot N{:}\dot{N}}$	$N_2 + [Et_2\ddot{C}]$	$13\cdot7$	$31\cdot9$	*d*
□⧹⧹N N	$N_2 + \left[□:\right]$	$13\cdot4$	$30\cdot5$	*d*
⬡⧹⧹N N	$N_2 + \left[\langle\rangle:\right]$	$13\cdot3$	$30\cdot9$	*d*

(cont.)

Table 7.18 (*cont.*)

Reactant	Products	$\log_{10} A_\infty/$ s^{-1}	$E_\infty/$ kcal mol^{-1}	Ref.
MeC̈Cl·N:N	$N_2 +$ [MeC̈Cl]	14·1	31·1	e
EtC̈Cl·N:N	$N_2 +$ [EtC̈Cl]	14·0	30·5	f
PrC̈Cl·N:N	$N_2 +$ [PrC̈Cl]	14·0	31·0	f
^{i}PrC̈Cl·N:N	$N_2 +$ [^{i}PrC̈Cl]	14·0	30·6	f
^{t}BuC̈Cl·N:N	$N_2 +$ [^{t}BuC̈Cl]	13·4	29·5	f
4-membered rings $(CH_2)_3O$	$C_2H_4 + CH_2O$	14·8^g	60^g	g
$CH_2OCH_2CMe_2$	$CH_2O + CH_2:CMe_2$	15·6	60·7	h
$CH_2CH_2O·CO$	$C_2H_4 + CO_2$	16·1	45·8	i
$CH_2CH_2CH_2SiMe_2$	$C_2H_4 +$ [$CH_2:SiMe_2$] $\rightarrow CH_2SiMe_2CH_2SiMe_2$	15·6	62·5	j
5-membered rings				
	$+ H_2$	12·7	48·5	k
	$+ H_2$	12·8	48·3	l
	$+ H_2$	12·3	44·6	m
	$+ H_2$	13·2^n	54·8^n	n
6-membered rings $(CH_2O)_3$	$3CH_2O$	15·0	47·4	o
$(MeCHO)_3$	$3MeCHO$	15·1	44·2	p
$(PrCHO)_3$	$3PrCHO$	14·4	42·0	q
$(^iPrCHO)_3$	3^iPrCHO	14·8	42·8	q

Table 7.18 (*cont.*)

Reactant	Products	$\log_{10} A_\infty/$ s^{-1}	$E_\infty/$ kcal mol^{-1}	Ref.
	$C_2H_4 + CH_2:CH\cdot CHO$	12·8	47·0	*r*
	$C_2H_4 + CH_2:CH\cdot CO\cdot Me$	14·5	51·2	*s*

[a] Ref. 42 (results in presence of C_3H_6); see also K. H. Mueller and W. D. Walters, *J. Amer. Chem. Soc.*, **73**, 1458 (1951); **76**, 330 (1954). [b] M. C. Flowers, R. M. Parker and M. A. Voisey, *J. Chem. Soc.* (*B*), 239 (1970). [c] H. M. Frey and I. D. R. Stevens, *J. Chem. Soc.*, 3865 (1962). [d] H. M. Frey and A. W. Scaplehorn, *J. Chem. Soc.* (*A*), 968 (1966). [e] M. R. Bridge, H. M. Frey and M. T. H. Liu, *J. Chem. Soc.* (*A*), 91 (1969). [f] H. M. Frey and M. T. H. Liu, *J. Chem. Soc.* (*A*), 1916 (1970). [g] D. A. Bittker and W. D. Walters, *J. Amer. Chem. Soc.*, **77**, 1429 (1955); results obtained in presence of nitric oxide. [h] G. F. Cohoe and W. D. Walters, *J. Phys. Chem.*, **71**, 2326 (1967). [i] T. L. James and C. A. Wellington, *J. Amer. Chem. Soc.*, **91**, 7743 (1969). [j] M. C. Flowers and L. E. Gusel'nikov, *J. Chem. Soc.* (*B*), 419, 1396 (1968). [k] C. A. Wellington and W. D. Walters, *J. Amer. Chem. Soc.*, **83**, 4888 (1961). [l] T. L. James and C. A. Wellington, *J. Chem. Soc.* (*A*), 2398 (1968). [m] A. C. Thomas and C. A. Wellington, *J. Chem. Soc.* (*A*), 2895 (1969). [n] C. A. Wellington, T. L. James and A. C. Thomas, *J. Chem. Soc.* (*A*), 2897 (1969) (preliminary results). [o] W. Hogg, D. M. McKinnon, A. F. Trotman-Dickenson and G. J. O. Verbeke, *J. Chem. Soc.*, 1403 (1961); see also R. LeG. Burnett and R. P. Bell, *Trans. Faraday Soc.*, **34**, 420 (1938). [p] C. C. Coffin, *Can. J. Res.* (*B*), **7**, 75 (1932). [q] C. C. Coffin, *Can. J. Res.* (*B*), **9**, 603 (1933). [r] D. G. Retzloff, B. M. Coull and J. Coull, *J. Phys. Chem.*, **74**, 2455 (1970). [s] C. S. Caton, *J. Amer. Chem. Soc.*, **91**, 7569 (1970).

7.1.7   Alkyl halides (Tables 7.19–7.23)

The elimination of hydrogen halides from alkyl halides to yield the corresponding olefins has been extensively studied, and has been recently reviewed by Maccoll.[3,4] The nature of the reaction which occurs upon pyrolysis of an alkyl halide, i.e. whether unimolecular or free radical, homogeneous or heterogeneous, depends upon the halogen involved and upon the structure of the compound concerned.

In many cases it is possible to isolate a unimolecular homogeneous mode of pyrolysis which is now widely accepted to involve a 4-centred activated complex. The reaction can be represented[3,4] by (7.23) although a

semi-ion-pair rather than an ion-pair transition state has also been suggested.[45]

Maccoll[46] has shown the analogy of these reactions to E1-type solvolytic reactions in solution; the evidence comes largely from the effect of substituents at the α- and β-carbon atoms (7.23) upon the rate of pyrolysis. The effects are largely on the activation energies and Table 7.19 shows

Table 7.19 Activation energies (kcal mol^{-1}) for the pyrolysis of substituted alkyl chlorides

$$R_1 \underset{R_2}{\overset{\beta}{\diagdown}} \underset{H}{\overset{}{C}} - \underset{Cl}{\overset{\alpha}{C}} \underset{R_4}{\overset{R_3}{\diagup}}$$

	β-methylation $\longrightarrow$ R_1, R_2		
R_3, R_4	H,H	H,CH$_3$	CH$_3$,CH$_3$
α-methylation			
H,H	58·4[a]	55·0[b]	56·9[c]
H,CH$_3$	51·0[d]	50·0[e]	—
CH$_3$,CH$_3$	45·0[f]	44·2[f]	42·3[f]

[a] Ref. 104. [b] D. H. R. Barton, A. J. Head and R. J. Williams, *J. Chem. Soc.*, 2039 (1951). [c] K. E. Howlett, *J. Chem. Soc.*, 4487 (1952). [d] D. H. R. Barton and A. J. Head, *Trans. Faraday Soc.*, **46**, 114 (1950). [e] A. Maccoll and R. H. Stone, *J. Chem. Soc.*, 2756 (1961). A. Maccoll and S. C. Wong, *J. Chem. Soc.* (*B*), 1492 (1968).

some values illustrating the effects of α- and β-methylation. It will be seen that the effect of α-methylation is to increase the rate considerably (by lowering the activation energy by 5–6 kcal mol^{-1} per methyl substitution) whereas β-methylation has a much smaller effect. This has been explained by the stabilization of the ion pair produced in (7.23) by hyperconjugation in the case of α-methylation, whereas β-methylation is considered to have a secondary effect upon the carbon–halogen bond. The extent of participation of the β-hydrogen atom in the activated complex is uncertain as some deuteriation experiments[47] appear to indicate complete breakage of the carbon–hydrogen bond. The effects of α- and β-chlorination are to increase and decrease the rate of pyrolysis respectively. These are smaller effects than in the case of α-methylation and can be explained qualitatively in terms of electronic effects.[3, 4] Quantitative interpretation is more difficult since the substitution of further chlorine atoms often results in radical chain modes of pyrolysis.

O'Neal and Benson[11, 48] have calculated the A-factors for a large number of alkyl halide pyrolyses from the absolute rate theory expression (7.24). The entropies of activation, $\Delta S^{\neq}$, were calculated by assigning bending, stretching and torsional frequencies to the activated complexes in a

$$A_\infty = (e\,kT/h)\exp(\Delta S^{\neq}/R) \qquad (7.24)$$

systematic way and also allowing for reaction path degeneracy. Agreement between calculated and experimental A-factors was obtained, generally within a factor of $\pm 0\cdot 3$ units in $\log_{10} A$. The calculated A-factors were then used with experimentally observed values of the rate-constants to obtain 'calculated' activation energies which were considered to be more self-consistent than the experimental values obtained from individual Arrhenius plots. It was noted that α-methylation produces an average lowering of about $5\cdot 7$ kcal mol^{-1} per methyl substitution whereas β-methylation lowers the activation energy by about $1\cdot 7$ kcal mol^{-1} per substitution, in general agreement with the trends observed by Maccoll (Table 7.19). Further discussion of the activated complex structure in alkyl halide pyrolysis will be found in sections 7.2.4 and 6.2.2.

Experimental data on *alkyl fluorides* have been obtained by chemical activation studies (see Chapter 8), and more recently (Table 7.20) by kinetic studies in conventional static systems[49] and in shock tubes.[50] The pyrolyses are reported to be clean unimolecular reactions in vessels seasoned with ethyl fluoride.[49]

Alkyl chloride pyrolyses (Table 7.21) are subject to surface effects and can be carried out heterogeneously at low temperatures (200–300 °C) on clean glass[51] and other[52] catalytic surfaces. In carbon-coated vessels these surface reactions are diminished although some carbon coatings, notably those derived from allyl bromide,[53] are found to be active catalysts in *cis–trans* isomerization reactions.[54] This must be borne in mind if the pyrolysis produces a mixture of olefins which can subsequently isomerize, as, for example, in the pyrolysis of 2-chlorobutane.[55] All the alkyl mono-chlorides (except methyl chloride) are thought to decompose thermally by unimolecular mechanisms in suitably coated vessels, the reactions being first-order and unaffected by inhibitors or accelerators. The decline of the first-order rate-constants with pressure has been observed in a number of cases and compared with unimolecular reaction rate theories (see section 7.2). When more than one chlorine atom is present, chain modes of decomposition may predominate. Of the chlorinated ethanes only 1,1-dichloroethane and ethyl chloride decompose by molecular mechanisms. The chain mechanism is found to be favoured if the radical formed by chlorine atom attack on the parent molecule can easily undergo a chain-propagating step producing olefin.[56, 57] For example, (7.25) occurs in the

$$ClCH_2\overset{\cdot}{C}HCl \longrightarrow CH_2{=}CHCl + Cl \qquad (7.25)$$
(from 1,2-dichloroethane)

chain decomposition of 1,2-dichloroethane whereas (7.26) cannot occur easily in a single step and 1,1-dichloroethane decomposes by a molecular reaction.

Similar arguments have been applied to *alkyl bromide pyrolyses* (Table 7.22) by Agius and Maccoll[58] and by Maccoll and Thomas[59] who have discussed schemes in terms of stopping radicals S [such as the bromine

$$Cl_2\overset{\bullet}{C}-CH_3 \xrightarrow{\;\;\times\;\;} CH_2=CHCl + Cl \qquad (7.26)$$

(from 1,1-dichloroethane)

analogue of that in (7.26)] and propagating radicals P [analogous to that shown in (7.25)]. For normal propyl bromide, for example, bromine atom attack appears to produce mainly the P-type radical $CH_3\overset{\bullet}{C}HCH_2Br$, and any S radicals formed are easily converted to P radicals by the step (7.27).

$$\overset{\bullet}{C}H_2CH_2CH_2Br + C_3H_7Br \longrightarrow CH_3\overset{\bullet}{C}HCH_2Br + C_3H_7Br \qquad (7.27)$$
$$(S) \qquad\qquad\qquad\qquad\qquad (P)$$

For isopropyl bromide, on the other hand, bromine atom attack is most likely to produce the S radical $CH_3—\overset{\bullet}{C}(CH_3)—Br$ which cannot be easily converted to a P radical by such a step as (7.27).

The direct unimolecular elimination of HBr is thus favoured over a chain mode of pyrolysis for isopropyl bromide, but the reverse is true for n-propyl bromide. The experimental facts concerning a number of bromide pyrolyses are summarized in Table 7.22. The pyrolyses of primary bromides occur by mixed chain and molecular mechanisms whereas secondary and tertiary compounds decompose predominantly by unimolecular mechanisms. Where both unimolecular and chain modes of pyrolysis are possible, information about the kinetic parameters of the unimolecular modes has sometimes been obtained from the pyrolyses fully inhibited by cyclohexene, propene or toluene.

In *alkyl iodide pyrolyses* (Table 7.23) the unimolecular elimination (7.28) is complicated by an iodine-atom-catalysed elimination which may

$$RCH_2CH_2I \longrightarrow RCH:CH_2 + HI \qquad (7.28)$$

well be a concerted process (7.29)[60] and by the iodine-atom-catalysed

$$RCH_2CH_2I + I \longrightarrow R-\overset{\overset{\displaystyle H}{|}}{\underset{\underset{\displaystyle I}{H}}{C}}\cdots CH_2 \longrightarrow \begin{array}{c} RCH=CH_2 + I \\ + HI \end{array} \qquad (7.29)$$

reduction by hydrogen iodide (7.30). The overall stoicheiometry of alkyl

$$\left. \begin{array}{l} I + RCH_2CH_2I \longrightarrow RCH_2CH_2 + I_2 \\ RCH_2CH_2 + HI \longrightarrow RCH_2CH_3 + I \end{array} \right\} \qquad (7.30)$$

iodide pyrolyses is thus represented by (7.31). Benson[60] has shown that

$$2RCH_2CH_2I \longrightarrow RCH=CH_2 + RCH_2CH_3 + I_2 \qquad (7.31)$$

the rate-limiting step is usually either (7.28) or (7.29) and is followed by the rapid reduction (7.30). The molecular elimination appears to be rate-controlling for ethyl, isopropyl and t-butyl iodides. For sec-butyl iodide, the iodine-atom-catalysed reaction (7.29) is important as well as (7.28) in determining the rate, although kinetic parameters relating to the molecular elimination have been derived[61] for the limited temperature range 290–330 °C.

Table 7.20 High-pressure Arrhenius parameters for the pyrolyses of alkyl fluorides

Reactant	Products	$\log_{10} A_\infty/$ s^{-1}	$E_\infty/$ kcal mol^{-1}	Ref.
ethyl fluoride	C_2H_4 (+HF)	13·3	58·2	49
		13·4	59·9	a
1-fluoropropane	propene (+HF)	13·3	58·3	a
2-fluoropropane	propene (+HF)	13·4	53·9	50
1-fluorobutane	butene (+HF)	13·3	56·8	50
1-fluoro-2-methylpropane (isobutyl fluoride)	isobutene (+HF)	13·3	58·6	50
2-fluoro-2-methylpropane (t-butyl fluoride)	isobutene (+HF)	13·4	51·5	50
1,1-difluoroethane	vinyl fluoride (+HF)	13·7	66·6	50
		13·9	61·9	b
1,1,1-trifluoroethane	$CH_2:CF_2$ (+HF)	13·5	71·1	50
		14·0	68·7	c
1,1,2,2-tetrafluoroethane	$CHF:CF_2$ (+HF)	13·4	70·1	d
1-chloro-2-fluoroethane	$CHCl:CH_2$ (+HF)	13·0	60·0	50

[a] P. Cadman, M. Day and A. F. Trotman-Dickenson, *J. Chem. Soc. (A)*, 2498 (1970).
[b] E. Tschuikow-Roux, W. J. Quiring and J. M. Simmie, *J. Phys. Chem.*, **74**, 2449 (1970).
[c] E. Tschuikow-Roux and W. J. Quiring, *J. Phys. Chem.*, **75**, 295 (1971). [d] G. E. Millward, R. Hartig and E. Tschuikow-Roux, *Chem. Commun.*, 465 (1971).

Table 7.21 High-pressure Arrhenius parameters for alkyl chloride pyrolysis

Reactant	Products	$\log_{10} A_\infty/$ s^{-1}	$E_\infty/$ kcal mol^{-1}	Ref.
ethyl chloride	ethylene + HCl	14·6	60·8	111, 112
		13·2	56·5	113
		13·5	56·6	114
		13·9	59·0	115
		13·5	56·6	116
		14·0	58·4	104
C_2D_5Cl	$C_2D_4 + DCl$	13·8	58·7	a
1,1-dichloroethane	CH_2:CHCl + HCl	11·7	48·3	105
		13·5	53·5	b
1-chloropropane	propene + HCl	13·5	55·0	c
2-chloropropane (isopropyl chloride)	propene + HCl	13·4	50·5	d,w
		13·6	51·1	113
		11·0	42·4	e
1,2-dichloropropane	chloropropenes + HCl	13·8	54·9	d
	cis-1-chloropropene + HCl	13·1	54·5	
	trans-1-chloropropene + HCl	13·3	56·1	f
	2-chloropropene + HCl	8·3	42·8	
	3-chloropropene + HCl	13·4	54·0	
1,1-dichloropropane	1-chloropropenes + HCl	12·8	51·2	g,h
2,2-dichloropropane	2-chloropropene + HCl	11·9	43·9	g
		14·5	51·4	i
1-chlorobutane	but-1-ene + HCl	14·0	57·0	c
		14·5	57·9	j
2-chlorobutane (sec-butyl chloride)	butenes + HCl	13·6	49·6	k
		14·0	50·6	l
1-chloro-2-methylpropane (isobutyl chloride)	isobutene + HCl	14·0	56·9	m
2-chloro-2-methylpropane (t-butyl chloride)	isobutene + HCl	14·3	46·0	n
		12·4	41·4	56
		13·7	44·9	o
		13·7	45·9	p
		14·4	46·6	q
		13·9	46·2	r
		14·6	46·0	s
		12·9	41·6	e
1,1-dichlorobutane	1-chlorobutenes + HCl	13·6	53·7	t
2,2-dichlorobutane	2-chlorobutenes + HCl	14·5	50·2	i
1,4-dichlorobutane	buta-1,3-diene + HCl	14·4	56·5	u
1-chloropentane	pent-1-ene + HCl	14·6	58·3	v
2-chloropentane	pentenes + HCl	14·1	50·7	w
3-chloropentane	pentenes + HCl	14·4	51·0	w,x

Table 7.21 (*cont.*)

Reactant	Products	$\log_{10} A_\infty/$ s^{-1}	$E_\infty/$ kcal mol^{-1}	Ref.
2-chloro-2-methylbutane	pentenes + HCl	13·8	44·2	*p*
1-chloro-2,2-dimethyl-propane	pentenes + HCl	13·3	60·0	*y*
cyclopentyl chloride	cyclopentene + HCl	13·5	48·3	*z,aa*
cyclohexyl chloride	cyclohexene + HCl	13·8	50·0	*aa, bb*
cycloheptyl chloride	cycloheptene + HCl	13·9	47·3	*aa*
cyclooctyl chloride	cyclooctene + HCl	13·2	45·0	*aa*
2-chloro-2,3-dimethyl-butane	hexenes + HCl	13·4	42·3	*p*
2-chloro-2,3,3-trimethyl-butane	heptenes + HCl	13·8	41·9	*p*
2-chlorooctane	octenes + HCl	13·5	48·7	*cc*
menthyl chloride	menthenes + HCl	12·6	45·0	*dd*
		11·8	42·1	*ee*
neomenthyl chloride	menthenes + HCl	10·7	40·1	*ee*
bornyl chloride	$C_{10}H_{16}$ hydrocarbons + HCl	13·8	50·6	*ff*
isobornyl chloride	$C_{10}H_{16}$ hydrocarbons + HCl	14·8	49·7	*gg*
$C_6H_5 \cdot CHCl \cdot CH_3$	$C_6H_5 \cdot CH:CH_2 + HCl$	12·6	44·8	*hh*
$p\text{-}F \cdot C_6H_4 \cdot CHCl \cdot CH_3$	$p\text{-}F \cdot C_6H_4 \cdot CH:CH_2 + HCl$	13·0	45·6	*hh*
$p\text{-}Cl \cdot C_6H_4 \cdot CHCl \cdot CH_3$	$p\text{-}Cl \cdot C_6H_4 \cdot CH:CH_2 + HCl$	12·2	44·1	*hh*
$p\text{-}Br \cdot C_6H_4 \cdot CHCl \cdot CH_3$	$p\text{-}Br \cdot C_6H_4 \cdot CH:CH_2 + HCl$	12·2	44·2	*hh*
$p\text{-}Me \cdot C_6H_4 \cdot CHCl \cdot CH_3$	$p\text{-}Me \cdot C_6H_4 \cdot CH:CH_2 + HCl$	14·1	47·7	*hh*
$m\text{-}Me \cdot C_6H_4 \cdot CHCl \cdot CH_3$	$m\text{-}Me \cdot C_6H_4 \cdot CH:CH_2 + HCl$	12·0	42·7	*hh*
$p\text{-}NC \cdot C_6H_4 \cdot CHCl \cdot CH_3$	$p\text{-}NC \cdot C_6H_4 \cdot CH:CH_2 + HCl$	12·5	47·0	*hh*
$CH_3CHClOMe$	$CH_2:CHOMe + HCl$	11·5	33·3	*ii*
$CH_3CHClOEt$	$CH_2:CHOEt + HCl$	10·5	30·3	*jj*
$CH_3CHCl \cdot CH:CH_2$	$CH_2:CH \cdot CH:CH_2 + HCl$	13·3	48·5	*kk*
$CH_2Cl \cdot CH_2 \cdot CH:CH_2$	$CH_2:CH \cdot CH:CH_2 + HCl$	13·7	55·0	*ll*
$CH_3CHCl \cdot CMe:CH_2$	$CH_2:CH \cdot CMe:CH_2 + HCl$	13·2	46·9	*kk*
$(CH_3)_2CCl \cdot CH:CH_2$	$CH_2:CMe \cdot CH:CH_2 + HCl$	13·3	42·6	*mm*
$CH_2Cl \cdot CH:C(CH_3)_2$	$CH_2:CH \cdot CMe:CH_2 + HCl$	12·0	38·3	*mm*
$CH_3 \cdot CHCl \cdot CO \cdot Me$	$CH_2:CH \cdot CO \cdot Me + HCl$	11·7	49·1	*nn*

(*cont.*)

Table 7.21 (*cont.*)

Reactant	Products	$\log_{10} A_\infty/$ s^{-1}	$E_\infty/$ kcal mol^{-1}	Ref.
MeCHCl·CH$_2$·CO·Me	MeCH:CH·CO·Me + HCl	12·2	44·5	*nn*
MeCH$_2$·CHCl·CO·Me	MeCH:CH·CO·Me + HCl	12·8	50·8	*nn*

a Ref. 106; see also section 9.5.2. *b* H. Hartmann, H. Heydtmann and G. Rinck, *Zeit. phys. Chem.* (*Frankfurt*), **28**, 71 (1961). *c* D. H. R. Barton, A. J. Head and R. J. Williams, *J. Chem. Soc.*, 2039 (1951). *d* D. H. R. Barton and A. J. Head, *Trans. Faraday Soc.*, **46**, 114 (1950). *e* M. Asahina and M. Onozuka, *J. Polym. Sci.* (*A*), **2**, 3505 (1964). *f* K. A. Holbrook and J. S. Palmer, *Trans. Faraday Soc.*, **67**, 80 (1971); see also G. J. Martens, M. Godfroid and L. Ramoisy, *Internat. J. Chem. Kinetics*, **2**, 123 (1970). *g* K. E. Howlett, *J. Chem. Soc.*, 945 (1953). *h* K. A. Holbrook and K. A. W. Parry, to be published. *i* B. C. Young and E. S. Swinbourne, *J. Chem. Soc.* (*B*), 1181 (1967). *j* H. Hartmann, H. Heydtmann and G. Rinck, *Zeit. phys. Chem.* (*Frankfurt*), **28**, 85 (1961). *k* A. Maccoll and R. H. Stone, *J. Chem. Soc.*, 2756 (1961). *l* H. Heydtmann and G. Rinck, *Zeit. phys. Chem.* (*Frankfurt*), **30**, 250 (1961). *m* K. E. Howlett, *J. Chem. Soc.*, 4487 (1952). *n* D. Brearley, G. B. Kistiakowsky and C. H. Stauffer, *J. Amer. Chem. Soc.*, **58**, 43 (1936). *o* R. L. Failes and V. R. Stimson, *Aust. J. Chem.*, **15**, 437 (1962). *p* A. Maccoll and S. C. Wong, *J. Chem. Soc.* (*B*), 1492 (1968). *q* B. Roberts, Ph.D. Thesis, University of London, 1961, quoted by A. Maccoll, ref. 4. *r* Ref. 62; this paper includes a useful summary of earlier work. *s* W. C. Herndon and J. Manion, *J. Miss. Acad. Sci.*, **10**, 159 (1964). *t* K. A. W. Parry, to be published. *u* R. J. Williams, *J. Chem. Soc.*, 113 (1953). *v* R. C. S. Grant and E. S. Swinbourne, *J. Chem. Soc.*, 4423 (1965). *w* V. Chytrý, B. Obereigner and D. Lím, *Chem. Ind.* (*Lond.*), 470 (1970). *x* See also F. Von Erbe, T. Grewer and K. Wehage, *Angew. Chem.*, **74**, 988 (1962). *y* J. S. Shapiro and E. S. Swinbourne, *Can. J. Chem.*, **46**, 1341 (1968). *z* E. S. Swinbourne, *J. Chem. Soc.*, 4668 (1960). *aa* J. M. Sullivan and W. C. Herndon, *J. Phys. Chem.*, **74**, 995 (1970). *bb* E. S. Swinbourne, *Aust. J. Chem.*, **11**, 314 (1958). *cc* C. J. Harding, A. Maccoll and R. A. Ross, *Chem. Commun.*, 289 (1967). *dd* D. H. R. Barton, A. J. Head and R. J. Williams, *J. Chem. Soc.*, 453 (1952). *ee* T. O. Bamkole and A. Maccoll, *J. Chem. Soc* (*B*), 1159 (1970). *ff* R. C. L. Bicknell and A. Maccoll, *Chem. Ind.* (*Lond.*), 1912 (1961). *gg* R. C. L. Bicknell, Ph.D. Thesis, University of London, 1962, quoted by A. Maccoll, ref. 4. *hh* M. R. Bridge, D. H. Davies, A. Maccoll, R. A. Ross and O. Banjoko, *J. Chem. Soc.* (*B*), 805 (1968). *ii* P. J. Thomas, *J. Chem. Soc.*, 136 (1961). *jj* R. L. Failes and V. R. Stimson, *Aust. J. Chem.*, **20**, 1553 (1967). *kk* P. J. Thomas, *J. Chem. Soc.* (*B*), 1238 (1967). *ll* P. Cadman, M. Day and A. F. Trotman-Dickenson, *J. Chem. Soc.* (*A*), 2058 (1970). *mm* C. J. Harding, A. Maccoll and R. A. Ross, *J. Chem. Soc.* (*B*), 634 (1969). *nn* M. Dakubu and A. Maccoll, *J. Chem. Soc.* (*B*), 1248 (1969).

Table 7.22 High-pressure Arrhenius parameters for alkyl bromide pyrolysis

Reactant	Products	$\log_{10} A_\infty/$ s^{-1}	$E_\infty/$ kcal mol^{-1}	Ref.
ethyl bromide	$C_2H_4 + HBr$	13·5	53·9	a
		12·9	52·3	b
C_2D_5Br	$C_2D_4 + DBr$	13·3	54·8	c
1-bromopropane	propene + HBr	12·8	50·7	e
		13·0	50·7	b
		13·2	51·9	f
2-bromopropane (isopropyl bromide)	propene + HBr	13·6	47·8	g
		13·6	47·7	b
		13·7	47·6	t
1-bromobutane	but-1-ene + HBr	13·2	50·9	e
2-bromobutane (sec-butyl bromide)	butenes + HBr	12·6	43·8	h
		13·0	45·5	i
		13·5	46·5	j
1-bromo-2-methyl-propane (isobutyl bromide)	isobutene + HBr	13·1	50·4	k
2-bromo-2-methyl-propane (t-butyl bromide)	isobutene + HBr	14·0	42·0	l
		13·3	40·5	h
		13·2	41·0	i
		13·5	41·5	d
1-bromopentane	pent-1-ene + HBr	13·1	50·5	m
2-bromo-2-methyl-pentane	pentenes + HBr	13·6	40·5	n
4-bromopent-1-ene	pentadienes + HBr	12·9	44·7	a
3-bromopentane	pent-2-ene + HBr	13·5	45·4	o
1-bromohexane	hex-1-ene + HBr	13·1	50·5	g
2-bromo-2,3-dimethyl-butane	hexenes + HBr	13·5	39·0	p
cyclopentylbromide	cyclopentene + HBr	12·8	43·7	q
cyclohexyl bromide	cyclohexene + HBr	13·5	46·1	r
$C_6H_5 \cdot CHBr \cdot CH_3$	$C_6H_5 \cdot CH:CH_2 + HBr$	12·2	38·8	s

[a] P. J. Thomas, *J. Chem. Soc.*, 1192 (1959). [b] A. T. Blades and G. W. Murphy, *J. Amer. Chem. Soc.*, **74**, 6219 (1952). [c] A. T. Blades, *Can. J. Chem.*, **36**, 1043 (1958). [d] Ref. 62; this paper includes a useful summary of earlier work. [e] Ref. 59; see also M. R. Bridge and J. L. Holmes, *J. Chem. Soc. (B)*, 1008 (1967). [f] J. T. D. Cross and V. R. Stimson, *Aust. J. Chem.*, **21**, 973 (1968). [g] A. Maccoll and P. J. Thomas, *J. Chem. Soc.*, 979 (1955). [h] A. Maccoll and P. J. Thomas, *J. Chem. Soc.*, 2455 (1955). [i] G. B. Sergeev, *Doklady Akad. Nauk S.S.S.R.*, **106**, 299 (1955). [j] M. N. Kale, A. Maccoll and P. J. Thomas, *J. Chem. Soc.*, 3016 (1958). [k] G. D. Harden and A. Maccoll, *J. Chem. Soc.*, 1197 (1959). [l] G. D. Harden and A. Maccoll, *J. Chem. Soc.*, 2454 (1955). [m] J. H. S. Green, A. Maccoll and P. J. Thomas, *J. Chem. Soc.*, 184 (1960). [n] G. D. Harden, *J. Chem. Soc.*, 5024 (1957). [o] N. Capon, A. Maccoll and R. A. Ross, *Trans. Faraday Soc.*, **63**, 1152 (1967). [p] G. D. Harden and A. Maccoll, *J. Chem. Soc.*, 5028 (1957). [q] M. N. Kale and A. Maccoll, *J. Chem. Soc.*, 5020 (1957). [r] J. H. S. Green and A. Maccoll, *J. Chem. Soc.*, 2449 (1955). [s] B. Stephenson, Ph.D. Thesis, University of London, 1957, quoted by A. Maccoll, ref. 4. [t] W. Tsang, *Internat. J. Chem. Kinetics.*, **2**, 311 (1970).

Table 7.23   High-pressure Arrhenius parameters for alkyl iodide pyrolysis

Reactant	Products	$\log_{10} A_\infty/$ s^{-1}	$E_\infty/$ kcal mol^{-1}	Ref.
ethyl iodide	$C_2H_4 + HI$	14·1	52·8	a
		13·7	50·0	b
		13·0	48·3	c
		14·2	51·0	d
2-iodopropane (isopropyl iodide)	propene + HI	14·5	48·2	e
		14·8	48·0	61
		13·0	43·5	f
		13·7	45·1	113
		13·2	42·9	g
2-iodobutane (sec-butyl iodide)	butenes + HI	15·2	47·9	61
2-iodo-2-methyl-propane (t-butyl iodide)	isobutene + HI	13·7	38·0	113
		13·5	38·3	d
		12·5	36·4	h
		11·1	32·0	i
$CH_3 \cdot CHI \cdot CO \cdot Me$	$CH_2 : CH \cdot CO \cdot Me + HI$	13·4	41·9	j

[a] J. H. Yang and D. C. Conway, *J. Chem. Phys.*, **43**, 1296 (1965). [b] S. W. Benson and A. N. Bose, *J. Chem. Phys.*, **37**, 2935 (1962). [c] R. A. Lee, M.Sc. Thesis, University of London, 1959, quoted by A. Maccoll, ref. 4. [d] G. Choudhary, quoted by A. Maccoll, ref. 4. [e] J. L. Holmes and A. Maccoll, *Proc. Chem. Soc.*, 175 (1957). [f] H. Teranishi and S. W. Benson, *J. Chem. Phys.*, **40**, 2946 (1964). [g] J. L. Jones and R. A. Ogg, *J. Amer. Chem. Soc.*, **59**, 1939 (1937). [h] S. W. Benson and A. N. Bose, *J. Chem. Phys.*, **38**, 878 (1963). [i] J. L. Holmes and G. Choudhary, *Symposium on Pyrolytic Reactions*, Chemical Institute of Canada, Ottawa (1964). [j] R. K. Solly, D. M. Golden and S. W. Benson, *Internat. J. Chem. Kinetics*, **2**, 393 (1970).

7.1.8   Esters (Tables 7.24–7.26)

Many simple esters decompose to give the corresponding carboxylic acid and one or more olefinic products. Esters of secondary and tertiary alcohols decompose by purely molecular mechanisms; for primary esters there is a concurrent chain reaction which can be suppressed by inhibitors. The Arrhenius parameters for some unimolecular ester pyrolyses of this type are listed in Tables 7.24 and 7.25.

For the unimolecular process, the six-centred transition state proposed by Hurd and Blunck[63] is generally accepted and is consistent with the observed A-factors which are usually in the range $10^{11.5\pm1.5}$ s^{-1}. The reaction may therefore be described by (7.32). The effects of α and β substitution upon the pyrolysis rate have been investigated by several groups of workers[64-66] and Maccoll has concluded that the pyrolyses of esters, like those of alkyl halides, are 'quasi-heterolytic' involving a polar transition state.[3, 67] The effect of α-methylation in particular is similar to, but smaller

$$R-C \overset{\overset{O}{\parallel}}{\underset{O-CR_2}{\diagup}} \overset{H \; \beta}{\underset{\alpha}{\diagdown}} CR_2 \longrightarrow R-C \overset{\overset{O\cdots H}{\diagup}}{\underset{O\cdots CR_2}{\diagdown}} CR_2 \longrightarrow RCO_2H + CR_2{=}CR_2 \quad (7.32)$$

than, that in the case of alkyl halide pyrolysis, and it is suggested that heterolysis is not as fully developed for esters as for alkyl halides.

For esters which are capable of decomposing in different ways, the predominant direction of elimination is given by the Hofmann rule, i.e. the favoured olefinic product is that bearing the fewest alkyl substituents around the double bond. Thus secondary butyl acetate yields 60% but-1-ene and 40% but-2-ene, whereas secondary butyl chloride yields 60% but-2-ene and 40% but-1-ene, more in accordance with the Saytzeff rule.

A number of special types of ester decompose or rearrange by different mechanisms to give products other than the simple olefin plus acid. The Arrhenius parameters for these are listed in Table 7.26. An example is the pyrolysis of *gem*-diesters, which appears to involve a six-membered cyclic activated complex, e.g. (7.33). Chloroformates and carbonates undergo

$$RCH \overset{O{\cdot}CO{\cdot}R'}{\underset{O{\cdot}CO{\cdot}R'}{\diagup}} \longrightarrow RCH \overset{\overset{O{=}C{-}R'}{\diagup}}{\underset{O\cdots CO}{\diagdown}} \longrightarrow RCHO + \overset{\overset{OC{-}R'}{\diagup}}{\underset{OC}{\diagdown}} O \quad (7.33)$$

formally similar fragmentation reactions illustrated by (7.34) for the former. In this case an alternative 'substitution' reaction (7.35) also occurs.

$$OC \overset{\overset{Cl}{\diagup}}{\underset{O-CH_2}{\diagdown}} \overset{H}{\underset{CHR}{\diagdown}} \longrightarrow OC \overset{\overset{Cl\cdots H}{\diagup}}{\underset{O\cdots CH_2}{\diagdown}} CHR \longrightarrow CO_2 + HCl + CH_2{:}CHR \quad (7.34)$$

$$OC \overset{Cl}{\underset{O}{\diagup}} CH_2CH_2R \longrightarrow CO_2 + ClCH_2CH_2R \quad (7.35)$$

The activated complex for the substitution reaction was concluded[68] to be more polar than that for the elimination reaction (7.34), but the measurements were made over a very restricted temperature range and the Arrhenius parameters must be of doubtful accuracy. The rearrangement of allyl esters is formally similar to the Cope rearrangement (section 7.1.5), but the Arrhenius parameters do not seem to be too well established.[2]

Table 7.24   High-pressure Arrhenius parameters for the pyrolysis of acetates

Reactant	Olefinic products (acetic acid formed in each case)	$\log_{10} A_\infty /$ s^{-1}	$E_\infty /$ kcal mol^{-1}	Ref.
ethyl acetate	C_2H_4	12·5	47·8	a
		11·6	44·0	b
ethyl-d_5 acetate	C_2D_4	12·7	49·5	c
isopropyl acetate	propene	13·0	45·0	a
		13·4	46·3	d
		12·1	42·9	e
		12·9	46·1	f
t-butyl acetate	isobutene	13·3	40·5	d
		13·5	42·1	f
sec-butyl acetate	[butenes]	13·3	46·6	d
t-amyl acetate	[pentenes]	13·4	40·3	d
cyclopentyl acetate	cyclopentene	12·9	44·1	g
cyclohexyl acetate	cyclohexene	12·6	45·2	g
1,2-diphenylethyl acetate	stilbene	13·0	43·3	65
bornyl acetate ⎫	bornylene, tricyclene,	12·0	45·3	h
isobornyl acetate ⎭	camphene	11·6	42·0	h
CH$_3$CHX·OAc				
X = (H, Me, Et, see above)				
CH$_2$:CH	CH$_2$:CH·CH:CH$_2$	11·6	41·8	f
CH$_3$CH:CH	CH$_3$·CH:CH·CH:CH$_2$	13·1	43·9	f
CH$_2$:CH·CH$_2$	1,3- and 1,4-pentadiene	11·5	41·8	f
		13·0	44·4	64
isopropyl	^{i}PrCH:CH$_2$ + Me$_2$C:CHMe	12·6	44·7	f
phenyl	PhCH:CH$_2$	12·3	43·0	f
		12·8	43·7	i
cyclopropyl	c-C$_3$H$_5$CH:CH$_2$ + cyclopentene	12·2	42·3	f
MeCO·CH$_2$	MeCO·CH:CHMe	11·9	37·4	64
(CH$_3$)$_2$CX·OAc				
X = (H, Me, see above)				
CH$_2$:CH	CH$_2$:CH·CMe:CH$_2$	13·6	40·6	f
CH$_2$:CH·CH$_2$	methylpentadienes	13·3	41·2	f
isopropyl	^{i}PrCMe:CH$_2$ + Me$_2$C:CMe$_2$	13·6	40·7	f
phenyl	PhCMe:CH$_2$	13·6	40·7	f
cyclopropyl	c-C$_3$H$_5$CMe:CH$_2$	13·9	40·8	f

^{a} A. T. Blades, *Can. J. Chem.*, **32**, 366 (1954). ^{b} M. M. Gil'burd and F. B. Moin, *Kinet. Katal.*, **8**, 261 (1967). ^{c} A. T. Blades and P. W. Gilderson, *Can. J. Chem.*, **38**, 1407 (1960). ^{d} E. U. Emovon and A. Maccoll, *J. Chem. Soc.*, 335 (1962). ^{e} B. S. Lennon and V. R. Stimson, *Aust. J. Chem.*, **21**, 1659 (1968). ^{f} K. K. Lum and G. G. Smith, *Internat. J. Chem. Kinetics.*, **1**, 401 (1969). ^{g} T. O. Bamkole and E. U. Emovon, *J. Chem. Soc. (B)*, 187 (1969). ^{h} E. U. Emovon, *J. Chem. Soc. (B)*, 588 (1966). ^{i} R. Taylor, G. G. Smith and W. H. Wetzel, *J. Amer. Chem. Soc.*, **84**, 4817 (1962).

Table 7.25   High-pressure Arrhenius parameters for the pyrolyses of other esters to give olefins plus carboxylic acids

Reactant	Products	$\log_{10} A_\infty/$ s^{-1}	$E_\infty/$ kcal mol^{-1}	Ref.
ethyl formate	ethylene + formic acid	11·3	44·1	a
propyl formate	propene + formic acid	9·4	39·7	b
isopropyl formate	propene + formic acid	12·4	44·2	b
		12·6	44·0	a
t-butyl formate	isobutene + formic acid	11·1	34·6	c
ethyl propionate	ethylene + propionic acid	12·7	48·5	d
t-butyl propionate	isobutene + propionic acid	13·2	40·0	e
		12·8	39·2	f
ethyl trimethylacetate	ethylene + trimethylacetic acid	11·2	44·0	g
isopropyl trimethylacetate	propylene + trimethylacetic acid	12·9	44·8	h
t-butyl chloroacetate	isobutene + chloroacetic acid	13·1	38·1	i
t-butyl dichloroacetate	isobutene + dichloroacetic acid	12·8	36·1	i
di(2-ethylhexyl) sebacate, $(CH_2)_8(CO_2CH_2CHEt^nBu)_2$	2-ethylhexene + acid + monoester	12·4	47·1	j
(−) menthyl benzoate	(+) p-menthenes + benzoic acid	11·0	38·1	k

^{a} A. T. Blades, *Can. J. Chem.*, **32**, 366 (1954). ^{b} R. B. Anderson and H. H. Rowley, *J. Phys. Chem.*, **47**, 454 (1943). ^{c} E. Gordon, S. J. W. Price and A. F. Trotman-Dickenson, *J. Chem. Soc.*, 2813 (1957). ^{d} A. T. Blades and P. W. Gilderson, *Can. J. Chem.*, **38**, 1412 (1960). ^{e} C. E. Rudy and P. Fugassi, *J. Phys. Chem.*, **52**, 357 (1948). ^{f} E. Warrick and P. Fugassi, *J. Phys. Chem.*, **52**, 1314 (1948). ^{g} J. T. D. Cross and V. R. Stimson, *Aust. J. Chem.*, **20**, 177 (1967). ^{h} B. S. Lennon and V. R. Stimson, *Aust. J. Chem.*, **21**, 1659 (1968). ^{i} E. U. Emovon, *J. Chem. Soc.*, 1246 (1963). ^{j} E. E. Sommers and T. I. Crowell, *J. Amer. Chem. Soc.*, **77**, 5443 (1955). ^{k} D. H. R. Barton, A. J. Head and R. J. Williams, *J. Chem. Soc.*, 1715 (1953).

Table 7.26   High-pressure Arrhenius parameters for the unimolecular decomposition of various esters to give different types of products

Reactant	Products	$\log_{10} A_\infty/$ s^{-1}	$E_\infty/$ kcal mol^{-1}	Ref.
Gem-diesters $[R_1CH(O \cdot CO \cdot R_2)_2]$				
methylene diacetate	$HCHO + (CH_3CO)_2O$	9·2	33·0	a
methylene dipropionate	$HCHO + (EtCO)_2O$	9·2	33·0	a
methylene dibutyrate	$HCHO + (PrCO)_2O$	9·2	33·0	a
ethylidene diacetate	$MeCHO + (CH_3CO)_2O$	10·3	32·9	b
ethylidene dipropionate	$MeCHO + (EtCO)_2O$	10·4	32·9	c
ethylidene dibutyrate	$MeCHO + (PrCO)_2O$	10·3	33·0	d
butylidene diacetate	$PrCHO + (CH_3CO)_2O$	10·5	32·9	c

(*cont.*)

Table 7.26 (*cont.*)

Reactant	Products	$\log_{10} A_\infty/$ s^{-1}	$E_\infty/$ kcal mol^{-1}	Ref.
heptylidene diacetate	$C_6H_{11}CHO + (CH_3CO)_2O$	10·5	33·0	d
furfurylidene diacetate	furfural + $(CH_3CO)_2O$	11·1	33·0	e
crotonylidene diacetate	$MeCH:CHCHO$ $+ (CH_3CO)_2O$	10·1	33·0	e
trichloroethylidene diacetate	$CCl_3CHO + (CH_3CO)_2O$	10·1	33·0	f
trichloroethylidene dibutyrate	$CCl_3CHO + (PrCO)_2O$	10·1	33·0	f

Chloroformates [$RO \cdot CO \cdot Cl$]

Reactant	Products	$\log_{10} A_\infty/$ s^{-1}	$E_\infty/$ kcal mol^{-1}	Ref.
isopropyl chloroformate	propene + HCl + CO_2	$\begin{cases} 12\cdot7^m \\ 9\cdot1 \end{cases}$	$\begin{matrix} 39\cdot2^m \\ 25\cdot8 \end{matrix}$	g, h
sec-butyl chloroformate	$\begin{cases} \text{but-1-ene} + HCl + CO_2 \\ \textit{cis}\text{-but-2-ene} + HCl + CO_2 \\ \textit{trans}\text{-but-2-ene} + HCl \\ \quad + CO_2 \end{cases}$	10·2 13·7 13·8	31·5 40·0 40·4	g
isobutyl chloroformate	isobutene + HCl + CO_2	$13\cdot0^n$	$40\cdot0^n$	h

Carbonates [$R_1O \cdot CO \cdot OR_2$]

Reactant	Products	$\log_{10} A_\infty/$ s^{-1}	$E_\infty/$ kcal mol^{-1}	Ref.
methyl ethyl carbonate	$C_2H_4 + CH_3OH + CO_2$	13·7	46·0	i
diethyl carbonate	$C_2H_4 + C_2H_5OH + CO_2$	13·9	46·0	i
$CH_3CHPh \cdot O \cdot CO \cdot OMe$	$CH_2:CHPh + CH_3OH$ $+ CO_2$	12·3	39·9	j

Allyl esters

Reactant	Products	$\log_{10} A_\infty/$ s^{-1}	$E_\infty/$ kcal mol^{-1}	Ref.
allyl formate	propene + CO_2	10·0	43·2	k
(structure)	(structure)	$11\cdot3^l$	40^l	l
(structure)	(structure)	$10\cdot3^l$	37^l	l

[a] C. C. Coffin and W. B. Beazley, *Can. J. Res.* (*B*), **15**, 229 (1937). [b] C. C. Coffin, *Can. J. Res.* (*B*), **5**, 636 (1931). [c] C. C. Coffin, *Can. J. Res.* (*B*), **6**, 417 (1932). [d] C. C. Coffin, J. R. Dacey and N. A. D. Parlee, *Can. J. Res.* (*B*), **15**, 247 (1937). [e] J. R. Dacey and C. C. Coffin, *Can. J. Res* (*B*), **15**, 260 (1937). [f] N. A. D. Parlee, J. R. Dacey and C. C. Coffin, *Can. J. Res.* (*B*), **15**, 254 (1937). [g] E. S. Lewis and W. C. Herndon, *J. Amer. Chem. Soc.*, **83**, 1955 (1961). [h] A. R. Choppin and E. L. Compere, *J. Amer. Chem. Soc.*, **70**, 3797 (1948). [i] A. S. Gordon and W. P. Norris, *J. Phys. Chem.*, **69**, 3013 (1965). [j] G. G. Smith and B. L. Yates, *J. Org. Chem.*, **30**, 434 (1965). [k] J. M. Vernon and D. J. Waddington, *Chem. Commun.*, 623 (1969). [l] E. S. Lewis, J. T. Hill and E. R. Newman, *J. Amer. Chem. Soc.*, **90**, 662 (1968); the Arrhenius parameters reported here for these and other similar reactions may be in error.[2] [m] Parameters calculated in ref. 11(*b*) from the data of ref. *g*. [n] Parameters calculated in ref. *h* from data of E. T. Lessig, *J. Phys. Chem.*, **36**, 2325 (1932).

7.1.9   Other unimolecular reactions (Tables 7.27–7.33)

Many other unimolecular reactions exist which fall outside the classifications employed in the preceding sections, and the high-pressure Arrhenius parameters for a number of these are listed in Tables 7.27–7.33.

Among these miscellaneous unimolecular reactions are the isomerizations of methyl and ethyl isocyanides (Table 7.27), which have proved to be of great importance in the development of unimolecular reaction rate theories. The isomerization can be regarded as a Wagner–Meerwein 1,2-shift and is assumed to occur via a cyclic activated complex (7.36).[69, 70]

$$R-N\overset{\text{\tiny +}}{\equiv}C \longrightarrow R\underset{C}{\overset{N}{\cdots}} \longrightarrow R-C\equiv N \qquad (7.36)$$

The reactions have been studied extensively at low pressures, and have yielded valuable data with which to test RRKM-based theories of isotope effects and collisional energy transfer (see Chapters 9 and 10). It now appears that the reactions may be more complicated than has been thought, for Yip and Pritchard have reported partial inhibition of the isomerization of methyl isocyanide by high pressures of added propane.[71] If the reaction proceeds partly by radical-chain processes, some revision of the very extensive data based on its study may be necessary.

Many of the other reactions quoted are believed to involve cyclic activated complexes. The pyrolyses of β-hydroxyolefins (Table 7.28), $\beta-\gamma$ unsaturated carboxylic acids (Table 7.29) and vinyl alkyl or allyl ethers (Table 7.30) provide examples of 6-centre complexes which have much in common with formally similar reactions quoted in previous sections. Some examples are shown below.

Four-centre complexes are postulated for a number of other unimolecular reactions such as the pyrolyses of chloroalkylsilanes[72] and fluoroalkylsilanes[73] [Table 7.31, e.g. (7.37)], and a selection of miscellane-

$$CHF_2 \cdot CH_2 \cdot SiF_3 \longrightarrow \begin{matrix} F \cdots\cdots SiF_3 \\ \vdots \quad \vdots \\ CHF = CH_2 \end{matrix} \longrightarrow \begin{matrix} SiF_4 \\ + \\ CHF:CH_2 \end{matrix} \qquad (7.37)$$

ous reactions (Table 7.32) including such processes as the elimination of water from t-butanol,[62] ammonia and methane from t-butylamine,[74] and isothiocyanic acid from isothiocyanates.

A few other unimolecular reactions are given in Table 7.33.

Table 7.27 High-pressure Arrhenius parameters for isomerization of alkyl isocyanides

Reactant	Product	$\log_{10} A_\infty/$ s^{-1}	$E_\infty/$ kcal mol^{-1}	Ref.
CH_3NC	CH_3CN	13·6	38·4	69
CD_3NC	CD_3CN	13·6	38·5	a
C_2H_5NC	C_2H_5CN	13·8	38·2	70
C_2D_5NC	C_2D_5CN	13·6	37·8	b

^{a} F. W. Schneider and B. S. Rabinovitch, *J. Amer. Chem. Soc.*, **85**, 2365 (1963).
^{b} K. M. Maloney, S. P. Pavlou and B. S. Rabinovitch, *J. Phys. Chem.*, **73**, 2756 (1969).

Table 7.28 High-pressure Arrhenius parameters for the pyrolyses of β-hydroxyolefins

Reactant	Products	$\log_{10} A_\infty/$ s^{-1}	$E_\infty/$ kcal mol^{-1}	Ref.
but-3-enol	$C_3H_6 + CH_2O$	11·7	41·0	a
pent-4-en-2-ol	$C_3H_6 + MeCHO$	11·9	40·9	a
2-methylpent-4-en-2-ol	$C_3H_6 + MeCO_2H$	12·1	40·7	a
3-phenylbut-3-enol	$PhCMe:CH_2 + CH_2O$	11·8	38·9	a
4-phenylbut-3-enol	$PhCH_2CH:CH_2 + CH_2O$	11·6	42·8	a
4-ethyl-1-phenylhexen-4-ol	$PhCH_2CH:CH_2 + Et_2CO$	12·1	41·8	b
$CH_2:CH \cdot CH_2CHOH \cdot C_6H_4X$	$C_3H_6 +$ $CHO \cdot C_6H_4X$			
X = H		10·8	36·2	c
p-Me		10·4	34·9	c
p-Cl		10·7	36·1	c
p-MeO		10·7	35·6	c
m-MeO		11·0	36·9	c
m-Me		10·4	34·9	c
o-Me		11·0	36·5	c
o-Cl		11·6	38·4	c
o-MeO		9·1	31·4	c

^{a} G. G. Smith and B. L. Yates, *J. Chem. Soc.*, 7242 (1965). ^{b} G. G. Smith and R. Taylor, *Chem. Ind. (Lond.)*, 949 (1961). ^{c} G. G. Smith and K. J. Voorhees, *J. Org. Chem.*, **35**, 2182 (1970).

Table 7.29 High-pressure Arrhenius parameters for decarboxylation of $\beta-\gamma$ unsaturated carboxylic acids

Reactant	Products	$\log_{10} A_\infty/$ s^{-1}	$E_\infty/$ kcal mol^{-1}	Ref.
but-3-enoic acid	propene + CO_2	11·7	40·3	a
2,2-dimethyl-4-phenyl-but-3-enoic acid	2-methyl-4-phenyl-but-2-ene + CO_2	10·9	38·8	b
2,2-dimethyl-3-phenyl-but-3-enoic acid	2-methyl-3-phenyl-but-2-ene + CO_2	10·2	30·6	c
2,2-dimethylbut-3-enoic acid	[olefin] + CO_2	11·1	36·6	d
2,2-dimethylpent-3-enoic acid	[olefin] + CO_2	11·8	40·9	d
2,2,3-trimethylbut-3-enoic acid	[olefin] + CO_2	10·9	32·9	d
2,2-dimethyl-3-ethyl-pent-3-enoic acid	[olefin] + CO_2	11·7	36·0	d
2-methyl-2-(cyclopent-1-enyl)-propionic acid	[olefin] + CO_2	9·0	29·9	d
2-methyl-2-(cyclohex-1-enyl)-propionic acid	[olefin] + CO_2	10·0	33·3	d
2-methyl-2-(cyclohept-1-enyl)-propionic acid	[olefin] + CO_2	8·3	28·6	d

[a] Rate-constants of G. G. Smith and S. E. Blau, *J. Phys. Chem.*, **68**, 1231 (1964); activation parameters re-evaluated by D. B. Bigley and R. W. May, ref. *d*. [b] D. B. Bigley and J. C. Thurman, *J. Chem. Soc.*, 6202 (1965). [c] D. B. Bigley and J. C. Thurman, *J. Chem. Soc. (B)*, 1076 (1966). [d] D. B. Bigley and R. W. May, *J. Chem. Soc. (B)*, 557 (1967); identification of the olefinic products was not discussed.

Table 7.30 High-pressure Arrhenius parameters for unimolecular reactions of ethers

Reactant	Products	$\log_{10} A_\infty/$ s^{-1}	$E_\infty/$ kcal mol^{-1}	Ref.
Alkyl ethers				
t-butyl ethyl ether	isobutene + EtOH	14·1	59·7	a
t-butyl isopropyl ether	$\left\{\begin{array}{l}\text{isobutene} + {}^{\text{i}}\text{PrOH} \\ \text{propene} + {}^{\text{t}}\text{BuOH}\end{array}\right.$	13·4 13·0	55·5 56·5	b
isopropyl ether	propene + ^{i}PrOH	14·6^e	63·5^e	c
Vinyl ethers				
vinyl ethyl ether [reaction (7.37)]	C_2H_4 + MeCHO	11·4	43·8	75
vinyl isopropyl ether	propene + MeCHO	$\left\{\begin{array}{l}12\cdot6 \\ 12\cdot2\end{array}\right.$	43·6 42·6	d e
vinyl butyl ether	but-1-ene + MeCHO	11·1	42·4	f
vinyl t-butyl ether	isobutene + MeCHO	10·9	36·2	e
vinyl 1-chloroethyl ether	CH_2:CHCl + MeCHO	11·5	44·8	e
Vinyl allyl ethers				
allyl vinyl ether	pent-4-enal	11·7	30·6	g
2-methylallyl vinyl ether	4-methylpent-4-enal^h	11·2	29·1	h
1-methylallyl vinyl ether	*trans*-hex-4-enal	11·3	27·9	i
allyl isopropenyl ether	hex-1-en-5-one	11·7	29·3	j

^{a} N. J. Daly and C. Wentrup, *Aust. J. Chem.*, **21**, 1535 (1968). ^{b} N. J. Daly and F. J. Ziolkowski, *Aust. J. Chem.*, **23**, 541 (1970). ^{c} N. J. Daly and V. R. Stimson, *Aust. J. Chem.*, **19**, 239 (1966); concurrent radical-chain reaction produces propane and acetone. ^{d} A. T. Blades, *Can. J. Chem.*, **31**, 418 (1953). ^{e} T. O. Bamkole and E. U. Emovon, *J. Chem. Soc. (B)*, 332 (1968). ^{f} T. O. Bamkole and E. U. Emovon, *J. Chem. Soc. (B)*, 523 (1967). ^{g} F. W. Schuler and G. W. Murphy, *J. Amer. Chem. Soc.*, **72**, 3155 (1950). ^{h} H. M. Frey and B. M. Pope, *J. Chem. Soc. (B)*, 209 (1966); identity of product as quoted in ref. 2. H. M. Frey and D. C. Montague, *Trans. Faraday Soc.*, **64**, 2369 (1968). ^{j} L. Stein and G. W. Murphy, *J. Amer. Chem. Soc.*, **74**, 1041 (1952).

Table 7.31 High-pressure Arrhenius parameters for decomposition of halogenoalkyl silicon compounds

Reactant	Products	$\log_{10} A_\infty/$ s^{-1}	$E_\infty/$ kcal mol^{-1}	Ref.
$CHF_2CH_2SiF_3$	$CHF:CH_2 + SiF_4$	12·3	32·7	a
$CHF_2CH_2SiF_2Me$	$CHF:CH_2 + MeSiF_3$	11·3	32·5	b
$CHF_2CH_2Si(OMe)_3$	$CHF:CH_2 + FSi(OMe)_3$	11·0	36·3	c
$CHF_2CH_2Si(OBu)_3$	$CHF:CH_2 + FSi(OBu)_3$	10·3	35·1	73
$CH_2ClCH_2SiCl_3$	$C_2H_4 + SiCl_4$	11·1[d]	45·0[d]	d
$CH_2Cl \cdot CH_2SiCl_2Et$	$C_2H_4 + EtSiCl_3$	12·1	46·2	e
$CH_2Cl \cdot CH_2 \cdot SiClEt_2$	$C_2H_4 + Et_2SiCl_2$	11·9	41·1	f
$CH_2Cl \cdot CH_2 \cdot SiEt_3$	$C_2H_4 + Et_3SiCl$	11·1	39·2	72
$CH_2Cl \cdot CH_2 \cdot SiMe_3$	$C_2H_4 + Me_3SiCl$	11·0	37·5	72
$CH_3 \cdot CHCl \cdot SiClEt_2$	$HCl + CH_2:CHSiClEt_2$	11·8	45·4	72

[a] R. N. Haszeldine, P. J. Robinson and R. F. Simmons, *J. Chem. Soc.*, 1890 (1964).
[b] D. Graham, R. N. Haszeldine and P. J. Robinson, *J. Chem. Soc.* (*B*), 652 (1969).
[c] D. Graham, R. N. Haszeldine and P. J. Robinson, *J. Chem. Soc.* (*B*), 611 (1971).
[d] I. M. T. Davidson, C. Eaborn and M. N. Lilly, *J. Chem. Soc.*, 2624 (1964), recalculated in ref. *f*; elimination of HCl and HSiCl$_3$ also occurs. [e] I. M. T. Davidson and C. J. L. Metcalfe, *J. Chem. Soc.*, 2630 (1964), recalculated in ref. *f*. [f] I. M. T. Davidson and M. R. Jones, *J. Chem. Soc.*, 5481 (1965).

Table 7.32 High-pressure Arrhenius parameters for some miscellaneous unimolecular reactions probably involving four-centre activated complexes

Reactant	Products	$\log_{10} A_\infty/$ s^{-1}	$E_\infty/$ kcal mol^{-1}	Ref.
$CH_2:CHF$	$C_2H_2 + HF$	14·4	81·4	a
		14·0	70·8	b
$CH_2:CF_2$	$C_2HF + HF$	14·4	86·0	c
$CH_2:CHCl$	$C_2H_2 + HCl$	13·8	72·1	a
$CH_2:CHBr$	$C_2H_2 + HBr$	13·0	62·6	a
t-butanol	isobutene + H_2O	13·4	61·6	62
t-butylamine	isobutene + NH_3	14·2	67·1	d
	'C_3NH_7' + CH_4	14·6	67·1	
^tBuSH	isobutene + H_2S	13·3	55·0	62
EtNCS	$[C_2H_4 + H(NCS)]$	12·4	45·3	e
^iPrNCS	[propene + H(NCS)]	13·0	42·9	e
^tBuNCS	[isobutene + H(NCS)]	13·0	39·4	e
	+ HCl	11·5	47·0	f
Et_3B	$C_2H_4 + Et_2BH$	11·6	33·7	g
$^iBu_3Al$	isobutene + $^iBu_2AlH$	11·2	26·6	h
$(Me_2CD \cdot CH_2)_3Al$	isobutene + $(^iBu\text{-}d_1)_2AlD$	11·1	27·2	i

[a] P. Cadman and W. J. Engelbrecht, *Chem. Commun.*, 453 (1970). [b] J. M. Simmie, W. J. Quiring and E. Tschuikow-Roux, *J. Phys. Chem.*, **74**, 992 (1970). [c] J. M. Simmie and E. Tschuikow-Roux, *J. Phys. Chem.*, **74**, 4075 (1970) [cf. *Chem. Commun.*, 773 (1970)]. [d] W. Tsang, *J. Chem. Phys.*, **40**, 1498 (1964), and earlier work cited therein. [e] N. Barroeta, A. Maccoll and A. Fava, *J. Chem. Soc. (B)*, 347 (1969). [f] A. G. Loudon, A. Maccoll and S. K. Wong, *J. Amer. Chem. Soc.*, **91**, 7577 (1969). [g] E. Albuin, J. Grotewold, E. A. Lissi and M. C. Vara, *J. Chem. Soc. (B)*, 1044 (1968). [h] K. W. Egger, *J. Amer. Chem. Soc.*, **91**, 2867 (1969). [i] K. W. Egger, *Internat. J. Chem. Kinetics*, **1**, 459 (1969).

Table 7.33   High-pressure Arrhenius parameters for some miscellaneous unimolecular reactions

Reactant	Products	$\log_{10} A_\infty/$ s^{-1}	$E_\infty/$ kcal mol^{-1}	Ref.
Me$_3$COCl	isobutene + CO + HCl	14·4	55·2	a
Me$_3$COBr	isobutene + CO + HBr	14·1	48·9	b
(CH$_3$CO)$_2$O	CH$_3$CO$_2$H + CH$_2$CO	12·1	34·5	c
(CO$_2$H)$_2$	HCO$_2$H + CO$_2$	11·9	30·0	d
EtCH(OCH$_2$CH:CH$_2$)$_2$	EtO$_2$CCH$_2$CH:CH$_2$ + propene	10·3	36·3	e
CH$_2$:CH·CH(OMe)$_2$	CH$_3$CH:CHOMe + CH$_2$O	9·2	34·8	e
Me$_2$CH·CH$_2$C $\equiv$ CH	propene + CH$_2$:C:CH$_2$	13·1	29·7	f
CH$_3$N$_3$	[CH$_3$N:] + N$_2$	14·4	40·5	g
CH$_2$N$_2$	[CH$_2$:] + N$_2$	12·1[h]	34·0[h]	h

[a] B. S. Lennon and V. R. Stimson, *J. Amer. Chem. Soc.*, **91**, 7562 (1969). [b] B. S. Lennon and V. R. Stimson, *Aust. J. Chem.*, **23**, 525 (1970). [c] J. Murawski and M. Szwarc, *Trans. Faraday Soc.*, **47**, 269 (1951). [d] G. Lapidus, D. Barton and P. E. Yankwich, *J. Phys. Chem.*, **68**, 1863 (1964). [e] F. Mutterer, P. Baumgartner and J.-P. Fleury, *Bull. Soc. Chim. France*, 1528 (1970). [f] W. Tsang, *Internat. J. Chem. Kinetics*, **2**, 23 (1970); an alternative decomposition to radicals also occurs. [g] M. S. O'Dell and B. deB. Darwent, *Can. J. Chem.*, **48**, 1140 (1970). [h] D. W. Setser and B. S. Rabinovitch, *Can. J. Chem.*, **40**, 1425 (1962); results at 25 Torr.

7.2 UNIMOLECULAR REACTIONS AT LOW PRESSURES

Whereas the main interest in unimolecular reactions in their high-pressure regions centres on the reaction mechanism, the interest in the low-pressure region is concerned with the change in order, the shape and position of the fall-off curve and in favourable cases with the second-order rate-constant. The fall-off characteristics such as $p_{\frac{1}{2}}$ and the shape of the fall-off curve are often used to test various theories of unimolecular reactions. Energy-transfer processes can also be studied by investigating the relative efficiencies of inert gases in restoring the rate-constant to its limiting high-pressure value.

Some reactions have been interpreted in terms of Kassel or Slater theory and values of the parameters s or n have been obtained. RRKM calculations have now been carried out for quite a number of unimolecular reactions. The fall-off curve in general tests both the assumptions inherent in the RRKM theory and the choice of model for the activated complex. Partly for this reason, the fall-off curve in itself is not a very sensitive test of any particular model for the activated complex. The low-pressure second-order rate-constant, on the other hand, is independent of the properties of the activated complex and depends only upon the properties of the energized molecule (see section 4.8).

Table 7.34 lists some of the experimental studies which have been made on unimolecular reactions in their fall-off regions. It is clearly useful to be able to predict the pressure at which the unimolecular rate-constant k_{uni} for a given reaction will begin to decline. All the theories predict that the pressure for a given degree of fall-off depends upon the number of modes of vibration involved. From Slater theory, for example, it is seen from (2.43)–(2.48) that the fall-off pressure $p_{\frac{1}{2}}$ is proportional to $(E_0/kT)^{-\frac{1}{2}(n-1)}$ if one assumes that the terms involving molecular frequencies and also k_2 are approximately constant for different molecules. It then follows that

$$\log p_{\frac{1}{2}} = constant - \tfrac{1}{2}(n-1)\log(E_0/kT)$$

Since E_0/kT is usually about 40 at temperatures which are convenient for conventional studies of thermal unimolecular reactions, one would expect an approximate correlation between $\log p_{\frac{1}{2}}$ and $(n-1)$. In Figure 7.1 some experimental fall-off pressures are plotted as $\log p_{\frac{1}{2}}$ vs the number of atoms (N) in the molecule. The result clearly suggests that $(n-1)$ is approximately a constant multiple of N. The dangers in assuming Kassel's s to be a constant fraction of the total number of degrees of freedom have been pointed out by Rabinovitch and co-workers.[76] Figure 7.1 shows, however, that the experimental data do conform approximately to this correlation, and this graph is clearly useful in a practical way for predicting the approximate fall-off pressure for any reaction. The experimental study of unimolecular reactions at very low pressures is complicated by the difficulties of measuring the extent of reaction in very small samples. In addition, when the mean free path is greater than the

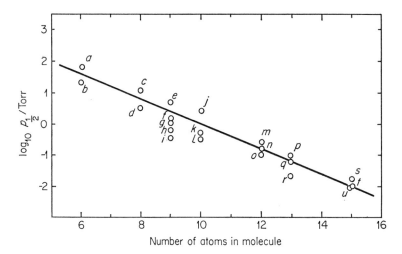

Figure 7.1 Dependence of transition pressure $p_{\frac{1}{2}}$ upon atomicity: (*a*) methyl isocyanide, (*b*) CD_3NC, (*c*) ethane, (*d*) ethyl chloride, (*e*) cyclopropane, (*f*) 1,1-dichlorocyclopropane, (*g*) cyclopropane-d_6, (*h*) ethyl isocyanide, (*i*) C_2D_5NC, (*j*) azomethane-d_6, (*k*) cyclobutene, (*l*) cyclobutene-d_6, (*m*) cyclobutane, (*n*) methylcyclopropane, (*o*) cyclobutane-d_8, (*p*) bicyclo[1.1.1]pentane, (*q*) 3-methylcyclobutene, (*r*) 1-methylcyclobutene, (*s*) methylcyclobutane, (*t*) 1,1-dimethylcyclopropane, (*u*) ethylcyclopropane. For references see Tables 7.34 and 7.36

diameter of the reaction vessel, energization will occur predominantly at the walls rather than in the gas phase; this complication has been treated theoretically by Maloney and Rabinovitch.[141] As a result of these difficulties, the fall-off regions are not easily experimentally accessible for molecules undergoing unimolecular reactions with activation energies from 40–60 kcal mol^{-1} at temperatures from 500–700 K unless the molecule contains fewer than ten to twelve atoms. This fact has restricted the number of examples of unimolecular reactions which have been studied over a wide extent of fall-off. The experimental data for a selected number of such reactions will now be examined in detail, and the calculations involving RRKM theory in particular will be described.

9

Table 7.34 Data for unimolecular reactions in the fall-off region (see also Tables 7.36 and 7.37)

Reactant	Products	Temperatures (°C)	Pressure range/Torr	$p_{1/2}$/Torr (obs.)	Exptl ref.	Theory applied[a] and refs.
cyclopropane	propene	490 ± 2	0.07–84	5	b	HL,[77] K,[77] S[77, c, d] RRKM[79, 83]
trans-cyclopropane-d_2	cis-cyclopropane-d_2	445	0.3–1000	6	81	K,[81] S[81]
cyclopropane-d_2	propene-d_2	445	0.05–1000	2	81	
cyclo-^{12}C$_2$^{13}CH$_6$	^{13}C-propene	450–519	1–760	—	e	I[e]
cyclopropane-d_6	propene-d_6	407–514	10^{-4}–1000	~1 (510 °C)	f, g	I[83, g, h]
methylcyclopropane	butenes	440–490	0.06–200	0.2 (490 °C)	85	RRKM[79, 83, 87]
1,1-dimethylcyclo-propane	methylbutenes	442–481	0.008–35	0.01 (442 °C)	88	K[88]
ethylcyclopropane	pentenes	454–484	0.05–84	~0.01 (468 °C)	86	RRKM[86]
1,1-dichlorocyclo-propane	2,3-dichloropropene	359–424	0.2–80	2 (424 °C)	90	RRKM[90]
cyclobutane	ethylene	448	0.06–100	0.2	93	HL,[93] K,[93, 94] S,[i] RRKM[79]
		449	2×10^{-4}–43	0.4	94	
		410–500	5×10^{-5}–20	0.2 (450 °C)	95	
		420–468	1–996	—	91	
cyclobutane-d_8	ethylene-d_4	449	0.005–100	~0.1	j, k	I[j, k]
methylcyclobutane	C$_2$H$_4$ + propene	400–450	0.003–0.45	0.02	97	K[96, 97]
	C$_2$H$_4$ + propene + butenes	409–489	10^{-4}–10	~0.02	96	
cis-1,2-dimethyl-cyclobutane	C$_2$H$_4$ + propene + butenes		2×10^{-4}–10		l	K[l]
trans-1,2-dimethyl-cyclobutane	C$_2$H$_4$ + propene + butenes		2×10^{-4}–10		l	K[l]
methylenecyclobutane	C$_2$H$_4$ + allene	450	0.5–35	<0.5	30	RRKM[143]

Table 7.34 (*cont.*)

Reactant	Products	Temperatures (°C)	Pressure range/ Torr	$p_{\frac{1}{2}}$/Torr (obs.)	Exptl ref.	Theory applied[a] and refs.
cyclobutene	buta-1,3-diene	130–175	0·02–23	0·5 (150 °C)	99	K,[99] S,[99] RRKM[100, 101, 143]
cyclobutene-d_6	butadiene-d_6	140–180	0·004–25	0·3 (150 °C)	103	RRKM[103]
1-methylcyclobutene	2-methylbuta-1,3-diene	150–175	0·001–1	0·02 (150 °C)	102	K,[102] RRKM[100]
3-methylcyclobutene	penta-1,*trans*-3-diene	124–149	0·01–45	0·06	m	K,[m] RRKM[100]
bicyclo[1.1.1]pentane	penta-1,4-diene	303	0·3–35	~0·1	n	
ethyl chloride	ethylene + HCl	402–521	0·2–120	2·8 (521 °C)	104	S,[104] K,[104]
		649	15–60	—	108	RRKM[79, 104, 106, 107]
		456	2–200	<8	105	
		439	0·1–30	0·15	106	
CHD_2CD_2Cl	C_2HD_3 + DCl / C_2D_4 + HCl	485–716	17–61	—	108	I[108]
CD_3CH_2Cl	$C_2D_2H_2$ + DCl	440–482	0·1–10	~0·2 (440 °C)	o	I[107]
C_2D_5Cl	C_2D_4 + DCl	439–480	0·2–7·5	0·6	106	I[106, 107]
1,1-dichloroethane	CH_2:CHCl + HCl	412–449	1–200	<17	105	K[105]
1,2-dichloroethane	CH_2:CHCl + HCl	—	—	—	—	RRKM[109]
1-chloropropane	propene + HCl	447	0·1–100	<1	105	
2-chloropropane	propene + HCl	407	0·5–10	<4	105	
isobutyl chloride	isobutene + HCl	362–475	0·1–100	<3	105	
t-butyl chloride	isobutene + HCl	339	0·1–100	<6	105, q	RRKM[p, 107]
2-bromopropane	propene + HBr	332	0·5–48	<4	r	HL,[s] K[r]
1,2-dibromoethane	CH_2:CHBr + HBr	—	—	—	—	RRKM[110]
1-bromo-2-chloroethane	CH_2:CHBr + HCl / CH_2:CHCl + HBr	—	—	—	—	RRKM[110]

Table 7.34 (*cont.*)

Reactant	Products	Temperatures (°C)	Pressure range/ Torr	$p_{\frac{1}{2}}$/Torr (obs.)	Exptl ref.	Theory applied[a] and refs
ethylene oxide	acetaldehyde	370–429	6–800	—	42	RRKM[43]
ethylene oxide-d_4	acetaldehyde-d_4	—	—	—	—	RRKM[43]
β-propiolactone	$C_2H_4 + CO_2$	259	0.01–100	~0.8	s	RRKM[s]
methyl isocyanide	methyl cyanide	200–260 / 279–282	0.02–10⁴ / 0.005–10	65 (230 °C) / —	69 / 117	RRKM[69, 117]
CD$_3$NC	CD$_3$CN	230	0.01–10⁴	30	118	I[118]
CH$_2$DNC	CH$_2$DCN	245	0.05–7000	55	119	I[119]
CH$_3$N^{13}C, ^{13}CH$_3$NC	CH$_3$^{13}CN, ^{13}CH$_3$CN	226	1–760	—	t	I[t]
ethyl isocyanide	ethyl cyanide	190–260	0.001–3000	0.6 (231 °C)	70	RRKM[70]
C$_2$D$_5$NC	C$_2$D$_5$CN	231	0.001–100	0.3 (231 °C)	120	I[120]
CH$_2$N$_2$	N$_2$ + [CH$_2$: → C$_2$H$_4$]	240–380	25–500	~400	u	RRKM[u]

[a] HL = Hinshelwood–Lindemann, S = Slater, K = Kassel (or RRK), RRKM = Rice–Ramsperger–Kassel–Marcus, I = isotope effect calculations based on RRKM (for fuller discussion of these see Chapter 9). [b] Ref. 77; see also L. W. Gay, Ph.D. Thesis, University of Oregon [cf. *Diss. Abs.* (B), **29**, 960 (1968–9)]. [c] R. C. Golike and E. W. Schlag, *J. Chem. Phys.*, **38**, 1886 (1963). [d] N. B. Slater, *Proc. Roy. Soc.* (A), **218**, 224 (1953). [e] L. B. Sims and P. E. Yankwich, *J. Phys. Chem.*, **71**, 3459 (1967). [f] A. T. Blades, *Can. J. Chem.*, **39**, 1401 (1961). [g] B. S. Rabinovitch, P. W. Gilderson and A. T. Blades, *J. Amer. Chem. Soc.*, **86**, 2994 (1964). [h] B. S. Rabinovitch, D. W. Setser and F. W. Schneider, *Can. J. Chem.*, **39**, 2609 (1961). [i] D. G. Retzloff, Ph.D. Thesis, University of Pittsburgh, 1967 [cf. *Diss. Abs.* (B), **28**, 4116 (1967–8)]. [j] J. Langrish and H. O. Pritchard, *J. Phys. Chem.*, **62**, 761 (1958). [k] R. W. Carr and W. D. Walters, *J. Amer. Chem. Soc.*, **88**, 884 (1966). [l] P. J. Conn, Ph.D. Thesis, University of Oregon, 1966 [cf. *Diss. Abs.* (B), **27**, 2311 (1966–7)]. [m] D. C. Marshall and H. M. Frey, *Trans. Faraday Soc.*, **61**, 1715 (1965). [n] R. Srinivasan, *J. Amer. Chem. Soc.*, **90**, 2752 (1968). [o] G. W. Völker and H. Heydtmann, *Zeit. Naturforsch.* (B), **23**, 1407 (1968). [p] H. Heydtmann, *Chem. Phys. Lett.*, **1**, 105 (1967). [q] B. Roberts, Ph.D. Thesis, University of London, 1961, quoted by A. Maccoll, ref. 4. [r] M. N. Kale and A. Maccoll, *J. Chem. Soc.*, 1513 (1964). [s] T. L. James and C. A. Wellington, *J. Amer. Chem. Soc.*, **91**, 7743 (1969). [t] J. F. Wettaw and L. B. Sims, *J. Phys. Chem.*, **72**, 3440 (1968). [u] M. B. Gabunia and A. A. Shteinman, *Zh. Fiz. Khim.*, **44**, 1682 (1970).

7.2.1 The structural and geometrical isomerizations of cyclopropane and its derivatives at low pressures

The most detailed experimental study of the structural isomerization of *cyclopropane* to propene at low pressures is that of Pritchard *et al.*[77] These workers studied the isomerization at pressures of 0·07–84 Torr and at $490 \pm 2\,°C$. Their results, together with those of Chambers and Kistiakowsky,[78] are shown in Figures 2.6 and 7.3. In Figure 2.6 the results were compared with the theoretical Hinshelwood–Lindemann, Kassel and Slater curves. The fall-off data were originally treated in terms of Slater theory (see section 2.4), but the agreement obtained is now generally regarded as fortuitous. RRKM theory was first applied to this reaction by Wieder and Marcus.[79, 80] These workers considered two 'rigid' activated complexes (Figure 7.2) which were selected by empirical procedures

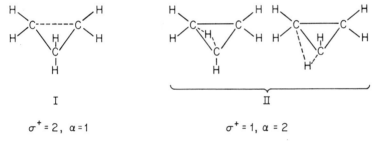

$$\sigma^+ = 2, \ \alpha = 1 \qquad\qquad \sigma^+ = 1, \ \alpha = 2$$

Figure 7.2 Activated complexes considered by Wieder and Marcus[79] for cyclopropane isomerization. The statistical factor

$$L^+ = \alpha\sigma/\sigma^+ = 1 \times 6/2 = 3$$

for Complex I, corresponding to breaking any of the three C—C bonds. For Complex II, $L^+ = 2 \times 6/1 = 12$; there are six H atoms, each of which can migrate to two carbon atoms. For further discussion of L^+ see section 4.9

similar to that described in section 6.2.3. Complex I is associated with a C—C stretching reaction coordinate and has C_{2v} symmetry. The vibration frequencies assigned to this complex were based on those for the normal molecule, with removal of a ring-deformation frequency and the lowering of two other frequencies to become torsional modes. Complex II corresponds to a C—H stretching coordinate with some ring deformation. It has no overall symmetry but exists as the two optical isomers shown in Figure 7.2. The frequencies for this complex were obtained by removal of a C—H frequency and lowering some ring-deformation frequencies. In some calculations, the semi-classical expression (5.30) was used for $N^*(E^*)$. In others, a more accurate treatment based on a quantum-mechanical treatment of the C—H stretching frequencies was used. The calculated curve for complex I is compared in Figure 7.3 with the experimental data. The calculated curve is displaced towards lower pressures, and it was found necessary to employ a collisional deactivation efficiency of $\lambda = 0.25$ in order to give agreement with the experimental curve.

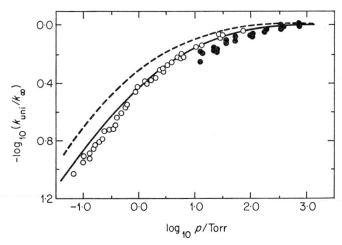

Figure 7.3 Comparison of some calculated and experimental fall-off curves for the isomerization of cyclopropane at 492 °C. Experimental points: ○, Pritchard and coworkers;[77] ●, Chambers and Kistiakowsky.[78] Calculated RRKM curves ($\lambda = 1$): – – – –, Wieder and Marcus;[79] ———, Lin and Laidler[83]

Complex II gave very similar results, the calculated curve differing from that for complex I only in a shift to lower pressure by 0.06 units in $\log_{10} p$. The temperature-dependence of the fall-off curve was calculated by Wieder and Marcus for complex II. A temperature decrease of 55 K shifted the curve towards lower pressures by 0.25 units. This can be compared with a value of 0.53 units calculated from the results of Schlag and Rabinovitch[81] at 445 °C and those of Pritchard and co-workers[77] corrected to 500 °C by Johnston and White.[82]

Lin and Laidler[83] considered the cyclopropane isomerization in terms of a mechanism involving a biradical intermediate (see the arguments in section 7.1.1). Using the high-pressure Arrhenius parameters of Falconer et al.[5] instead of those of Chambers and Kistiakowsky,[78] much better agreement with experiment was obtained than in the calculations of Wieder and Marcus[79] (see Figure 7.3). Although the Lin and Laidler curve was calculated for a biradical mechanism, similar results could be obtained assuming a concerted mechanism. The improved agreement was attributed largely to the use of different Arrhenius parameters (cf. section 6.5.2).

The geometrical isomerization of *trans-cyclopropane-d_2* to the *cis*-isomer was studied by Schlag and Rabinovitch[81] who found a fall-off curve similar to that for the structural isomerization. The data correspond to a Slater *n* value of 13·5 to 14·0. Simons and Rabinovitch[84] have made some RRKM calculations based on the values $A_\infty = 10^{16.1}$ s^{-1}, $E_\infty = 64.1$ kcal mol^{-1} and using a ring-deformation mode as reaction coordinate. The shape of the theoretical curve fits the experimental data (see Figure 7.4)

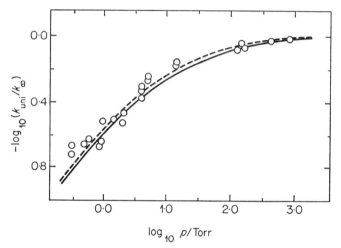

Figure 7.4   Fall-off curves for the geometrical isomerization of *trans*-cyclopropane-d_2 at 445 °C. Experimental points from Schlag and Rabinovitch.[81] Calculated RRKM curves: ———, Lin and Laidler;[83] – – – –, Simons and Rabinovitch[84] (displaced by 0·2 log units)

but is displaced by 0·2 units along the $\log_{10} p$ axis. Also shown in Figure 7.4 is the curve calculated by Lin and Laidler[83] based on a biradical intermediate. Their calculations used the values $A_\infty = 10^{16.1}$ s^{-1}, $E_\infty = 65.1$ kcal mol^{-1} for the ring-opening step and gave good agreement with experiment.

Methylcyclopropane undergoes four parallel unimolecular reactions to give the isomeric butenes. The fall-off curves were obtained at 490·4 and 446·9 °C by Chesick[85] and discussed in terms of Kassel and Slater theories. The fall-off curves for methylcyclopropane[85] and ethylcyclopropane[86] are displaced towards lower pressures compared with the cyclopropane curve (Figure 7.5), corresponding to a change in the Kassel *s* value from cyclopropane (12–13) to methylcyclopropane (19) and ethylcyclopropane

(21–23). Chesick found that the activation energies for but-1-ene and but-2-ene production were approximately equal, and less than that for isobutene production by about 2–3 kcal mol^{-1}. Ignoring isobutene production and treating the other reactions as a single unimolecular

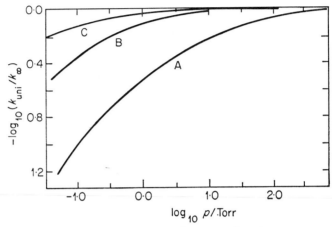

Figure 7.5 Experimental fall-off curves for the structural isomerizations of cyclopropane (A)[77, 78] and methylcyclopropane (B)[85] at 490 °C, and for ethylcyclopropane (C)[86] at 468 °C

reaction, Wieder and Marcus[79] applied RRKM theory to calculate the fall-off curve at 490 °C. The reaction coordinate was chosen to be a partial H-atom migration combined with an unsymmetrical ring deformation (Figure 4.4). A collision efficiency of $\lambda = 0.22$ was found necessary to produce agreement with experiment.

Lin and Laidler[83] have applied RRKM theory separately to the production of but-1-ene, but-2-ene (*cis + trans*) and isobutene from methylcyclopropane using Arrhenius parameters derived from Chesick's[85] data and also some more recent values of Setser and Rabinovitch.[87] The calculated rates gave poor fits for all three reactions and λ values of 0.17, 0.16 and 0.028 were found necessary to give agreement with experiment.

RRKM calculations were carried out by Halberstadt and Chesick[86] for the production of 1- and 2-pentene (treated as a single unimolecular reaction) from *ethylcyclopropane*. Although the experimental data are rather scattered and do not extend to $p_{\frac{1}{2}}$, these authors concluded that a low value of λ (0.04 − 0.09) gave the best fit for the calculated curve. Significantly, it was pointed out that an activation energy error of only 1.4 kcal mol^{-1} would account for this apparently low value of λ compared with that found for analogous reactions (cf. section 6.5.2).

The decline in rate-constant with pressure for the isomerization of *1,1-dimethylcyclopropane* to 2-methylbut-2-ene and 3-methylbut-1-ene at 442 and 481 °C was studied by Flowers and Frey.[88] Fall-off commences at about 16 Torr at 442 °C, a similar value to that observed for the structural isomerization of 1,2-dimethylcyclopropane.[89] The Kassel equation was used to fit the data with the value of $s = 23$.

The isomerization of *1,1-dichlorocyclopropane* to 2,3-dichloropropene was studied at initial pressures down to 0·2 Torr and at 359 and 424 °C by Holbrook et al.[90] The activated complex for this reaction was assumed to involve chlorine-atom migration with possibly some ring deformation. The application of RRKM theory to this reaction has been treated in detail in Chapter 6. If the high-pressure Arrhenius parameters of Parry and Robinson[14] are used, a collisional deactivation efficiency of $\lambda = 0·4$ is necessary to give agreement with the more extensive experimental results at 424 °C. Alternatively an increase in E_∞ by 1–2 kcal mol^{-1} with the corresponding adjustment of A_∞ would also produce the desired agreement.

7.2.2 The reactions of cyclobutane and its derivatives at low pressures

The decomposition of *cyclobutane* itself to two molecules of ethylene has been studied at low pressures by Walters and co-workers,[91, 92] by Pritchard et al.,[93] by Butler and Ogawa[94] and by Vreeland and Swinehart.[95] The last pair of workers carried out an extensive study of the reaction using mass spectrometric analysis over the temperature range 410–500 °C and a pressure range of 5×10^{-5} to 20 Torr. The data were fitted by the Kassel equation using $s = 18$ and $E_0 = 63·2$ kcal mol^{-1}. Rate-constants below 10^{-3} Torr were higher than those predicted theoretically and it was suggested that this was due to the occurrence of two competing unimolecular mechanisms. Butler and Ogawa[94] discussed a possible explanation of the low-pressure discrepancy in terms of a wall effect, but some experiments of Vreeland and Swinehart using packed vessels appeared to invalidate this. Small amounts of propene and but-1-ene are also formed by zero-order reactions at low pressures and it is clear that the role of the surface in this reaction is not properly understood.

RRKM calculations have been made by Wieder and Marcus[79] for cyclobutane decomposition using a reaction coordinate involving the simultaneous expansion of two opposite C—C bonds in the ring. The resulting theoretical curve with $\lambda = 0·20$ was in fair agreement with the experimental data of Pritchard and co-workers.[93]

Lin and Laidler[83] have carried out RRKM calculations based on a scheme (7.38) involving the tetramethylene biradical and also incorporating the back-reaction in which energized species revert to reactants. Assuming that $k_{-2E} = k_{3E}$ and taking the high-pressure Arrhenius parameters of Vreeland and Swinehart,[95] Lin and Laidler[83] obtained good agreement

$$\square \xrightarrow[k_{-1}[M]]{k_1[M]} \square^* \xrightarrow[k_{-2E}]{k_{2E}} \left[\begin{smallmatrix}\bullet\\\bullet\end{smallmatrix}\right]^* \xrightarrow{k_{3E}} 2C_2H_4{}^* \xrightarrow{k_4[M]} 2C_2H_4$$

(7.38)

between theory and experiment without the use of a low collisional efficiency. They concluded, however, that because of the insensitivity of calculated rates to the structure of the activated complex, the agreement obtained did not provide positive evidence for the participation of a biradical intermediate in the cyclobutane system. Their curves are compared with experimental data in Figure 7.6.

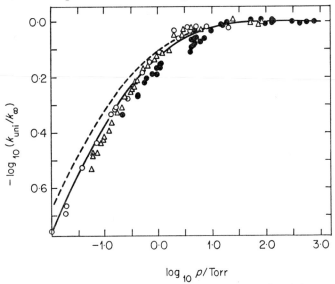

Figure 7.6 Fall-off curves for the decomposition of cyclobutane at 449 °C. Experimental points: △, Pritchard and coworkers;[93] ○, Vreeland and Swinehart;[95] ●, Walters and coworkers.[91, 92] Calculated curves of Lin and Laidler[83] using

$$\log_{10} k_\infty/\text{s}^{-1} = 15.85 - 63.2/\theta \quad (\text{———})$$

and

$$\log_{10} k_\infty/\text{s}^{-1} = 15.62 - 62.5/\theta \quad (\text{- - - - -})$$

Methylcyclobutane undergoes two unimolecular reactions in the region of 400–450 °C, a decomposition to ethylene and propene and an isomerization to pent-1-ene (7.39). Thomas *et al.* found[96] that the ratio k_i/k_d

$$\begin{array}{c}CH_3\\ \backslash\\ CH-CH_2\\ |\quad\quad|\\ CH_2-CH_2\end{array}\begin{array}{l}\xrightarrow{k_d} CH_3-CH=CH_2 + C_2H_4\\ \\ \xrightarrow{k_i} CH_3-CH_2-CH_2-CH=CH_2\end{array}$$

(7.39)

decreases with initial pressure. The fall-off curve (for the total rate-constant $k = k_i + k_d$) was found to level out at pressures below 10^{-3} Torr, and packed vessel experiments indicated that surface activation was important at these pressures. The decomposition reaction has also been studied at low pressures by Patarrachia and Walters[97] who found that the fall-off curve is displaced to lower pressures by 1·2 units in $\log_{10} p$ compared with Vreeland and Swinehart's data for cyclobutane. Quantum and classical forms of Kassel theory were applied. No RRKM calculations appear to have been carried out at the present time for this molecule.

Although *cis–trans* isomerizations of suitable cyclobutane derivatives have been reported[98] there do not appear to be any extensive studies of such reactions at low pressures.

7.2.3   The isomerizations of cyclobutene and its derivatives at low pressures

The isomerization of *cyclobutene* to buta-1,3-diene was studied by Hauser and Walters[99] at 150 °C and from 0·015 Torr to 23 Torr. The fall-off region was found to be similar to that for cyclobutane. The fall-off curve was fitted by the classical Kassel curve with $s = 10$, the quantum Kassel curve with $s = 13$ and the Slater curve with $n = 15$ or 16.

RRKM theory has been applied to the isomerizations of cyclobutene, *1-methylcyclobutene* and *3-methylcyclobutene* by Elliott and Frey,[100] and to cyclobutene by Lin and Laidler[101] and by Engelbrecht and DeVries.[143] Elliott and Frey introduced a λ factor of 0·3 in order to produce agreement with the experimental data for cyclobutene and 3-methylcyclobutene. A more satisfactory treatment[101, 143] uses $\lambda = 1$ but makes allowance for the back reaction of energized product molecules (see section 6.5.2). The experimental fall-off curve for 1-methylcyclobutene occurs at about one-third the pressure of that for 3-methylcyclobutene at a common temperature, but is almost identical at a common value of (E_0/kT).[102]

The isomerization of *perdeuteriocyclobutene* to butadiene-d_6 has recently been studied[103] at pressures from 0·004–25 Torr at 150 °C. Compared with the fall-off curve for 'light' cyclobutene there is a cross-over point, since the deuterium-substituted compound reacts faster at low pressures but slower at high pressures (see section 9.5.2). Although conventional RRKM calculations were insensitive to the model used for the activated complex, the coordinates of the predicted cross-over point enabled a distinction between the various models to be made. The use of isotopic substitution in this way appears to be a useful advance in the application of RRKM theory to experimental data. The experimental fall-off curves for C_4H_6 and C_4D_6 are shown in Figure 7.7.

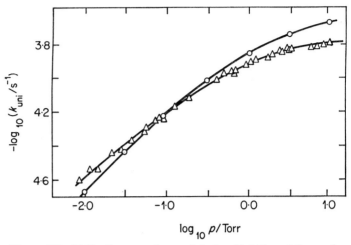

Figure 7.7   Fall-off curves for cyclobutene[99] ($\bigcirc$) and for cyclo-butene-d_6[103] ($\triangle$) at 150·5 °C

7.2.4   The pyrolysis of alkyl halides at low pressures

Apart from some work on chemically activated alkyl fluorides, and a few isolated studies of bromide and iodide decompositions at low pressures, most of the low-pressure studies of alkyl halide pyrolysis concern alkyl chlorides. In particular the pyrolysis of *ethyl chloride* has attracted much attention.[104–108] Although there is good agreement between various workers in the values of measured rate-constants at a particular temperature, exact comparison of the fall-off curves is made difficult by the uncertainty in the high-pressure Arrhenius parameters (Table 7.35). For

Table 7.35   Comparison of Arrhenius parameters for ethyl chloride pyrolysis at high pressures

	$\log_{10} A_\infty/\text{s}^{-1}$	$E_\infty/\text{kcal mol}^{-1}$	Ref.
Experimental values			
Barton and Howlett	14·60	60·80	111
(recalculated by Howlett[112])			
Tsang	13·16	56·46	113
Capon and Ross	13·46	56·62	114
Shilov and Sabirova	13·90	59·00	115
Hartmann, Bosche and Heydtmann	13·51	56·61	116
Holbrook and Marsh	14·03	58·43	104
Values assumed by other workers in calculations			
Benson and Bose	13·4	57·1	45
Wieder and Marcus	14·0	57·5	79
Hassler and Setser	13·4	56·5	109

this reason we have selected k_∞ values from the rate expression preferred by Hassler and Setser[109] (which agrees well with the measured k_∞ values of Blades and co-workers[108] at 922·5 K and also of Heydtmann and Völker[106] at 712·5 K) and replotted the fall-off data of Heydtmann and Völker,[106] Howlett[105] and Holbrook and Marsh[104] as $\log(k_{uni}/k_\infty)$ vs $\log$ *pressure* (Figure 7.8). It will be seen that the Heydtmann and Völker values correspond to a value of $p_{\frac{1}{2}}$ which is ten times lower than those observed by the other workers.

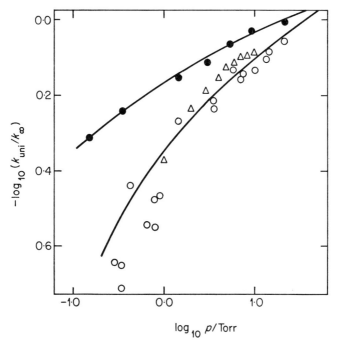

Figure 7.8 Experimental fall-off curves for ethyl chloride decomposition (see text). ●,Heydtmann and Völker,[106] 712·5 K; ○, Holbrook and Marsh,[104] 730·0 K; △, Howlett,[105] 729·2 K

RRKM theory calculations have been carried out for this reaction by Wieder and Marcus,[79] Holbrook and Marsh,[104] Heydtmann and Völker[106] and by Setser and co-workers.[109, 110] Wieder and Marcus considered a rigid cyclic activated complex. One C—H stretching frequency was removed to serve as the reaction coordinate and the remaining frequencies were grouped in multiples of the lowest frequency and adjusted to fit the experimental entropy of activation. The Arrhenius parameters used by Wieder and Marcus are shown in Table 7.35. The calculated fall-off curve was displaced to low pressures compared with Howlett's data,[105] and a collisional efficiency of 0·25 would be necessary to give agreement at $p_{\frac{1}{2}}$. Holbrook and Marsh[104] came to essentially the same conclusion from a more detailed study at four different temperatures. Hassler and Setser[109] used several different models of the activated complex for this reaction, assigning frequencies by analogy with the parent molecule and by vibrational analysis of the 4-membered ring in the complex as described in section 6.2.2. The results of taking one of the external rotations to be active rather than adiabatic were also discussed. In a subsequent publication, Johnson and Setser[110] reported some slight

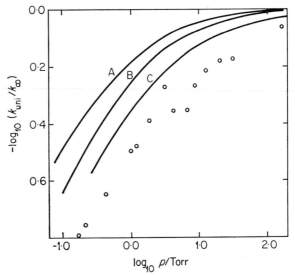

Figure 7.9   Comparison of calculated fall-off curves for ethyl chloride decomposition at 794 K. Calculated curves ($\lambda = 1$): (A) Hassler and Setser[109] model I; (B) and (C) Holbrook and Marsh [104] with $E_\infty = 58·4$ and 59·9 kcal mol⁻¹ respectively. Circles are experimental points[104]

changes to the calculations but concluded that the fall-off remained essentially unchanged. Calculations have been repeated by the present authors using Setser's model I for the activated complex but a temperature of 794 K in order to facilitate comparison with the experimental data of Holbrook and Marsh. The comparison between the various calculations is shown in Figure 7.9 and confirms the fact (reported by Johnson and Setser[110]) that there is a discrepancy of 1·0 log units on the pressure axis between Setser's calculated curve and the experimental data. This discrepancy is at least in part due to the different Arrhenius parameters taken by Hassler and Setser, since (as also shown in Figure 7.9) a change in E_∞ by $1 \cdot 5\,\text{kcal}\,\text{mol}^{-1}$ results in a displacement of the calculated curve by $0 \cdot 3$ log units (compare curves B and C). Similar conclusions were reached in section 6.5.2.

The problem of the correct statistical factor L^+ for 4-centre eliminations has been discussed by a number of workers (see particularly ref. 109). For ethyl chloride, values of 1, 2, 3 and 6 have been used, and not always correctly. Various values will arise depending on (a) whether there is free internal rotation ($\sigma = 3$) or merely torsional vibration ($\sigma = 1$) in the normal molecule, and (b) the geometry of the activated complex. Of the three conceivable complexes I–III the first is not valid because it can lead to two different sets of products (see section 4.9).

I	II	III
$\sigma^+ = 1, \alpha = 1$	$\sigma^+ = 1, \alpha = 1$	$\sigma^+ = 1, \alpha = 2$

The second will be invalid for the same reason if the molecule has a torsional vibration, but might in principle be valid for the free rotor model, in which case $L^+ = \alpha\sigma/\sigma^+ = 3$ (see section 4.9). The most likely complex seems, in fact, to be III, which exists as optical isomers since the 4-centre ring is twisted. In this case $L^+ = 6$ for the freely rotating molecule or 2 for the torsional model, these being the values used by Hassler and Setser.[109]

7.2.5 The isomerization of isocyanides at low pressures

The thermal isomerization of *methyl isocyanide* to methyl cyanide has been thoroughly investigated by Rabinovitch and co-workers.[69, 117] This reaction is especially favourable for study from the point of view of unimolecular reaction rate theory since the methyl isocyanide molecule is relatively small and symmetrical (C_{3v} symmetry) and the reaction has a low activation energy. Fall-off curves are therefore accessible at low temperatures and can be studied over a wide pressure range. The reaction has long been thought to be an uncomplicated unimolecular process with no heterogeneous or radical contribution, although it can be sensitized by t-butyl peroxide, and recent evidence suggests that there may be a minor radical chain contribution (see section 7.1.9 and ref. 71).

RRKM calculations have been carried out[69, 117] for several 'rigid' models of the activated complex, and some of the results are compared with the experimental points in Figure 7.10. The agreement obtained is very

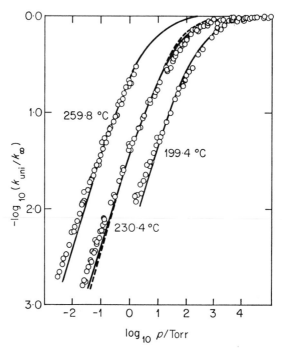

Figure 7.10 Fall-off curves for the isomerization of methyl isocyanide, from Schneider and Rabinovitch.[69] In this figure the 259·8 °C curve is arbitrarily displaced by one log unit to the left and the 199·4 °C curve by one log unit to the right. The lines are theoretical curves for two different models

satisfactory and the authors conclude that calculation of a definite k_{uni}/k_∞ value is possible within a factor of two. Good agreement has also been obtained for the temperature displacement of the fall-off curves; a 60 K increase in temperature gives $\Delta \log_{10} p$(obs.) = 0·12 at $k_{uni}/k_\infty = 0·1$, cf. $\Delta \log_{10} p$(calc.) = 0·11. The activation energy was found to decline at low pressures and the variation was quantitatively in agreement with the theoretical prediction: for example $(E_\infty - E_{bim})$(obs.) = 2·1 kcal mol⁻¹, cf. $(E_\infty - E_{bim})$(calc.) = 1·8 kcal mol⁻¹.

Attempts were also made in these calculations to improve the estimation of the state density $N^*(E^*)$ for the energized molecule by taking an-harmonicity into account. The vibrations were still treated as independent, i.e. cross-terms in the potential energy expression were ignored. The energy levels for stretching vibrations were described by Morse functions, and bending vibrations were also assumed to obey a Morse function with force constants chosen empirically so that the dissociation limits occurred at much higher energies than for stretching vibrations. The results of anharmonicity corrections are most apparent at low pressures, where the second-order rate-constant is increased due to the increase in $N^*(E^*)$. In this case, calculated values of k_{bim} were found to increase by about 30% on including anharmonicity corrections. The fall-off curves were little affected, and k_∞ was only changed by a few per cent since anharmonicity corrections for the energized molecule and activated complex tend to cancel out at the high-pressure limit.

The isomerizations of isocyanides have also been important in two further aspects of unimolecular reactions, both of which are treated in more detail in later chapters. The first of these is the study of the relative efficiencies of added gases in increasing the rate of a unimolecular reaction in its second-order region. The thermal isomerization of methyl isocyanide has been particularly useful in this respect and has been studied in the presence of over 100 added gases (see section 10.4). Secondly, the effects of varying the molecular structure of the isocyanide have been examined by studying the fall-off characteristics of CH_3NC,[69] C_2H_5NC[70] and the isotopically substituted molecules CD_3NC,[118] CH_2DNC[119] and C_2D_5NC.[120] Deuterium substitution is found to move the fall-off curve to lower pressures; for example, the $p_{\frac{1}{2}}$ values for CH_3NC and CD_3NC are about 65 Torr and 30 Torr respectively at 230 °C. The fall-off curves (Figure 9.8) for all these molecules are well reproduced by RRKM calculations and the observed isotope effects are in good agreement with theory (see section 9.5.2).

7.3 UNIMOLECULAR REACTIONS PRODUCING FREE RADICALS

Because it is difficult to study radical reactions in isolation, information about unimolecular reactions involving free radicals has often to be deduced from complex reaction systems. For this reason it is convenient to separate these reactions from unimolecular reactions involving only stable molecules.

Table 7.36 summarizes some of the experimental work done on decomposition reactions producing free radicals or atoms. An example of

Table 7.36 Data for some unimolecular reactions producing free radicals and atoms†

Reaction	Temperatures (°C)	Pressure range/Torr	Kinetic data^a	$p_{\frac{1}{2}}$/Torr (obs.)	Exptl ref.	Theory applied^b and refs.
$M + O_3 \rightarrow O_2 + O + M$	100	20-95	$\log k_{bim} = 12\cdot7 - 24\cdot0/\theta$ (M = O_3)	—	c	K,^{d,e} S,^{d,e} RRKM^{79}
$M + N_2O \rightarrow N_2 + O + M$	568-779	13-3 × 10⁴	$\log k_{bim} = 12\cdot7 - 59\cdot2/\theta$ (M = N_2O)	1·3 × 10⁴ (615 °C)	f	K,^{g} S,^{g} RRKM^{79}
$M + H_2O_2 \rightarrow 2OH + M$	300-600 / 241-659	0·2-20 / 1·6-7	$\log k_{bim} = 15\cdot4 - 48\cdot0/\theta$ (M = H_2O_2)	—	h / i	K,^{k} S,^{k} RRKM^{79}
$M + NO_2Cl \rightarrow NO_2 + Cl + M$	432 / 180-250	6-22 / 0·6-3	[$\log k_{bim} = 16\cdot0 - 48\cdot1/\theta$] / $\log k_{bim} = 13\cdot8 - 27\cdot5/\theta$ (M = NO_2Cl)	—	j / l	S,^{n} RRKM^{79}
$NOCl \rightarrow NO + Cl$	203 / 200-400	3-9	$\log k_{bim} = 13\cdot8 - 35\cdot4/\theta$ (M = NOCl)	—	m / ff	RRKM^x
$M + CH_3Cl \rightarrow CH_3 + Cl + M$	790-870	10-35	$\log k_{bim} = 1\cdot27$ at 844 °C (M = CH_3Cl) (ref. 122)	—	121	S,^{122} RRKM$^{123, 124}$
$M + N_2O_5 \rightarrow NO_2 + NO_3 + M$	27-71	0·05-10	$\log k_{bim} = 16\cdot1 - 19\cdot3/\theta$ (M = NO)	290 (27 °C)	o	K,^{d} S,^{d} RRKM^{79}
$(M+)C_2H_6 \rightarrow 2CH_3(+ M)$	550-620 / 640-726	40-600 / 1-200	$\log k_\infty = 16\cdot0 - 86\cdot0/\theta$ / $\log k_{bim} = 18\cdot3 - 72\cdot5/\theta$ (M = C_2H_6 or CO_2)	20 (685 °C)	125	K,^{d} S,^{d} RRKM$^{79, 129, 130, 131, 144}$
	566-600	0·4-300	—	6·3 (600 °C)	126	
$C_2F_6 \rightarrow 2CF_3$	1027-1327	3000-3300	$\log k_\infty = 17\cdot6 - 94\cdot4/\theta$	—	132	RRKM$^{130, 133}$
$N_2F_4 \rightarrow 2NF_2$	71-137	456-4560	$\log k_\infty = 15\cdot0 - 19\cdot4/\theta$	~2300	p	—
$M + F_2O \rightarrow FO + F + M$	250-270	50-800	$\log k_{bim} = 14\cdot7 - 39\cdot0/\theta$ (M = F_2O, O_2, SiF_4)	—	q	—
$M + N_2O_4 \rightarrow 2NO_2 + M$	−20-28	380-5320	$\log k_{bim} = 14\cdot3 - 11\cdot0/\theta$ (M = N_2 or CO_2)	~450	r	RRKM^{79}
$N_2H_4 \rightarrow 2NH_2$	—	—	—	1400 (27 °C) (calc.)	—	RRKM^s
$M + CH_3CHO \rightarrow CH_3 + CHO + M$	502-536	100-400	$\log k_{bim} = 17\cdot0 - 73\cdot8/\theta$ (M = CH_3CHO)	7 (27 °C) (calc.)	u	RRKM^t

Reaction	Temperatures (°C)	Pressure range/Torr	Kinetic data[a]	$p_{\frac{1}{2}}$/Torr (obs.)	Exptl ref.	Theory applied[b] and refs.
$CHF_3 \rightarrow CF_2 + HF$	927–1327	2000–3000	$\log k_\infty = 12 \cdot 1 - 63 \cdot 0/\theta$	—	v	—
$(CH_3)_2N_2 \rightarrow 2CH_3 + N_2$ (in presence of NO)	250–320	1–160	$\log k_\infty = 17 \cdot 3 - 55 \cdot 5/\theta$	—	134	RRKM[135]
$(CD_3)_2N_2 \rightarrow 2CD_3 + N_2$ (in presence of NO)	256–297	1–150	$\log k_\infty = 15 \cdot 5 - 50 \cdot 7/\theta$	2–3	136	RRKM[136]
$(CF_3)_2N_2 \rightarrow 2CF_3 + N_2$ (in presence of NO)	328–378		$\log k_\infty = 15 \cdot 3 - 52 \cdot 8/\theta$	—	w	—
$^tBu\text{-}^tBu \rightarrow 2^tBu$ related reactions[z, aa]	712–868	530–1270	$\log k_\infty = 16 \cdot 3 - 68 \cdot 5/\theta$	—	y	—
$^tBu\text{—}CH_2CH\!:\!CH_2 \rightarrow$ $^tBu + $ allyl related reactions[z, aa]	727–857	520–1100	$\log k_\infty = 15 \cdot 8 - 65 \cdot 5/\theta$	—	bb	—
$C_6H_5CH_2CH_3 \rightarrow$ $C_6H_5CH_2 + CH_3$ related reactions[z, dd]	603–727		$\log k_\infty = 14 \cdot 6 - 70 \cdot 1/\theta$	—	cc	—
$ZnMe_2 \rightarrow Me + ZnMe$ related reactions[dd]	573–827		$\log k_{16\ \mathrm{Torr}} = 11 \cdot 3 - 47 \cdot 2/\theta$	—	ee	—

[a] 'log k_{bim}' $\equiv \log_{10}(k_{\mathrm{bim}}/\mathrm{dm^3\ mol^{-1}\ s^{-1}})$ and 'log k_∞' $\equiv \log_{10}(k_\infty/\mathrm{s^{-1}})$. [b] For key to abbreviations see footnote a to Table 7.34. [c] S. W. Benson and A. E. Axworthy, J. Chem. Phys., **26**, 1718 (1957). [d] E. K. Gill and K. J. Laidler, Proc. Roy. Soc. (A), **250**, 121 (1959). [e] E. K. Gill and K. J. Laidler, Trans. Faraday Soc., **55**, 753 (1959). [f] H. S. Johnston, J. Chem. Phys., **19**, 663 (1951). [g] E. K. Gill and K. J. Laidler, Can. J. Chem., **36**, 1570 (1958). [h] P. A. Giguere and I. D. Liu, Can. J. Chem., **35**, 283 (1957). [i] D. E. Hoare, J. B. Protheroe and A. D. Walsh, Trans. Faraday Soc., **55**, 548 (1959). [j] W. Forst, Can. J. Chem., **36**, 1308 (1958). [k] E. K. Gill and K. J. Laidler, Proc. Roy. Soc. (A), **251**, 66 (1959). [l] H. F. Cordes and H. S. Johnston, J. Amer. Chem. Soc., **76**, 4264 (1954). [m] M. Volpe and H. S. Johnston, J. Amer. Chem. Soc., **78**, 3903 (1956). [n] N. B. Slater, Theory of Unimolecular Reactions, Methuen, 1959, p. 175. [o] H. S. Johnston and R. L. Perrine, J. Amer. Chem. Soc., **73**, 4782 (1951). [p] L. Brown and B. deB. Darwent, J. Chem. Phys., **42**, 2158 (1965). [q] W. Koblitz and H. J. Schumacher, Zeit. phys. Chem. (Leipzig) (B), **25**, 283 (1934). [r] T. Carrington and N. Davidson, J. Phys. Chem., **57**, 418 (1953). [s] D. W. Setser and W. C. Richardson, Can. J. Chem., **47**, 2593 (1969). [t] D. W. Setser, J. Phys. Chem., **70**, 826 (1966). [u] A. B. Trenwith, J. Chem. Soc., 4426 (1963). [v] E. Tschuikow-Roux and J. E. Marte, J. Chem. Phys., **42**, 2049 (1965). [w] T. H. McGee and C. E. Waring, J. Phys. Chem., **73**, 2838 (1969), and earlier work cited therein. [x] W. Forst and P. St. Laurent, J. Chim. Phys. Physicochim. Biol., **67**, 1018 (1970). [y] W. Tsang, J. Chem. Phys., **44**, 4283 (1965). [z] Data for other similar reactions have been reviewed by H. M. Frey and R. Walsh, ref. 2. [aa] See also W. Tsang, Internat. J. Chem. Kinetics, **2**, 23, 311 (1970), and references cited therein. [bb] W. Tsang, J. Chem. Phys., **46**, 2817 (1967). [cc] G. L. Esteban, J. A. Kerr and A. F. Trotman-Dickenson, J. Chem. Soc., 3873 (1963). [dd] Data for other similar reactions have been reviewed by J. A. Kerr, Chem. Rev., **66**, 465 (1966); see also R. J. Kominar and S. J. W. Price, Can. J. Chem., **47**, 991 (1969); A. N. Dunlop and S. J. W. Price, Can. J. Chem., **48**, 3205 (1970); S. J. W. Price and J. P. Richard, Can. J. Chem., **48**, 3209 (1970). [ee] S. J. W. Price and A. F. Trotman-Dickenson, Trans. Faraday Soc., **53**, 1208 (1957); the reaction is in the fall-off region at 16 Torr. [ff] P. G. Ashmore and M. S. Spencer, Trans. Faraday Soc., **55**, 1868 (1959); P. G. Ashmore and M. G. Burnett, Trans. Faraday Soc., **57**, 1315 (1961); **58**, 1801 (1962).

such a decomposition reaction which has been treated theoretically is the decomposition of methyl chloride. The pyrolysis of methyl chloride was studied by Shilov and Sabirova[121] who postulated a non-chain radical mechanism:

$$CH_3Cl \longrightarrow CH_3 + Cl \qquad (a)$$

$$CH_3 + CH_3Cl \longrightarrow CH_4 + CH_2Cl \qquad (b)$$

$$Cl + CH_3Cl \longrightarrow HCl + CH_2Cl \qquad (c)$$

$$CH_2Cl + CH_3Cl \longrightarrow CH_2Cl_2 + CH_3 \qquad (d)$$

$$2CH_2Cl \longrightarrow CH_2ClCH_2Cl \qquad (e)$$

$$C_2H_4Cl_2 \longrightarrow HCl + CH_2:CHCl \qquad (f)$$

$$CH_2:CHCl \longrightarrow HCl + C_2H_2 \qquad (g)$$

At 800 °C steps (e) and (f) are more important than (d) and so the slow step (a) is followed by the rapid production of HCl by steps (c), (f) and (g). Since each chlorine atom produced in step (a) leads to the formation of three molecules of HCl, the rate of step (a) can be measured as one-third of the rate of production of HCl. From an analysis of the kinetic data, Holbrook[122] deduced a second-order rate-constant

$$k_{\rm bim} = (1 \cdot 9 \pm 0 \cdot 2) \times 10^4 \ {\rm cm^3 \, mol^{-1} \, s^{-1}}$$

for the experimental temperature of 1117 K and pressures in the range 10–35 Torr. Slater theory was used to calculate the second-order rate-constant using two different reaction coordinates, a symmetrical C—Cl stretch and an equally weighted combination of C—Cl stretch plus methyl group deformation. Since both these reaction coordinates are of A_1 symmetry, only the three A_1 vibrations out of the total nine modes of vibration can contribute on the basis of Slater's harmonic theory. It is therefore not surprising that a large discrepancy was found between the experimental second-order rate of energization and the values calculated in this way; the calculated values were too low by a factor of 10^5. Better agreement has been obtained by Forst and St. Laurent[123] using RRKM theory. The calculation for $k_{\rm bim}$ does not involve any properties of the activated complex but is largely a matter of evaluating $N^*(E^*)$ for the energized molecule (see section 4.8). Using the Whitten–Rabinovitch approximation for $N^*(E^*)$ (see section 5.4.3), Forst and St. Laurent calculated a value of $k_{\rm bim} = 5 \cdot 8 \times 10^2 \ {\rm cm^3 \, mol^{-1} s^{-1}}$, still too low by a factor of 32. Allowance for anharmonicity[123, 124] and the inclusion of the figure axis rotation as an active degree of freedom[124] brought the agreement with experiment within a factor of 2.

An important reaction from the theoretical point of view is the dissociation of ethane into two methyl radicals. Experimental studies[125, 126] have shown that the rate-constant at 600 °C begins to decline at about 40 Torr and that $p_{\frac{1}{2}}$ is about 1 Torr at this temperature. Lin and Back found the rate expression (7.40). A high A-factor was predicted by

$$\log_{10} k_\infty/s^{-1} = 16 \cdot 0 - 86 \cdot 0/\theta \qquad (7.40)$$

Marcus[127] who showed that the recombination of methyl radicals by the reverse process is very efficient. Steel and Laidler[128] have discussed the high A-factor and its implications of a loose activated complex leading to a high entropy of activation. Loose complexes have been assumed in the RRKM calculations which have been carried out for this reaction.[9, 129, 144] Lin and Laidler[129] took a number of different models, in some of which the activated complex had an active free internal rotation of the two methyl groups against each other. In one model the overall rotation about the C—C bond was also taken to be active in the activated complex. Rate-constants were then calculated from the RRKM equation appropriate when active rotations are involved [equation (5.20)]. Differences between the various models were slight, but those involving active rotations were best. Corrections were also made for the centrifugal effect of adiabatic rotations, but the details of these corrections were erroneous.[130] Waage and Rabinovitch have recently reviewed all the data on ethane decomposition and methyl radical recombination and given a detailed discussion of the theoretical treatment of these reactions. They note, however, that unresolved discrepancies still remain.[131]

An analogous reaction is the decomposition of hexafluoroethane (7.41)

$$C_2F_6 \longrightarrow 2CF_3 \qquad (7.41)$$

for which Tschuikow-Roux[132] has carried out shock-tube studies and measured the high-pressure dissociation rate-constant at 1300–1600 K:

$$\log_{10} k_\infty/s^{-1} = 17 \cdot 6 - 94 \cdot 4/\theta$$

This reaction also has an abnormally high A-factor implying a loose activated complex with some free rotation. RRKM calculations[133] have been used to predict the fall-off behaviour on the basis of a model employing five active internal rotations. Centrifugal corrections were also found to be important but the calculations were based on an erroneous equation.[130] Another interesting reaction of this type is the unimolecular decomposition of azomethane (7.42) to produce two methyl radicals and a molecule of

$$(CH_3)_2N_2 \longrightarrow 2CH_3 + N_2 \qquad (7.42)$$

nitrogen. The overall reaction is undoubtedly complex, but there is evidence that the reaction inhibited by nitric oxide is rate-controlled by the unimolecular reaction (7.42), for which the high-pressure rate expression

$$\log_{10} k_\infty/\text{s}^{-1} = 17 \cdot 3 - 55 \cdot 5/\theta \qquad (7.43)$$

(7.43) has been obtained.[134] RRKM calculations by Forst[135] reproduced the high A-factor by using a loose activated complex whose rotational degrees of freedom are of paramount importance. A recent paper on the thermal decomposition of azomethane-d_6 has shown some unexpected differences between this molecule and azomethane itself.[136] The authors conclude that the experimental complications are fewer in the case of $(CD_3)_2N_2$ than $(CH_3)_2N_2$ and that the decomposition of this molecule is therefore a better reaction with which to test the theory.

7.4 UNIMOLECULAR DECOMPOSITION REACTIONS OF FREE RADICALS

Radical decompositions, of the type

$$R \longrightarrow R' + M$$

(where R' is a hydrogen atom or small radical and M is a stable molecule) are known to occur as steps in complex free radical chain reactions. They are unimolecular processes which are characterized by low activation energies due to the low bond strengths involved, and consequently they have relatively high transition pressures $[p_{\frac{1}{2}} \propto (E_0/kT)^{-\frac{1}{2}(n-1)}$ on Slater theory, see section 7.2].

A great deal of kinetic information is now available for thermal radical decompositions, these having been studied both in pyrolysis systems and by other preparative methods such as photolysis and mercury photosensitization. The rate-constants for such decompositions have usually to be obtained by a relative method in which they are compared with those for radical–radical recombination or radical abstraction from the parent molecule. Since these latter rate-constants are not always themselves well known, the kinetic data, although extensive, are not always reliable. The difficulties in measuring rate-constants for radical decompositions have been outlined by Kerr and Lloyd in a recent review[137] to which the reader is referred for more detail.

Here we shall concern ourselves only with those few radical decomposition reactions which have been observed to show fall-off behaviour over a sufficient pressure range to allow the high- and low-pressure Arrhenius parameters to be determined with reasonable certainty. Table 7.37 lists the major examples of such reactions known at present and summarizes the kinetic information. Kerr and Lloyd have pointed out that the pressure range is rather limited for most studies of radical decompositions and that

Table 7.37 Unimolecular decomposition reactions of some free radicals

Reaction	Reaction system	Temperatures (°C)	Pressure range/Torr	E_c^∞/kcal mol⁻¹	$\log_{10} A^\infty$/s⁻¹	E_{bim}/kcal mol⁻¹	$\log_{10}(A_{\text{bim}}/\text{dm}^3\ \text{mol}^{-1}\ \text{s}^{-1})$	$p_{\frac{1}{2}}$/Torr	Theory applied and refs.
$C_2H_5 \rightarrow C_2H_4 + H$	ethane pyrolysis[125]	640–726	1–200	38·0	13·6	32·4	15·3	—	RRKM[129,144]
	Hg-photosens. C_2H_6 pyrolysis[138]	400–500	4–650	40·9	14·4	31·8	14·8	—	
$CH_3OCH_2 \rightarrow CH_3 + CH_2O$	Hg-photosens. Me_2O pyrolysis[a]	200–300	3–600	25·5	13·2	18·1	13·5	—	RRKM[129]
$CH_3CO \rightarrow CH_3 + CO$	Photol. $Me_2CO + HI$[b]	235–295	20–160	15·0	10·3	12·0	11·5	130 (270 °C)	HL[b]
$C_2H_5CO \rightarrow C_2H_5 + CO$	Photol. $Et_2N_2 + EtCHO$[c]	30–80	20–140	14·7	13·3	10·5	12·6	110 (42 °C)	HL[c]
$(CH_3)_2CHO \rightarrow$ $CH_3 + CH_3CHO$	Pyrol. $^iPrONO + NO$[d]	160–200	20–230	17·3	11·8	8·3	10·1	39 (160 °C) 101 (200 °C)	—
$EtMeCHO \rightarrow$ $Et + MeCHO$	Pyrol. ^sBuONO[e]	150–190	12–200	17·5	13·4	10·6	11·5	110 (190 °C)	—
$Me_3CO \rightarrow Me + Me_2CO$	Pyrol. $^tBu_2O_2 + NO$[f]	125–163	10–60	22·8	14·7	13·4	12·7	112 (163 °C)	—
	Pyrol. $^tBu_2O_2 + PhOH$[g]	209–274	4–32	—	—	—	—	80 (209 °C)	—

[a] L. F. Loucks and K. J. Laidler, *Can. J. Chem.*, **45**, 2767 (1967). [b] H. E. O'Neal and S. W. Benson, *J. Chem. Phys.*, **36**, 2196 (1962). [c] J. A. Kerr and A. C. Lloyd, *Trans. Faraday Soc.*, **63**, 2480 (1967). [d] D. L. Cox, R. A. Livermore and L. Phillips, *J. Chem. Soc. (B)*, 245 (1966). [e] R. L. East and L. Phillips, *J. Chem. Soc. (A)*, 1939 (1967). [f] M. J. Yee Quee and J. C. J. Thynne, *Trans. Faraday Soc.*, **63**, 2970 (1967). [g] M. F. R. Mulcahy and D. J. Williams, *Aust. J. Chem.*, **17**, 1291 (1964).

the data can often be treated by a Hinshelwood–Lindemann extrapolation method to obtain the limiting values of the Arrhenius parameters. Since this procedure is known to be inadequate (see Chapter 1), the results of such extrapolations must be treated with caution.

An important example of a radical decomposition which has been studied over a large pressure range is the decomposition of the ethyl radical (7.44) to which RRKM theory has been applied by Lin and

$$C_2H_5 \longrightarrow C_2H_4 + H \qquad (7.44)$$

Laidler.[129] The experimental data of Loucks and Laidler[138] give the high-pressure rate-constant as (7.45) assuming the rate-constant for ethyl

$$\log_{10} k_\infty/s^{-1} = 14\cdot4 - 40\cdot9/\theta \qquad (7.45)$$

radical recombination to be $10^{13\cdot3}$ cm^3 mol^{-1} s^{-1}. Two models were considered for the ethyl radical decomposition reaction. In model I, no active rotations were assumed for either the energized radical or the

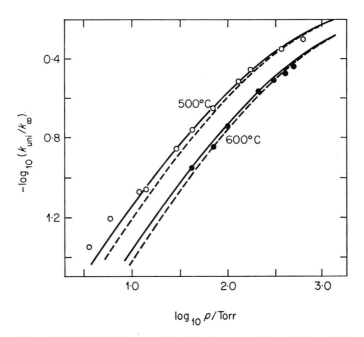

$\log_{10} p/\text{Torr}$

Figure 7.11 Fall-off curves for the decomposition of the ethyl radical. Experimental points: ○, Loucks and Laidler;[138] ●, Lin and Back.[125] Calculated curves:[129] – – –, model I; ——, model II (see text)

activated complex (A). The methyl torsional frequency in the radical was

$$
\underset{H}{\overset{H}{\diagdown}}C\!=\!\!\!\underset{H}{\overset{\overset{\displaystyle H}{|}}{C}}\!\!\underset{H}{\overset{H}{\diagup}} \quad (A)
$$

taken to be 275 cm^{-1} as in the ethane molecule[139] and other frequencies were taken from Purnell and Quinn.[140] The vibration frequencies of the activated complex were taken to be the geometric mean of frequencies for the radical and for the ethylene molecule. The C—H bending frequencies were adjusted to fit the observed entropy of activation at 450 °C.

Model II treats the torsional vibration in the energized radical as a free internal rotation which is removed in the activated complex owing to partial double bond formation. The overall (external) rotation about the C—C bond axis was taken to be active in both the radical and the complex. The collision diameter was taken to be 2·75 Å in all calculations. The results are compared with experimental data at 500 and 600 °C in Figure 7.11. It will be seen that both models give good agreement with the results, although the authors preferred model II in view of its better prediction of the extrapolated experimental second-order rate-constant k_{bim}.

Lin and Laidler also applied RRKM theory to the decomposition of the methoxymethyl radical (7.46) and obtained good agreement with the

$$
CH_3OCH_2 \longrightarrow CH_3 + CH_2O \qquad (7.46)
$$

experimental results for a model incorporating two active rotations in both the activated complex and the energized radical.

References

1. H. M. Frey, *Advan. Phys. Org. Chem.*, **4**, 148 (1966).
2. H. M. Frey and R. Walsh, *Chem. Rev.*, **69**, 103 (1969).
3. A. Maccoll, *Advan. Phys. Org. Chem.*, **3**, 91 (1965).
4. A. Maccoll, *Chem. Rev.*, **69**, 33 (1969).
5. W. E. Falconer, T. F. Hunter and A. F. Trotman-Dickenson, *J. Chem. Soc.*, 609 (1961).
6. M. C. Flowers and H. M. Frey, *J. Chem. Soc.*, 2758 (1960).
7. S. W. Benson and P. S. Nangia, *J. Chem. Phys.*, **38**, 18 (1963).
8. R. J. Crawford and T. R. Lynch, *Can. J. Chem.*, **46**, 1457 (1968); J. A. Berson and J. M. Balquist, *J. Amer. Chem. Soc.*, **90**, 7343 (1968); W. L. Carter and R. G. Bergmann, *J. Amer. Chem. Soc.*, **90**, 7344 (1968).
9. R. G. Bergmann and W. L. Carter, *J. Amer. Chem. Soc.*, **91**, 7411 (1969).
10. M. R. Willcott and V. H. Cargle, *J. Amer. Chem. Soc.*, **91**, 4310 (1969).
11. (*a*) S. W. Benson, *Thermochemical Kinetics*, Wiley, New York, 1968. (*b*) S. W. Benson and H. E. O'Neal, *Kinetic Data on Gas-phase Unimolecular Reactions*, Nat. Bur. Standards, NSRDS-NBS 21, 1970.
12. H. E. O'Neal and S. W. Benson, *J. Phys. Chem.*, **72**, 1866 (1968).

13. R. C. S. Grant and E. S. Swinbourne, *Chem. Commun.*, 620 (1966).
14. K. A. W. Parry and P. J. Robinson, *J. Chem. Soc.* (*B*), 49 (1969).
15. K. A. Holbrook and K. A. W. Parry, *J. Chem. Soc.* (*B*), 1019 (1970).
16. M. C. Flowers and H. M. Frey, *J. Chem. Soc.*, 3547 (1961); see also C. A. Wellington, *J. Phys. Chem.*, **66**, 1671 (1962); D. G. Retzloff, B. M. Coull and J. Coull, *J. Phys. Chem.*, **74**, 2455 (1970).
17. J. P. Chesick, *J. Amer. Chem. Soc.*, **85**, 2720 (1963).
18. R. J. Ellis and H. M. Frey, *J. Chem. Soc.*, 5578 (1964).
19. K. A. W. Parry and P. J. Robinson, *Chem. Commun.*, 1377 (1968); finalized Arrhenius parameters quoted in Table 7.4.
20. B. Atkinson and D. McKeagan, *Chem. Commun.*, 189 (1966).
21. R. A. Mitsch and E. W. Neuvar, *J. Phys. Chem.*, **70**, 546 (1966).
22. A. D. Ketley, A. J. Berlin, E. Gorman and L. P. Fisher, *J. Org. Chem.*, **31**, 305 (1966).
23. R. Hoffmann and R. B. Woodward, *Acc. of Chem. Res.*, **1**, 17 (1968); R. B. Woodward and R. Hoffmann, *Angew. Chem. Internat. Ed.*, **8**, 781 (1969). See also J. J. Vollmer and K. L. Servis, *J. Chem. Educ.*, **45**, 214 (1968); **47**, 491 (1970).
24. A. J. Cocks, H. M. Frey and I. D. R. Stevens, *Chem. Commun.*, 458 (1969).
25. J. E. Baldwin and P. W. Ford, *J. Amer. Chem. Soc.*, **91**, 7192 (1969).
26. R. J. Ellis and H. M. Frey, *Trans. Faraday Soc.*, **59**, 2076 (1963).
27. J. J. Gajewski and C. N. Shih, *J. Amer. Chem. Soc.*, **89**, 4532 (1967); **91**, 5900 (1969); W. von E. Doering and W. R. Dolbier, *J. Amer. Chem. Soc.*, **89**, 4534 (1967).
28. R. Hoffmann, S. Swaminathan, B. G. Odell and R. Gleiter, *J. Amer. Chem. Soc.*, **92**, 7091 (1970); R. Hoffmann, B. G. Odell and A. Imamura, *J. Chem. Soc.* (*B*), 1675 (1970).
29. J. P. Chesick, *J. Phys. Chem.*, **65**, 2170 (1961); R. L. Brandauer, B. Short and S. M. E. Kellner, *J. Phys. Chem.*, **65**, 2269 (1961).
30. W. J. Engelbrecht and M. J. DeVries, *J. S. Afr. Chem. Inst.*, **23**, 163, 172 (1970).
31. R. W. Carr and W. D. Walters, *J. Phys. Chem.*, **69**, 1073 (1965).
32. S. W. Benson, *Advan. Photochem.*, **2**, 14 (1964).
33. H. M. Frey, A. M. Lamont and R. Walsh, *Chem. Commun.*, 1583 (1970).
34. G. R. Branton, H. M. Frey, D. C. Montague and I. D. R. Stevens, *Trans. Faraday Soc.*, **62**, 659 (1966).
35. H. M. Frey and I. D. R. Stevens, *Trans. Faraday Soc.*, **61**, 90 (1965).
36. R. B. Cundall, *Prog. React. Kinetics*, **2**, 165 (1964).
37. M. C. Lin and K. J. Laidler, *Can. J. Chem.*, **46**, 973 (1968).
38. J. E. Douglas, B. S. Rabinovitch and F. S. Looney, *J. Chem. Phys.*, **23**, 315 (1955).
39. K. E. Lewis and H. Steiner, *J. Chem. Soc.*, 3080 (1964).
40. H. M. Frey and R. J. Ellis, *J. Chem. Soc.*, 4770 (1965).
41. H. M. Frey and R. K. Solly, *Trans. Faraday Soc.*, **64**, 1858 (1968); A. Amano and M. Uchiyama, *J. Phys. Chem.*, **69**, 1278 (1965).
42. M. L. Neufeld and A. T. Blades, *Can. J. Chem.*, **41**, 2956 (1963).
43. D. W. Setser, *J. Phys. Chem.*, **70**, 826 (1966).
44. M. Lenzi and A. Mele, *J. Chem. Phys.*, **43**, 1974 (1965).
45. S. W. Benson and A. N. Bose, *J. Chem. Phys.*, **39**, 3463 (1963).

46. A. Maccoll, in 'Theoretical Organic Chemistry' (Kekulé Symposium of the Chemical Society, 1958), Butterworths, London, 1959.
47. A. T. Blades, *Can. J. Chem.*, **40**, 1533 (1962).
48. H. E. O'Neal and S. W. Benson, *J. Phys. Chem.*, **71**, 2903 (1967).
49. M. Day and A. F. Trotman-Dickenson, *J. Chem. Soc. (A)*, 233 (1969).
50. P. Cadman, M. Day, A. W. Kirk and A. F. Trotman-Dickenson, *Chem. Commun.*, 203 (1970); P. Cadman, M. Day and A. F. Trotman-Dickenson, *J. Chem. Soc. (A)*, 248, 1356 (1971).
51. J. S. Shapiro, E. S. Swinbourne and B. C. Young, *Aust. J. Chem.*, **17**, 1217 (1964).
52. G. M. Schwab and H. Noller, *Zeit. Elektrochem.*, **58**, 762 (1954).
53. A. Maccoll, *J. Chem. Soc.*, 965 (1955).
54. J. L. Holmes and L. S. M. Ruo, *J. Chem. Soc. (A)*, 1231 (1968).
55. K. A. Holbrook and J. J. Rooney, *J. Chem. Soc.*, 247 (1965).
56. D. H. R. Barton and P. F. Onyon, *Trans. Faraday Soc.*, **45**, 725 (1949).
57. K. E. Howlett, *Trans. Faraday Soc.*, **48**, 25 (1952).
58. P. J. Agius and A. Maccoll, *J. Chem. Soc.*, 973 (1955).
59. A. Maccoll and P. J. Thomas, *J. Chem. Soc.*, 5033 (1957).
60. S. W. Benson, *J. Chem. Phys.*, **38**, 1945 (1963).
61. J. L. Holmes and A. Maccoll, *J. Chem. Soc.*, 5919 (1963).
62. W. Tsang, *J. Chem. Phys.*, **40**, 1498 (1964).
63. C. D. Hurd and F. H. Blunck, *J. Amer. Chem. Soc.*, **60**, 2419 (1938).
64. E. U. Emovon and A. Maccoll, *J. Chem. Soc.*, 227 (1964).
65. G. G. Smith, F. D. Bagley and R. Taylor, *J. Amer. Chem. Soc.*, **83**, 3647 (1961).
66. A. T. Blades and P. W. Gilderson, *Can. J. Chem.*, **38**, 1401 (1960).
67. A. Maccoll, *J. Chem. Soc.*, 3398 (1958).
68. E. S. Lewis and W. C. Herndon, *J. Amer. Chem. Soc.*, **83**, 1961 (1961).
69. F. W. Schneider and B. S. Rabinovitch, *J. Amer. Chem. Soc.*, **84**, 4215 (1962).
70. K. M. Maloney and B. S. Rabinovitch, *J. Phys. Chem.*, **73**, 1652 (1969).
71. C. K. Yip and H. O. Pritchard, *Can. J. Chem.*, **48**, 2942 (1970).
72. I. M. T. Davidson, M. R. Jones and C. Pett, *J. Chem. Soc. (B)*, 937 (1967), and references cited therein.
73. R. N. Haszeldine, P. J. Robinson and J. A. Walsh, *J. Chem. Soc. (B)*, 578 (1970), and references cited therein.
74. H. O. Pritchard, R. G. Sowden and A. F. Trotman-Dickenson, *J. Chem. Soc.*, 546 (1954).
75. A. T. Blades and G. W. Murphy, *J. Amer. Chem. Soc.*, **74**, 1039 (1952).
76. D. W. Placzek, B. S. Rabinovitch, G. Z. Whitten and E. Tschuikow-Roux, *J. Chem. Phys.*, **43**, 4071 (1965).
77. H. O. Pritchard, R. G. Sowden and A. F. Trotman-Dickenson, *Proc. Roy. Soc. (A)*, **217**, 563 (1953).
78. T. S. Chambers and G. B. Kistiakowsky, *J. Amer. Chem. Soc.*, **56**, 399 (1934).
79. G. M. Wieder and R. A. Marcus, *J. Chem. Phys.*, **37**, 1835 (1962).
80. G. M. Wieder, Ph.D. Thesis, Polytechnic Institute of Brooklyn, 1961.
81. E. W. Schlag and B. S. Rabinovitch, *J. Amer. Chem. Soc.*, **82**, 5996 (1960). The observed A-factor of $10^{16.4}$ s^{-1} refers to $k = k_f + k_r$.[83, 84]
82. H. S. Johnston and J. R. White, *J. Chem. Phys.*, **22**, 1969 (1954).

83. M. C. Lin and K. J. Laidler, *Trans. Faraday Soc.*, **64**, 927 (1968).
84. J. W. Simons and B. S. Rabinovitch, *J. Phys. Chem.*, **68**, 1322 (1964).
85. J. P. Chesick, *J. Amer. Chem. Soc.*, **82**, 3277 (1960).
86. M. L. Halberstadt and J. P. Chesick, *J. Phys. Chem.*, **69**, 429 (1965).
87. D. W. Setser and B. S. Rabinovitch, *J. Amer. Chem. Soc.*, **86**, 564 (1964).
88. M. C. Flowers and H. M. Frey, *J. Chem. Soc.*, 1157 (1962).
89. M. C. Flowers and H. M. Frey, *Proc. Roy. Soc. (A)*, **257**, 122 (1960).
90. K. A. Holbrook, J. S. Palmer, K. A. W. Parry and P. J. Robinson, *Trans. Faraday Soc.*, **66**, 869 (1970).
91. C. T. Genaux, F. Kern and W. D. Walters, *J. Amer. Chem. Soc.*, **75**, 6196 (1953); R. W. Carr and W. D. Walters, *J. Phys. Chem.*, **67**, 1370 (1963).
92. F. Kern and W. D. Walters, *Proc. Nat. Acad. Sci.*, **38**, 937 (1952).
93. H. O. Pritchard, R. G. Sowden and A. F. Trotman-Dickenson, *Proc. Roy. Soc. (A)*, **218**, 416 (1953).
94. J. N. Butler and R. B. Ogawa, *J. Amer. Chem. Soc.*, **85**, 3346 (1963).
95. R. W. Vreeland and D. F. Swinehart, *J. Amer. Chem. Soc.*, **85**, 3349 (1963).
96. T. F. Thomas, P. J. Conn and D. F. Swinehart, *J. Amer. Chem. Soc.*, **91**, 7611 (1969).
97. A. F. Patarrachia and W. D. Walters, *J. Phys. Chem.*, **68**, 3894 (1964).
98. H. R. Gerberich and W. D. Walters, *J. Amer. Chem. Soc.*, **83**, 4884 (1961).
99. W. P. Hauser and W. D. Walters, *J. Phys. Chem.*, **67**, 1328 (1963).
100. C. S. Elliott and H. M. Frey, *Trans. Faraday Soc.*, **62**, 895 (1966).
101. M. C. Lin and K. J. Laidler, *Trans. Faraday Soc.*, **64**, 94 (1968).
102. M. K. Knecht, *J. Amer. Chem. Soc.*, **91**, 7667 (1969).
103. H. M. Frey and B. M. Pope, *Trans. Faraday Soc.*, **65**, 441 (1969).
104. K. A. Holbrook and A. R. W. Marsh, *Trans. Faraday Soc.*, **63**, 643 (1967).
105. K. E. Howlett, *J. Chem. Soc.*, 3695, 4487 (1952).
106. H. Heydtmann and G. W. Völker, *Zeit. Phys. Chem. (Frankfurt)*, **55**, 296 (1967).
107. H. Heydtmann, *Ber. Bunsenges Phys. Chem.*, **72**, 1009 (1968).
108. A. T. Blades, P. W. Gilderson and M. G. H. Wallbridge, *Can. J. Chem.*, **40**, 1526 (1962).
109. J. C. Hassler and D. W. Setser, *J. Chem. Phys.*, **45**, 3246 (1966).
110. R. L. Johnson and D. W. Setser, *J. Phys. Chem.*, **71**, 4366 (1967).
111. D. H. R. Barton and K. E. Howlett, *J. Chem. Soc.*, 165 (1949).
112. K. E. Howlett, quoted by D. H. R. Barton and A. J. Head, *Trans. Faraday Soc.*, **46**, 114 (1950).
113. W. Tsang, *J. Chem. Phys.*, **41**, 2487 (1964).
114. N. Capon and R. A. Ross, *Trans. Faraday Soc.*, **62**, 1560 (1966).
115. A. E. Shilov and R. D. Sabirova, *Kinet. Katal.*, **5**, 40 (1964).
116. H. Hartmann, H. G. Bosche and H. Heydtmann, *Zeit. Phys. Chem. (Frankfurt)*, **42**, 329 (1964).
117. F. J. Fletcher, B. S. Rabinovitch, K. W. Watkins and D. J. Locker, *J. Phys. Chem.*, **70**, 2823 (1966).
118. F. W. Schneider and B. S. Rabinovitch, *J. Amer. Chem. Soc.*, **85**, 2365 (1963).
119. B. S. Rabinovitch, P. W. Gilderson and F. W. Schneider, *J. Amer. Chem. Soc.*, **87**, 158 (1965).
120. K. M. Maloney, S. P. Pavlou and B. S. Rabinovitch, *J. Phys. Chem.*, **73**, 2756 (1969).

121. A. E. Shilov and R. D. Sabirova, *Zhur. Fiz. Khim.*, **33**, 1365 (1959).
122. K. A. Holbrook, *Trans. Faraday Soc.*, **57**, 2151 (1961).
123. W. Forst and P. St. Laurent, *Can. J. Chem.*, **43**, 3052 (1965).
124. W. Forst and P. St. Laurent, *Can. J. Chem.*, **45**, 3169 (1967).
125. M. C. Lin and M. H. Back, *Can. J. Chem.*, **44**, 505, 2357 (1966).
126. A. B. Trenwith, *Trans. Faraday Soc.*, **62**, 1538 (1966).
127. R. A. Marcus, *J. Chem. Phys.*, **20**, 364 (1952).
128. C. Steel and K. J. Laidler, *J. Chem. Phys.*, **34**, 1827 (1961).
129. M. C. Lin and K. J. Laidler, *Trans. Faraday Soc.*, **64**, 79 (1968).
130. E. V. Waage and B. S. Rabinovitch, *Chem. Rev.*, **70**, 377 (1970).
131. E. V. Waage and B. S. Rabinovitch, *Internat. J. Chem. Kinetics*, **3**, 105 (1971).
132. E. Tschuikow-Roux, *J. Chem. Phys.*, **43**, 2251 (1965).
133. E. Tschuikow-Roux, *J. Chem. Phys.*, **49**, 3115 (1968).
134. W. Forst and O. K. Rice, *Can. J. Chem.*, **41**, 562 (1963).
135. W. Forst, *J. Chem. Phys.*, **44**, 2349 (1966).
136. D. Chang and O. K. Rice, *Internat. J. Chem. Kinetics*, **1**, 171 (1969).
137. J. A. Kerr and A. C. Lloyd, *Quart. Rev.*, **22**, 549 (1968).
138. L. F. Loucks and K. J. Laidler, *Can. J. Chem.*, **45**, 2795 (1967).
139. G. Herzberg, *Infrared and Raman Spectra of Polyatomic Molecules*, van Nostrand, Princeton, 1945.
140. J. H. Purnell and C. P. Quinn, *J. Chem. Soc.*, 4090 (1964).
141. K. M. Maloney and B. S. Rabinovitch, *J. Phys. Chem.*, **72**, 4483 (1968).
142. H. E. O'Neal and S. W. Benson, *Internat. J. Chem. Kinetics*, **2**, 424 (1970).
143. W. J. Engelbrecht and M. J. DeVries, *J. S. Afr. Chem. Inst.*, **24**, 46 (1971).
144. B. S. Rabinovitch and D. W. Setser, *Advan. Photochem.*, **3**, 1 (1964).

8 Chemical Activation

As was shown in Chapter 1, molecules undergo thermal unimolecular reactions as a result of energization by molecular collisions. Collisions between molecules at a given temperature produce energized molecules with an equilibrium distribution of energy (the Maxwell–Boltzmann distribution) which enables the fraction of molecules energized into a particular energy range to be calculated [equation (4.5)]. The average energy of the reacting molecules can also be calculated in various ways as is discussed later (section 8.1.3).

Methods of energization other than by molecular collisions may produce a non-equilibrium situation in which molecules can acquire energies far in excess of the average thermal energy. This excess energy may be dissipated by molecular collisions or may result in further chemical reaction if a suitable reaction path is available. One such method of energization is by absorption of radiation (i.e. photoactivation) and this type of study is discussed in Chapter 10. When the energization occurs by virtue of the energy changes in a chemical reaction producing the molecules, the process is known as *chemical activation*, which is the subject of the present chapter.

Consider, for example, the potential energy profiles in Figures 8.1 and 8.2; these refer to the general reaction scheme:

$$\text{Reactant(s)} \xrightarrow[\text{(reaction 1)}]{\substack{\text{Activating} \\ \text{reaction}}} \underset{\substack{\text{(energized for reaction 2)}}}{\substack{\text{Chemically activated} \\ \text{molecule}}} \xrightarrow[\substack{\text{or isomerization} \\ \text{(reaction 2)}}]{\substack{\text{Unimolecular} \\ \text{decomposition}}} \text{Product(s)}$$

From these figures it is clearly seen that if the critical energy for reaction 1 is $(E_0)_1$ and the difference in zero-point levels of reactants and products is ΔE_0 (equal to the heat of reaction at 0 K) then the chemically activated molecules are energized by at least an amount $E_{\min} = (E_0)_1 - \Delta E_0$ where ΔE_0 is negative or positive depending upon whether the reaction is exothermic or endothermic. In addition, if the reactants possess average thermal energy $\langle E_{\text{th}} \rangle$ at the reaction temperature, then the average total energy possessed by the chemically activated molecules is given by (8.1);

$$\langle E \rangle = (E_0)_1 - \Delta E_0 + \langle E_{\text{th}} \rangle = E_{\min} + \langle E_{\text{th}} \rangle \qquad (8.1)$$

in this chapter the non-fixed energy of an energized molecule will be denoted by E rather than E^*. If the critical energy required for a subsequent reaction, $(E_0)_2$, is less than E_{min}, then the molecules are able to undergo

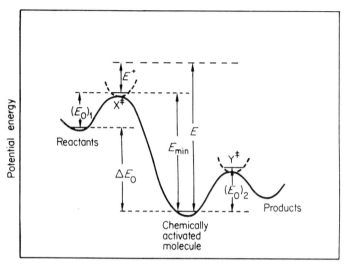

Figure 8.1 Chemical activation by an exothermic reaction; $X^{\ddagger}$ represents the activated complex for the activating reaction, and $Y^{\ddagger}$ the activated complex for subsequent reaction of the chemically activated molecule

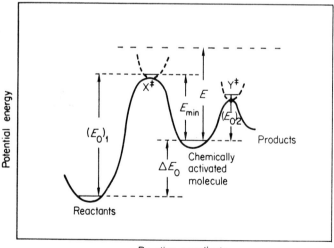

Figure 8.2 Chemical activation by an endothermic reaction

such reaction unless they are stabilized (i.e. their excess energy is removed) by molecular collisions. From the relationship between $(E_0)_1$, ΔE_0 and the activation energy of the back reaction, it will be seen that chemically activated molecules are always sufficiently energized to undergo the back reaction. This may be important in reducing the net efficiency of some unimolecular isomerization reactions[1] but for sufficiently exothermic reactions the back reaction is usually unimportant compared with collisional stabilization of the product (see section 6.5.2).

The most important examples of chemically activated systems which are relevant to unimolecular rate theories are bimolecular association reactions. The best known of these are the reactions between the methylene radical and olefins to produce vibrationally excited cyclopropanes, e.g. (8.2), and the reactions between atoms (principally H atoms) and olefins

$$\ddot{C}H_2 + C_2H_4 \longrightarrow \triangle^* \begin{array}{c} \overset{k_a(E)}{\nearrow} \text{propene} \\ \underset{k_S}{\searrow} \overset{+M}{} \\ \triangle \end{array} \qquad (8.2)$$

to produce vibrationally excited radicals, e.g. (8.3). The competition between decomposition of the vibrationally excited species [rate-constant $k_a(E)$] and collisional stabilization (rate-constant k_S) is reflected in the

$$H + C_4H_8 \longrightarrow \text{sec-}C_4H_9^* \begin{array}{c} \overset{k_a(E)}{\nearrow} CH_3 + \text{propene} \\ \underset{k_S}{\searrow} \overset{+M}{} \\ \text{sec-}C_4H_9 \end{array} \qquad (8.3)$$

experimentally measured ratio (D/S) of decomposition to stabilization products obtained. For each system studied, the overall reaction mechanism must be established to enable this ratio to be determined from the experimental product analysis. If it is assumed that deactivation occurs upon every collision, the rate-constant k_S can be equated with the collision frequency ω which can be calculated from the kinetic theory expression (6.21). It is then possible to derive an experimental value of $k_a(E)$ which will in fact be an average value for the range of energized molecules involved.

The possibility of generating vibrationally excited species of widely different energies, by using different reactions for their production, enables $k_a(E)$ to be measured as a function of their energy content. This has provided an extremely valuable test of the RRK and RRKM expressions for the rate-constant $k_a(E)$ as a function of energy. Chemical activation experiments can alternatively provide a test of the strong collision assumption if it is accepted that the RRKM expression for $k_a(E)$ is correct. Such aspects of chemical activation are discussed in Chapter 10. For a comprehensive account of chemical activation, the reader is referred to the review by Rabinovitch and Flowers[2] and for more specific detail on rate formulations and energy expressions to that by Rabinovitch and Setser.[3] In this work, section 8.1 deals with the basic principles and section 8.2 with experimental data for some examples of major interest for unimolecular reaction rate theories.

8.1 BASIC PRINCIPLES

8.1.1 The average rate-constant $\langle k_a \rangle$

As an example of the treatment of chemical activation systems, consider the formation of a vibrationally excited molecule A* by an association of two radicals according to the scheme (8.4). According to this scheme,

$$\text{R} + \text{R}' \underset{k_a'(E)}{\overset{k_a(E)}{\rightleftarrows}} \text{A*} \begin{array}{c} \nearrow \text{Decomposition} \\ \text{products (D)} \\ \searrow_{\omega} \\ \text{A (S)} \end{array} \tag{8.4}$$

energized molecules A* at energy E can reform reactants with a rate-constant $k_a'(E)$, form decomposition products with a rate-constant $k_a(E)$ or be de-energized to stable molecules A. On the strong-collision assumption the first-order rate-constant for de-energization is equal to the collision frequency, $\omega = Zp$ where p is the total pressure and Z is given by (6.21). In some cases de-energization by collision with the walls should be included[4] but this complication has usually been ignored. Now suppose that the fraction of molecules which are energized per unit time into the energy range between E and $E + \delta E$ is $f(E) \delta E$. The fraction of A* decomposing by path D compared with those stabilized by path S is $k_a(E)/[k_a(E) + \omega]$, and if the back reaction reforming reactants can be ignored, the fraction of molecules in the energy range E to $E + \delta E$ decomposing to products is therefore $\{k_a(E)/[k_a(E) + \omega]\} f(E) \delta E$. The total number of molecules decomposing per unit time (D), at all energies above the critical energy E_0, is therefore given by (8.5). The total rate of stabilization (S) is given by a similar integral with $k_a(E)$ in the numerator replaced by ω.

$$D = \int_{E_0}^{\infty} \frac{k_a(E)}{k_a(E) + \omega} f(E) \, dE \tag{8.5}$$

In a strictly monoenergetic system, the experimental ratio D/S is equal to $k_a/\omega(E)$ [cf. (8.4)]. Where there is a distribution of energies, an average

10

rate-constant $\langle k_a \rangle$ for all energies above E_0 is similarly defined by (8.6).

$$\frac{\langle k_a \rangle}{\omega} = \frac{D}{S} = \frac{\text{No. of molecules decomposing per unit time}}{\text{No. of molecules being stabilized per unit time}} \quad (8.6)$$

Hence utilizing (8.5) and the equivalent expression for stabilization, the result (8.7) is obtained for $\langle k_a \rangle$.

$$\langle k_a \rangle = \omega \frac{\displaystyle\int_{E_0}^{\infty} \{k_a(E)/[k_a(E)+\omega]\} f(E)\, \mathrm{d}E}{\displaystyle\int_{E_0}^{\infty} \{\omega/[k_a(E)+\omega]\} f(E)\, \mathrm{d}E} \quad (8.7)$$

At high pressures $\omega \gg k_a(E)$, and writing the average rate-constant as $\langle k_a \rangle_\infty$, we obtain (8.8):

$$\langle k_a \rangle_\infty = \omega \frac{\displaystyle\int_{E_0}^{\infty} [k_a(E)/\omega] f(E)\, \mathrm{d}E}{\displaystyle\int_{E_0}^{\infty} f(E)\, \mathrm{d}E} = \omega \langle k_a(E)/\omega \rangle = \langle k_a(E) \rangle \quad (8.8)$$

Similarly at low pressures $\omega \ll k_a(E)$, and the corresponding rate-constant $\langle k_a \rangle_0$ is given by (8.9):

$$\langle k_a \rangle_0 = \omega \frac{\displaystyle\int_{E_0}^{\infty} f(E)\, \mathrm{d}E}{\displaystyle\int_{E_0}^{\infty} [\omega/k_a(E)] f(E)\, \mathrm{d}E} = \omega \langle \omega/k_a(E) \rangle^{-1} = \langle 1/k_a(E) \rangle^{-1}$$

$$(8.9)$$

From (8.8) and (8.9),

$$\langle k_a \rangle_\infty / \langle k_a \rangle_0 = \langle k_a(E) \rangle \langle 1/k_a(E) \rangle \quad (8.10)$$

If the vibrationally excited molecules are produced in a virtually mono-energetic manner, the ratio $\langle k_a \rangle_\infty / \langle k_a \rangle_0$ will approximate to unity. If there is a spread of energies, the less-excited molecules contribute more heavily to (8.9) so that $\langle k_a \rangle$ falls as the pressure is reduced and $\langle k_a \rangle_\infty / \langle k_a \rangle_0 > 1$. This ratio is therefore a measure of the spread in energies of the species A* and may be related to the energy distribution function $f(E)$. In some experimental systems $\langle k_a \rangle$ is found to increase again at sufficiently low pressures, and this indicates failure of the strong collision assumption (see section 10.3).

8.1.2   The form of the distribution function $f(E)$

In thermal energization systems, the distribution function $f(E)$ is simply the thermal quantum Boltzmann distribution $K(E)$ and the rate of energization into the energy range E to $E + \delta E$ is given by (8.11) [which appeared earlier as (4.5), since $K(E)\,\delta E = \delta k_1/k_2$]. For the chemically

$$K(E)\,\delta E = \frac{N(E)\exp(-E/kT)\,\delta E}{Q_2} \qquad (8.11)$$

activated system described here, the distribution function is given[8a] by (8.12). This can be derived as follows by using the principle of detailed

$$f(E)\,\delta E = \frac{k_a'(E)\,K(E)\,\delta E}{\displaystyle\int_{E_{\min}}^{\infty} k_a'(E)\,K(E)\,\mathrm{d}E} \qquad (8.12)$$

balancing applied to the reverse process of reforming radicals R and R′ with a rate-constant $k_a'(E)$. Consider a situation in which the processes D and S can be ignored and equilibrium is established between R, R′ and A* in the scheme (8.4). At equilibrium, the fraction of molecules with energy between E and $E + \delta E$ is the Boltzmann equilibrium population $K(E)\,\delta E$. The rate of dissociation to R + R′ is then $k_a'(E)\,K(E)\,\delta E$, and by the principle of detailed balancing this also gives the rate of combination of R and R′ to give A* in this energy range. Thus by arguments similar to those used earlier for thermal reactions, the rate of production of these A* is $k_a'(E)\,K(E)\,\delta E$ even when some of them are removed by processes D and S. The total rate of energization to all levels above the minimum energy $E_{\min}$ is therefore given by (8.13) and the fraction of molecules energized to energies between E and $E + \delta E$ is thus (8.12).

$$\text{Total rate of energization} = \int_{E_{\min}}^{\infty} k_a'(E)\,K(E)\,\mathrm{d}E \qquad (8.13)$$

In order to evaluate (8.12) and hence (8.7), equation (8.11) is used for $K(E)\,\delta E$, and this necessitates a knowledge of the energy density $N(E)$ for the energized molecules; suitable methods for the calculation of this quantity are discussed in Chapter 5. It is also necessary to know the rate-constant $k_a'(E)$ for the reverse dissociation of the energized molecule and this, like $k_a(E)$, is usually calculated from RRKM theory. The rate-constant $k_a'(E)$ can be evaluated from the basic RRKM equation (8.14)

$$k_a'(E) = \frac{1}{hN^*(E)} \sum_{E_{vr}^+=0}^{E^+} P'(E_{vr}^+) \qquad (8.14)$$

[cf. (4.15)], in which $N^*(E)$ is the energy density of the energized molecule and $\sum P'(E_{vr}^+)$ refers to the activated complex for the reverse dissociation into radicals (shown as X⁺ in Figures 8.1 and 8.2). The treatment therefore requires the construction of a model for this activated complex as well as the activated complex for the decomposition reaction (Y⁺ in Figures 8.1

and 8.2). Statistical factors and centrifugal effects can obviously be included if necessary.

Substitution of (8.11) and (8.14) into (8.12) gives the final expression (8.15) for the distribution function, where the integration limits in the

$$f(E)\,\delta E = \frac{\{\sum P'(E_{vr}^+)\}\exp\left(-E^+/kT\right)\delta E^+}{\displaystyle\int_{E^+=0}^{\infty}\{\sum P'(E_{vr}^+)\}\exp\left(-E^+/kT\right)\mathrm{d}E^+} \tag{8.15}$$

denominator correspond to all possible energies of the activated complex. Calculation of the distribution function by (8.15) is hence possible if $\sum P'(E_{vr}^+)$ can be evaluated. The results of such calculations show that the distribution function $f(E)$ produces a very narrow distribution of energies in which often 80–90% of the excited molecules have energies within ten per cent of the average energy. This is illustrated in Figure 8.3.

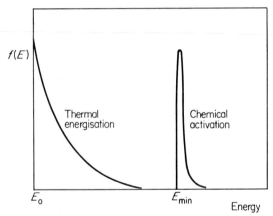

Figure 8.3 Schematic plot of the distribution functions for thermally and chemically activated molecules

8.1.3 The average energy

The final result for the average rate-constant $\langle k_a \rangle$ is obtained upon substituting from (8.15) and the equivalent of (8.14) for $k_a(E)$ into (8.7) [or (8.8) and (8.9) for $\langle k_a \rangle_\infty$ and $\langle k_a \rangle_0$ respectively]. The calculated values of these rate-constants can then be compared with those obtained experimentally.

It is also useful to calculate the average energy of the *reacting molecules* from the distribution function $\phi(E)$ which is appropriate to the system under consideration. Use is made of the general equation (8.16), in which the integration is carried out over all energies of interest. Some expressions

$$\langle E \rangle = \frac{\int E\phi(E)\,\mathrm{d}E}{\int \phi(E)\,\mathrm{d}E} \tag{8.16}$$

for the distribution functions and average energies of the reacting molecules in thermal reactions and in chemical activation systems are compared in Table 8.1.

The last section of Table 8.1 refers to the important special case where $k_a(E) = 0$ and only the back reaction to reform radicals has to be considered. Practically, this refers to calculations of the rates of radical recombination in the general-pressure region according to the scheme:

$$R + R' \underset{k_{a'}(E)}{\overset{\longrightarrow}{\longleftarrow}} A^* \xrightarrow{k_S} A$$

Thus radical recombination can be treated by the general methods described in this chapter; the methyl radical–ethane system has been the subject of much discussion and forms a good illustration which is discussed in detail in ref. 5. An interesting point is that the average energy of the A^* decomposing in the chemically activated system at low pressure is the same as that of the decomposing molecules in the thermal dissociation of A to $R + R'$ at high pressure.

8.2 EXPERIMENTAL STUDIES

Although it has long been realized that chemical activation occurs, for example in chemiluminescent reactions and in atom recombination reactions,[6] the use of chemical activation as a test of unimolecular rate theories is relatively recent.

One of the most comprehensive investigations has been that of Rabinovitch and co-workers[7] on the reactions of hydrogen or deuterium atoms with butenes. This system produces vibrationally excited sec-butyl radicals which can be stabilized or can decompose to propene and methyl radicals as shown in the scheme (8.3). The possibility of using H or D atoms with any of the three butene isomers leads to six possible minimum energies $E_{\min}$ of the butyl radicals formed. The use of two different temperatures and high and low pressures also gives 4 possible values of the thermal energy and hence a possible total of 24 different average energies $\langle E \rangle$ for the decomposing radicals. Table 8.2 shows some selected values of the observed rate-constants $\langle k_a \rangle_0$ and $\langle k_a \rangle_\infty$ derived from the experimental results, and also the values calculated theoretically on the basis outlined in section 8.1.1. It will be seen that the agreement is quite good

Table 8.1 Distribution functions and average energies for reacting molecules in the scheme (8.4)

Fraction of molecules energized into range E to $E+\delta E$	Fraction of molecules reacting from range E to $E+\delta E$, $\phi(E)\,\delta E$	Average energy of reacting molecules
		$\langle E\rangle = \int_{E_0}^{\infty} E\phi(E)\,dE \Big/ \int_{E_0}^{\infty} \phi(E)\,dE$

Thermal activation

$K(E)\,\delta E$
(Boltzmann)

$$\left[\frac{k_a(E)}{k_a(E)+\omega}\right] K(E)\,\delta E$$

low pressure $\longrightarrow K(E)\,\delta E$

high pressure $\longrightarrow [k_a(E)/\omega]\,K(E)\,\delta E$

$$\int_{E_0}^{\infty} EK(E)\,dE \Big/ \int_{E_0}^{\infty} K(E)\,dE$$
(Classical value $E_0 + kT$)

$$\int_{E_0}^{\infty} Ek_a(E)\,K(E)\,dE \Big/ \int_{E_0}^{\infty} k_a(E)\,K(E)\,dE$$
(Classical value $E_0 + skT$)

Chemical activation ⎰ *General case*
[System described
by (8.4)]

$$\left[\frac{k_a(E)}{k_a(E)+k'_a(E)+\omega}\right] k'_a(E)\,K(E)\,\delta E$$

low pressure $\longrightarrow \left[\dfrac{k_a(E)}{k_a(E)+k'_a(E)}\right] \times k'_a(E)\,K(E)\,\delta E$

high pressure $\longrightarrow \left[\dfrac{k_a(E)}{\omega}\right] \times k'_a(E)\,K(E)\,\delta E$

$k'_a(E)\,K(E)\,\delta E$

General expressions are cumbersome. Often $k'_a(E) \ll k_a(E)$ and the following simpler forms result:

$$\int_{E_0}^{\infty} Ek'_a(E)\,K(E)\,dE \Big/ \int_{E_0}^{\infty} k'_a(E)\,K(E)\,dE$$

$$\int_{E_0}^{\infty} Ek_a(E)\,k'_a(E)\,K(E)\,dE \Big/ \int_{E_0}^{\infty} k_a(E)\,k'_a(E)\,K(E)\,dE$$

For reversible recombination of radicals, where $k_a(E) = 0$

$$\left[\frac{k'_a(E)}{k'_a(E)+\omega}\right] k'_a(E)\,K(E)\,\delta E$$

low pressure $\longrightarrow k'_a(E)\,K(E)\,\delta E$

high pressure $\longrightarrow \dfrac{[k'_a(E)]^2}{\omega}\,K(E)\,\delta E$

$$\int_{E_0}^{\infty} Ek'_a(E)\,K(E)\,dE \Big/ \int_{E_0}^{\infty} k'_a(E)\,K(E)\,dE$$

$$\int_{E_0}^{\infty} E[k'_a(E)]^2\,K(E)\,dE \Big/ \int_{E_0}^{\infty} [k'_a(E)]^2\,K(E)\,dE$$

considering the possible inaccuracies in the thermochemical data and the choice of models for the activated complex, upon which the theoretical predictions are based. The ratio $\langle k_a \rangle_\infty / \langle k_a \rangle_0$ is seen to vary in the range

ble 8.2 Observed rate-constants $\langle k_a \rangle$ for some butyl radical decompositions compared h calculated values given in parentheses. (See ref. 7 for details of other similar results)

Reaction	$\langle E \rangle_0/$ kcal mol^{-1}	$10^{-7}\langle k_a \rangle_0/$s^{-1}	$\langle E \rangle_\infty/$ kcal mol^{-1}	$10^{-7}\langle k_a \rangle_\infty/$s^{-1}	$\langle k_a \rangle_\infty/\langle k_a \rangle_0$
cis-but-2-ene 195 K	8·45	>0·42, 0·61 (0·62)	8·96	0·73, 0·77 (0·75)	1·26 (1·21)
cis-but-2-ene 298 K	9·88	0·90 (1·01)	11·41	>1·4, 1·5 (1·67)	1·63 (1·65)
but-1-ene 298 K	10·98	1·30 (1·65)	12·40	2·4, 2·2 (2·57)	1·69, 1·85 (1·56)
cis-but-2-ene 298 K	11·78	1·35, 2·2 (1·79)	13·19	2·2, 3·3 (2·72)	1·50 (1·52)
but-1-ene 298 K	12·88	2·5, 3·1, 3·2 (2·79)	14·19	4·0, 4·7, 4·2 (4·05)	1·52, 1·60 (1·45)

1·2–1·8, which indicates a narrow spread of energies. The comparable ratio for thermal energization systems may be a factor of ten higher.[8] The average energy of the decomposing radicals was calculated as shown in section 8.1.3, and Figure 8.4 shows both the experimental and calculated

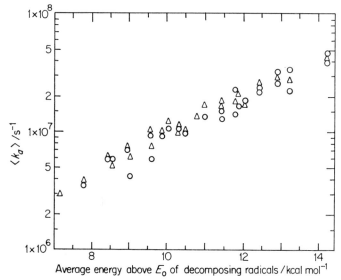

Figure 8.4 The rate constant $\langle k_a \rangle$ as a function of average energy for the decomposition of chemically activated sec-butyl radicals:[7] ○, experimental values; △, calculated values

rate-constants as a function of this energy. Again, the agreement is strikingly good and provides strong support for the postulated model in which all modes of vibration and all internal rotations were taken to be active. The strong collision assumption was made and is seen to be valid for these calculations which refer to the case where most collisions of the activated radicals were with butene molecules. Experiments were also carried out with the reactants highly diluted by gases less efficient in energy transfer. The calculations for this case are considerably more complicated and are described in Chapter 10.

The study of chemically activated radicals formed from H atoms plus olefins has also been extended to radicals which decompose by loss of ethyl, propyl, isopropyl, t-butyl and neo-pentyl radicals rather than methyl radicals, and to competitive systems in which the chemically activated radicals decompose by more than one process. The results can be interpreted satisfactorily in terms of RRKM theory.[9]

One of the first demonstrations of the efficiency of intramolecular vibrational energy transfer was that of Butler and Kistiakowsky[10] who produced vibrationally excited methylcyclopropane by the reaction of methylene radicals with either cyclopropane or propene. The lifetimes of the excited methylcyclopropanes were found to be dependent upon the reaction used and upon the source of the methylene radicals (photolysis of ketene or diazomethane), but the composition of the isomeric butene mixture formed was found to be invariant. This was interpreted as evidence for the rapid exchange of energy among the vibrational modes of the hot molecule. Similar but more extensive studies of the reactions of methylene with ethylene to give hot cyclopropane and with cis-but-2-ene to give hot dimethylcyclopropane have been carried out by Rabinovitch and co-workers.[11] The methylene was produced by both the photolytic and the thermal decomposition of diazomethane and also by the photolysis of ketene. The minimum energy E_{min} of the hot molecules was calculated from the known thermochemistry to be about 100 kcal mol^{-1} compared with critical energies for isomerization of about 60 kcal mol^{-1}. The average energy of the hot molecules was estimated (ca 110 kcal mol^{-1}) and hence values of $k_a(E)$ were obtained which were in agreement with the experimental $\langle k_a \rangle$ values. Comparison between $k_a(E)$ and $\langle k_a \rangle$ is adequately valid since the energy dispersion is small relative to the excess energy present. There is, in fact, some difficulty in the estimation of absolute energies in these systems, although the differences in energy between different cases are known with better accuracy. It is therefore useful to compare the observed and calculated ratios of $k_a(E)$ (or $\langle k_a \rangle$) for two cases. Such a comparison is made in Table 8.3 and it will be seen that the values in the last two columns of this table are in good agreement.

The study of these systems has also been extended to weak-collision cases where information about collisional energy transfer efficiency can be obtained (see section 10.3).

Dorer and Rabinovitch[12] similarly studied chemically activated $cyclo$-$C_3H_5(CH_2)_nCH_3$ and $cyclo$-$C_3H_5(CH_2)_nCF_3$ ($n = 0, 1, 2$), and obtained

Table 8.3 Variation of rate-constants for the isomerization of chemically activated cyclopropane and 1,2-dimethylcyclopropane with the energy of the excited molecules[11]

System[a]	$(E_2 - E_1)$/kcal mol^{-1}	$\langle k_a \rangle_2 / \langle k_a \rangle_1$ Observed	$\langle k_a \rangle_2 / \langle k_a \rangle_1$ Calculated
Cyclopropane			
1. C_2H_4 + DM, 325 °C	2·7	2·0	1·4
2. C_2H_4 + DM, 450 °C			
Cyclopropane-d_2			
1. CHD:CHD + ketene, 332 nm	'3'	1·8	1·6
2. CHD:CHD + ketene, 320 nm			
1,2-Dimethylcyclopropane			
1. CB + DM, 250 °C	5·4	2·3	2·1
2. CB + DM, 400 °C			

[a] DM = diazomethane, CB = cis-but-2-ene, CHD:CHD was *trans*.

rate-constants in good agreement with theory, as regards both the variation of $\langle k_a \rangle$ along each homologous series, and also the statistical-weight effect resulting from replacement of CH_3 by CF_3 (cf. statistical-weight isotope effects, see Chapter 9). Methylene radicals have also been used to produce chemically activated molecules by insertion into C—H bonds in alkanes and Si—H bonds in organosilanes, the results being in good agreement with RRKM calculations.[13]

The third and last main class of chemical-activation unimolecular reaction systems of current interest are the elimination reactions of hydrogen halides from chemically activated alkyl halide molecules. Chemically activated alkyl chlorides and alkyl bromides have been studied by Setser and co-workers[14] by utilizing the abstraction reactions between methylene radicals and chloro- or bromomethanes. These reactions generate halo-methyl radicals which recombine to form vibrationally excited alkyl halide molecules, with energies typically in the range 85–95 kcal mol^{-1} compared with critical energies of 50–55 kcal mol^{-1}

for the elimination reactions. When methylene (generated by the photolysis of ketene or diazomethane, or by the pyrolysis of diazomethane) reacts with methyl chloride, the major reactions at high pressure (> 1000 Torr) are (8.17)–(8.20):

$$CH_2 + CH_3Cl \longrightarrow CH_3 + CH_2Cl \qquad (8.17)$$

$$2CH_3 \longrightarrow C_2H_6^* \qquad (8.18)$$

$$CH_3 + CH_2Cl \longrightarrow CH_3CH_2Cl^* \qquad (8.19)$$

$$CH_2Cl + CH_2Cl \longrightarrow ClCH_2CH_2Cl^* \qquad (8.20)$$

The vibrationally excited chloroethanes are then either collisionally de-energized or undergo elimination reactions according to (8.21)–(8.24):

$$C_2H_5Cl^* \quad \begin{array}{c} \overset{\omega}{\nearrow} \; C_2H_5Cl \qquad (S_1) \qquad (8.21) \\ \underset{\langle k_{a1} \rangle}{\searrow} \; C_2H_4 + HCl \qquad (D_1) \qquad (8.22) \end{array}$$

$$C_2H_4Cl_2^* \quad \begin{array}{c} \overset{\omega}{\nearrow} \; C_2H_4Cl_2 \qquad (S_2) \qquad (8.23) \\ \underset{\langle k_{a2} \rangle}{\searrow} \; C_2H_3Cl + HCl \qquad (D_2) \qquad (8.24) \end{array}$$

In the case of 1,2-dichloroethane the rate-constant for elimination, $\langle k_{a2} \rangle$, was obtained from the slope of a plot of D_2/S_2 vs $1/pressure$. On the strong collision assumption, equation (8.6) predicts that such a plot should be linear. The results in Figure 8.5 show that linear plots are obtained at both

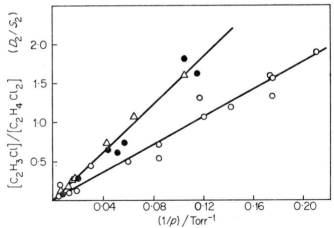

Figure 8.5 Plot of [vinyl chloride]/[1,2-dichloroethane] against $1/pressure$ for the CH_3/CH_2Cl radical system.[14a] ○, CH_2N_2 at 300 °C (the $1/p$ scale should be divided by 10 for these points); △, CH_2N_2 at 25 °C; ●, CH_2CO at 25 °C

25 °C and 300 °C and that the plot at 25 °C is independent of the methylene source, thus demonstrating that the methylene radicals are at thermal equilibrium at this temperature.

Determination of the rate-constant $\langle k_{a1} \rangle$ is complicated by the fact that ethylene is also produced by the reaction of methylene radicals with ketene. However, steady-state treatment of the reaction scheme shows that $\langle k_{a1} \rangle$ can be obtained from the slope of a plot of

$$[C_2H_4Cl_2 + C_2H_3Cl]/[C_2H_5Cl] \quad vs \quad 1/pressure$$

Calculations using RRKM theory have been made for these systems by Setser and co-workers.[15, 16] Models were constructed for the activated complexes for the radical association and the elimination reactions. The best agreement with the experimental data was found using a 4-centred activated complex for the elimination reaction having bond orders of 0·2, 0·2, 0·8 and 1·8 for the C—H, H—Cl, C—Cl and C—C bonds respectively. All internal degrees of freedom were taken to be active. The results of some of these calculations are compared with experimental values for the average elimination rate-constant in Table 8.4. Of particular interest are

Table 8.4 Calculated and observed rate-constants[16] for the elimination reactions of chemically activated alkyl halides at 300 K

Reactant molecule	Molecule eliminated	$E_0/$ kcal mol^{-1}	$\langle E \rangle_\infty/$ kcal mol^{-1}	$10^{-9}\langle k_a \rangle/s^{-1}$ Observed	Calculated
CH_3CH_2Cl	HCl	55·0	91·1	2·6	2·5
CH_3CH_2Br	HBr	52·0	91·4	6·1	5·8
CH_2BrCH_2Br	HBr	52·0	88·5	1·7	1·5
CH_2BrCH_2Cl	⎧HCl	(55·0)	88·4	0·12	0·42
	⎨HBr	(52·0)	88·4	0·30	1·1
	⎩(ratio HBr/HCl	—	—	2·5	2·6)
CH_2ClCH_2Cl	HCl	55·0	88·4	0·18	0·83

the calculations for the competitive rates of elimination of HCl and HBr from 1-bromo-2-chloroethane. Although the *absolute* values of $\langle k_a \rangle$ for HCl and HBr elimination are calculated to be three to four times larger than the observed values it will be noted that the agreement is good for the *relative* rates of these reactions. The calculated rate-constants are sensitive to the critical energies E_0 and unfortunately these parameters are not readily accessible for either 1-bromo-2-chloroethane or 1,2-dichloroethane, both of which decompose thermally by predominantly free radical chain reactions.[17, 18] Some support for the proposed models for these elimination reactions has come from work[19] on the isotopic pairs C_2H_5Cl/C_2D_5Cl and

$C_2H_4Cl_2/C_2D_4Cl_2$. Calculations based upon the 4-centred activated complex proposed previously give chemical activation isotopic rate ratios of 3·0 for C_2H_5Cl/C_2D_5Cl (expt. $3·3 \pm 0·4$) and 2·8 for $C_2H_4Cl_2/C_2D_4Cl_2$ (expt. $3·5 \pm 0·1$).

Vibrationally excited alkyl fluorides have been produced by radical recombination reactions occurring during the photolysis of fluorinated aldehydes and ketones. Benson and Haugen[20] first analysed the data for 1,2-difluoroethane which reacts according to (8.25). Taking (D/S) (8.6) as

$$2CH_2F \rightleftharpoons CH_2FCH_2F^* \begin{array}{c} \xrightarrow{k_a(E)} CH_2{=}CHF + HF \\ \xrightarrow{\omega} CH_2F{-}CH_2F \end{array} \qquad (8.25)$$

the ratio of the quantum yields ϕ of vinyl fluoride and 1,2-difluoroethane, and using the RRK theory for $k_a(E)$ gives (8.26), in which E is the average

$$\frac{\phi(C_2H_4F_2)}{\phi(C_2H_3F)} = \frac{\omega}{\langle k_a \rangle} = \frac{\lambda Zp}{A_\infty}\left(\frac{E}{E-E_0}\right)^{s-1} \qquad (8.26)$$

energy of the energized molecules and the other symbols have their usual significance. E can be estimated from the known enthalpy of the combination reaction and the specific heats of the radicals concerned. Benson and Haugen concluded that for 1,2-difluoroethane the best fit corresponded to values of $s = 12$, $\lambda = 0·1$–$0·2$ and $E_0 = 62\ \text{kcal mol}^{-1}$.

Trotman-Dickenson and co-workers[21] studied a number of chemically activated alkyl fluorides and similarly analysed the results in terms of the RRK theory by equation (8.26). For a series of compounds, s was assigned by analogy with cyclopropane derivatives, λ was assumed to be unity, A_∞ was taken to be $10^{13·5}\ \text{s}^{-1}$ and the results were used to calculate E_0 values for the reactions under consideration. The resulting E_0 values are given in Table 8.5. Of particular interest is the study of chemically activated ethyl fluoride[22] which has been made at three different levels of excitation by using reactions (8.27)–(8.29). The average energies of the molecules

$$C_2H_5 + F_2 \longrightarrow C_2H_5F^* + F \qquad (8.27)$$

$$CH_3 + CH_2F \longrightarrow C_2H_5F^* \qquad (8.28)$$

$$CH_2 + CH_3F \longrightarrow C_2H_5F^* \qquad (8.29)$$

produced in (8.27), (8.28) and (8.29) were calculated to be 69, 90 and ca 109 kcal mol^{-1}. Average rate-constants $\langle k_a \rangle$ were derived from experimental measurements of the relative amounts of ethylene and ethyl fluoride formed in the three systems; strong collisions were assumed and an average collision frequency was taken to be $10^7\ \text{Torr}^{-1}\ \text{s}^{-1}$. These rate-constants were then compared with values of $k_a(E)$ calculated from RRKM theory. The calculations were based on those of Hassler and

Table 8.5 Critical energies for HF elimination from alkyl fluorides, estimated from chemical activation experiments on the basis of RRK theory

Alkyl fluoride	Source	E_0/kcal mol^{-1} (estimated)	Ref.
C_2H_5F	$\left\{ \begin{array}{l} \left\{ \begin{array}{l} CH_3 + CH_2F \\ C_2H_5 + F_2 \\ CH_2 + CH_3F \end{array} \right\} \\ CH_3 + CH_2F \end{array} \right.$	51 59	a b
CH_2FCH_2F	$2\ CH_2F$	$\left\{ \begin{array}{l} 52 \\ 62 \pm 3 \end{array} \right.$	c b, d
CH_3CHF_2	$\left\{ \begin{array}{l} CH_3 + CHF_2 \\ CH_2 + CH_2F_2 \end{array} \right\}$	47–48	21
$CH_3CH_2CHF_2$	$CH_2 + CH_3CHF_2$	54	21
$CH_3CF_2CH_3$	$CH_2 + CH_3CHF_2$	48–49	21

[a] J. A. Kerr, A. W. Kirk, B. V. O'Grady and A. F. Trotman-Dickenson, *Chem. Commun.*, 365 (1967). [b] G. O. Pritchard and R. L. Thommarson, *J. Phys. Chem.*, **71**, 1674 (1967). [c] J. A. Kerr, A. W. Kirk, B. V. O'Grady, D. C. Phillips and A. F. Trotman-Dickenson, *Disc. Faraday Soc.*, **44**, 263 (1967). [d] Data of G. O. Pritchard, M. Venugopalan and T. F. Graham, *J. Phys. Chem.*, **68**, 1786 (1964), evaluated by S. W. Benson and G. Haugen, ref. 20; RRKM treatment applied by H. W. Chang and D. W. Setser, *J. Amer. Chem. Soc.*, **91**, 7648 (1969).

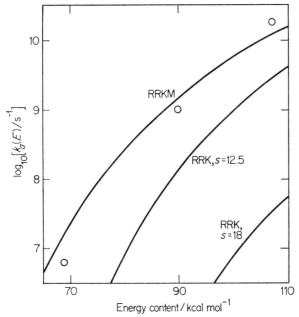

Figure 8.6 Calculated RRKM and RRK curves for the rate-constant $k_a(E)$ against energy content for chemically activated ethyl fluoride;[22] the circles represent experimental values

Setser[15] for chloroethanes and bromoethanes except that vibrational frequencies were assigned to the activated complex for the elimination reaction according to the method of Benson and O'Neal.[23] The calculated variation of $k_a(E)$ with energy is compared with the experimental values of $\langle k_a \rangle$ plotted against energy in Figure 8.6. Also shown are some calculations based on the classical RRK theory. It will be seen that the agreement with the RRKM curve is quite good despite the assumptions made in equating $k_a(E)$ and $\langle k_a \rangle$. The RRK curves, on the other hand, predict much lower values of $\langle k_a \rangle$ than those observed, unless the strong collision assumption is abandoned, for which there seems little justification in this case.

References

1. M. C. Lin and K. J. Laidler, *Trans. Faraday Soc.*, **64**, 94 (1968).
2. B. S. Rabinovitch and M. C. Flowers, *Quart. Rev.*, **18**, 122 (1964).
3. B. S. Rabinovitch and D. W. Setser, *Advan. Photochem.*, **3**, 1 (1964).
4. K. M. Maloney, *J. Phys. Chem.*, **74**, 4177 (1970).
5. E. V. Waage and B. S. Rabinovitch, *Internat. J. Chem. Kinetics*, **3**, 105 (1971).
6. M. Trautz, *Zeit. Phys. Chem.*, **53**, 1 (1905); Symposium on Inelastic Collisions of Atoms and Simple Molecules, *Disc. Faraday Soc.*, **33** (1962).
7. B. S. Rabinovitch, R. F. Kubin and R. E. Harrington, *J. Chem. Phys.*, **38**, 405 (1963), and references cited therein; additional data are illustrated in Figure 9.2.
8. (a) B. S. Rabinovitch and R. W. Diesen, *J. Chem. Phys.*, **30**, 735 (1959). (b) H. S. Johnston, *J. Chem. Phys.*, **20**, 1103 (1952); **22**, 1969 (1954); *Gasphase Reaction Rate Theory*, Ronald, New York, 1966, Chap. 15B.
9. C. W. Larson and B. S. Rabinovitch, *J. Chem. Phys.*, **52**, 5181 (1970), and references cited therein.
10. J. N. Butler and G. B. Kistiakowsky, *J. Amer. Chem. Soc.*, **82**, 759 (1960).
11. D. W. Setser and B. S. Rabinovitch, *Can. J. Chem.*, **40**, 1425 (1962); see also D. W. Setser, B. S. Rabinovitch and J. W. Simons, *J. Chem. Phys.*, **40**, 1751 (1964); J. W. Simons, B. S. Rabinovitch and D. W. Setser, *J. Chem. Phys.*, **41**, 800 (1964); J. D. Rynbrandt and B. S. Rabinovitch, *J. Phys. Chem.*, **74**, 1679 (1970).
12. F. H. Dorer and B. S. Rabinovitch, *J. Phys. Chem.*, **69**, 1952, 1964, 1973 (1965).
13. R. L. Johnson, W. L. Hase and J. W. Simons, *J. Chem. Phys.*, **52**, 3911 (1970); W. L. Hase and J. W. Simons, *J. Chem. Phys.*, **52**, 4004 (1970), **54**, 1277 (1971), and references cited therein.
14. (a) J. C. Hassler, D. W. Setser and R. L. Johnson, *J. Chem. Phys.*, **45**, 3231 (1966); (b) J. C. Hassler and D. W. Setser, *J. Chem. Phys.*, **45**, 3237 (1966).
15. J. C. Hassler and D. W. Setser, *J. Chem. Phys.*, **45**, 3246 (1966).
16. R. L. Johnson and D. W. Setser, *J. Phys. Chem.*, **71**, 4366 (1967).
17. K. D. King and E. S. Swinbourne, *J. Chem. Soc. (B)*, 687 (1970).
18. K. A. Holbrook, R. W. Walker and W. R. Watson, *J. Chem. Soc. (B)*, 577 (1971).

19. K. Dees and D. W. Setser, *J. Chem. Phys.*, **49**, 1193 (1968).
20. S. W. Benson and G. Haugen, *J. Phys. Chem.*, **69**, 3898 (1965).
21. J. A. Kerr, D. C. Phillips and A. F. Trotman-Dickenson, *J. Chem. Soc. (A)*, 1806 (1968), and references cited therein.
22. A. W. Kirk, A. F. Trotman-Dickenson and B. L. Trus, *J. Chem. Soc. (A)*, 3058 (1968).
23. S. W. Benson and H. E. O'Neal, *J. Phys. Chem.*, **71**, 2903 (1967).

9 Kinetic Isotope Effects in Unimolecular Reactions

The rate of a unimolecular reaction is in general altered by isotopic substitution in the reactant molecule. The study of isotope effects is important since it can often lead to a more detailed understanding of the reaction process than can be obtained by study of the unsubstituted compound alone. Isotopic substitution can have several simultaneous effects, and this complication sometimes produces results which are at first sight curious. For example, a deuterium-substituted compound may isomerize more slowly than the parent compound at high pressures, but more rapidly at low pressures. The general theory of kinetic isotope effects has been adequately presented elsewhere[1] and the present discussion is limited to a review of the principles involved and some detail of their application to unimolecular reactions. In this context the review by Rabinovitch and Setser[2] is useful, and many of the illustrative calculations given later are taken from that source. For simplicity the discussion is in terms of hydrogen–deuterium isotope effects, but the principles obviously apply to other systems as well. Some of the terminology used in connection with isotope effects is summarized in Table 9.1.

9.1 GENERAL DISCUSSION OF ISOTOPE EFFECTS

The origin of all kinetic isotope effects lies in the changes in the quantized molecular energy levels which occur when the vibration frequencies and moments of inertia of a molecule are modified by isotopic substitution. The changes are purely mass effects, since isotopic substitution has no effect on the electron distribution or potential-energy surface for a molecule, and hence no effect on the ground-state geometry or the force constants for vibration of the molecule. The changes in moments of inertia are generally less important than those in vibration frequency and will be neglected here. In the case of a simple harmonic oscillator the vibration frequency is given by (9.1), and isotopic substitution alters the

Table 9.1   Terminology used for kinetic isotope effects

Intermolecular	Comparison of rates between two different molecules [e.g. reactions (9.3) and (9.4)]
Intramolecular	Comparison of competitive rates of two reactions of one molecule [e.g. reactions (9.14)]
Primary	Atom(s) substituted are directly involved in the reaction [e.g. reactions (9.3) and (9.4)]
Secondary	Atom(s) substituted are not directly involved in the reaction [e.g. reactions (9.5) and (9.6)][a]
Normal (regular)	$k(\text{light}) > k(\text{heavy})$
Inverse	$k(\text{heavy}) > k(\text{light})$
Statistical-weight	Arises from changes in the distribution of energy levels as opposed to changes in critical energy. Usually predominant in secondary isotope effects
Non-equilibrium	Effects found when activated complexes are not in equilibrium with reactants, e.g. unimolecular reactions in the fall-off region, or chemically activated systems

[a] More rigorously, the reaction coordinate is orthogonal to all vibrational modes involving the atom(s) substituted; the normal coordinates for these modes have effectively zero weighting in the reaction coordinate.

vibration frequency by changing the reduced mass μ with a constant value

$$\nu = \frac{1}{2\pi}\sqrt{\frac{k}{\mu}} \tag{9.1}$$

of the force constant k. A simple illustration of this result is the observation that the stretching vibration frequencies of the bonds X—H and X—D, where X is a relatively heavy group such as Cl, Br, alkyl, etc., are related by $\nu_{\text{XH}}/\nu_{\text{XD}} \approx \sqrt{2}$ (it will be noted that $\mu_{\text{XH}}/\mu_{\text{XD}} \approx \frac{1}{2}$ if X is heavy). For the more general case where the molecular distortions must be considered in terms of normal vibrations which are not highly localized (see Chapter 2 and ref. 3), the properties of the isotopic forms are still related in an overall sense by the Teller–Redlich Product Rule.[3] For the case of a non-linear molecule this is (9.2), in which I denotes moment of inertia, M

$$\prod_{i=1}^{3N-6} \frac{\nu_i}{\nu_i'} = \left(\frac{I_A I_B I_C}{I_A' I_B' I_C'}\right)\left(\frac{M}{M'}\right)^{\frac{3}{2}} \prod_{i=1}^{N} \left(\frac{m_i'}{m_i}\right)^{\frac{3}{2}} \tag{9.2}$$

molecular weight, m_i the masses of the N atoms in the molecule and primed quantities refer to the isotopically substituted molecule. The product rule is a useful relationship when it is desired to set up consistent models of a molecule and its isotopically substituted form, particularly in the case of activated complexes where experimental measurement is not possible even in principle.[4]

The crucial effect of the changes in vibration frequencies is to modify the vibrational energy levels of the molecule (for example the spacing of the levels is $h\nu$ for a simple harmonic oscillator) and these changes can have essentially two consequences. Firstly, there are *critical energy effects*; the zero-point energies ($\frac{1}{2}h\nu$) of the affected vibrations will change on isotopic substitution, and if these vibrations contribute substantially to the reaction coordinate the critical energy E_0 will be changed. The compounds XH and XD again form a simple example, the relevant energy diagrams being shown in Figure 9.1. The group X is here regarded as

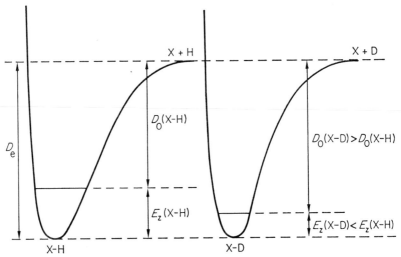

Figure 9.1   Potential energy curves for the molecules X—H and X—D, showing difference in dissociation energy

essentially a point mass, so that the unimolecular reaction is a pure bond fission and there is no zero-point energy in the activated complex. Fission of the X—H or X—D bond thus requires a critical energy which is equal to $D_e - E_z$ and may differ substantially for the two compounds. For R_3C—H and R_3C—D, for example, the zero-point energies are $\frac{1}{2}h\nu_0 \approx 4.1$ and $2.9 \text{ kcal mol}^{-1}$ respectively, giving a difference in the critical energies for C—D and C—H fission of

$$E_0(R_3C\text{—}D) - E_0(R_3C\text{—}H) = \Delta E_0 \approx 1.2 \text{ kcal mol}^{-1}$$

This difference would give a rate ratio of

$$k(R_3C\text{—}H)/k(R_3C\text{—}D) = \exp(\Delta E_0/kT) = 2.7 \text{ at } 600 \text{ K}$$

which is a substantial difference, and when this effect is present it is frequently predominant. The reaction involving the lighter isotope is generally favoured (as in the above case), and this is known as a *normal kinetic isotope effect* [k(light) > k(heavy)]. The rate ratio will not always be as large as above since the reaction may be such that the hydrogen atom is incompletely removed in the activated complex. In this case there will still be some zero-point energy of the corresponding vibration(s) and the change in E_0 on isotopic substitution will be less than before. If the reaction involves transfer of an atom from one site to another it is even possible for the atom to be bound more tightly in the activated complex than in the molecule. In this case the result can be that k(heavy) > k(light), an *inverse kinetic isotope effect*. Inverse effects can also arise for other reasons discussed below.

These examples have referred to the case where the atom to be isotopically substituted actually takes part in the reaction, i.e. the reaction coordinate involves the vibration of which the frequency is altered. Examples would be reactions (9.3) and (9.4). Such effects are known as *primary kinetic isotope effects*, and a major contribution to such effects is

$$C_2H_5-H \longrightarrow C_2H_5 + H \qquad (9.3)$$

$$C_2H_5-D \longrightarrow C_2H_5 + D \qquad (9.4)$$

the change in critical energy discussed above. Statistical-weight effects (see below) may also be important, however, and can sometimes predominate.

The second consequence of changes in the energy level spacing on isotopic substitution is a change in the state densities of reactant and activated complex, usually referred to as a *statistical-weight effect*. Even when isotopic substitution occurs at a site remote from the reaction centre the reaction rate may be affected. Reactions (9.5) and (9.6) comprise an example, and any difference in rate is due to a *secondary kinetic isotope effect*. To a good approximation the critical energy is unaffected by such

$$CH_3CH_2-CH_2CH_3 \longrightarrow 2CH_3CH_2 \qquad (9.5)$$

$$CD_3CH_2-CH_2CD_3 \longrightarrow 2CD_3CH_2 \qquad (9.6)$$

changes, and secondary isotope effects arise mainly from changes in the quantities $\sum P(E_{vr}^+)$ and $N^*(E^*)$ which (with E_0) principally affect the results of the RRKM calculation. These quantities are certain to change on isotopic substitution because of the changes in energy level spacing; the effects are clearly illustrated in the Whitten–Rabinovitch equations (5.32) and (5.38) and a specific example is given in Table 9.2 and Figure 9.5. Such effects are known as *statistical-weight isotope effects*, since they arise from changes in the numbers of quantum states at various energy levels, i.e. the statistical weights of these levels. Primary isotope effects invariably include statistical-weight effects as well as critical-energy effects, and the statistical-weight effects are not always of smaller magnitude.

Isotopic substitution at sites not directly involved in the reaction can also produce small changes in critical energy, due to small contributions to the reaction coordinate from vibrational modes involving the substituted atoms. Isotope effects in such systems are sometimes called secondary effects, although they are not pure secondary effects in the sense used here; the terms 'conventional secondary isotope effects' and 'mixed primary-secondary isotope effects' have been used.

9.2 BASIS OF APPLICATION TO UNIMOLECULAR REACTIONS

The calculation of isotope effects for unimolecular reactions is relatively simple for chemical activation or photochemical activation experiments, in which the energized molecules are produced in a relatively narrow band of energies. There is in practice still a spread of energies which cannot be ignored (see section 8.1.2), but ideal experiments involving monoenergetic excitation would give $k_a(E^*)$ for the isotopic reactions directly. The rate-constants are each calculable from RRKM theory [equation (4.16)] and the isotopic rate ratio is thus given by (9.7), in which statistical factors and centrifugal effects have been omitted for simplicity. This equation applies equally to intermolecular isotope effects and (with a simplification

$$\frac{k_{aH}(E_H^*)}{k_{aD}(E_D^*)} = \frac{\sum\limits^{E_H^+} P_H(E_{vr}^+)\Big/\sum\limits^{E_D^+} P_D(E_{vr}^+)}{N_H^*(E_H^*)/N_D^*(E_D^*)} \tag{9.7}$$

described later) to intramolecular isotope effects. The expression becomes relatively simple for a secondary isotope effect, for which a natural comparison can be made for the case where $E_H^* = E_D^*$ and $E_H^+ = E_D^+$. For primary isotope effects the situation is more complicated since E_0 is different for the two compounds, and therefore $E_H^* \neq E_D^*$ and/or $E_H^+ \neq E_D^+$.

In thermal activation experiments, the formulation is complicated even for a secondary isotope effect because of the simultaneous changes in k_a and in the energy distribution function. The fractional concentration $f_H'(E^*)\,\delta E^*$ of energized molecules in the energy range E^* to $E^* + \delta E^*$ in the steady state is given by (9.8), and the corresponding quantity $f_D'(E^*)\,\delta E^*$ is given by a similar equation. The isotopic rate ratio is thus (9.9). The general equation is clearly very complex, but the high- and low-pressure

$$f'_H(E^*)\,\delta E^* = \left(\frac{[A_{H(E^* \to E^* + \delta E^*)}]}{[A]}\right)_{\text{steady state}}$$

$$= \frac{N_H^*(E^*)\exp(-E^*/kT)/Q_{2H}}{1+k_{aH}(E^*)/k_2[M]}\,\delta E^* \qquad (9.8)$$

$$\frac{(k_{\text{uni}})_H}{(k_{\text{uni}})_D} = \frac{\displaystyle\int_{E^*=(E_0)_H}^{\infty} f'_H(E^*)\,k_{aH}(E^*)\,dE^*}{\displaystyle\int_{E^*=(E_0)_D}^{\infty} f'_D(E^*)\,k_{aD}(E^*)\,dE^*} \qquad (9.9)$$

limits (9.10) and (9.11) are relatively simple and will be useful in the discussion. These equations may be obtained from (9.9) or more simply

$$(p \to \infty) \qquad \frac{(k_\infty)_H}{(k_\infty)_D} = \frac{(Q_2^+/Q_2)_H}{(Q_2^+/Q_2)_D}\exp\{[(E_0)_D-(E_0)_H]/kT\} \qquad (9.10)$$

$$(p \to 0) \qquad \frac{(k_{\text{bim}})_H}{(k_{\text{bim}})_D} = \frac{(Q_2^{*'}/Q_2)_H}{(Q_2^{*'}/Q_2)_D}\exp\{[(E_0)_D-(E_0)_H]/kT\} \qquad (9.11)$$

from (4.21) and (4.24) respectively. In (9.11) the reasonable approximation has been made that $k_{2H} = k_{2D}$, and it will be recalled that $Q_2^{*'}$ is the partition function for the active degrees of freedom in the energized molecule with the energy zero taken as the ground state of A^+, i.e. the lowest energy an A^* can have.

The type and size of isotope effects in unimolecular reactions can now be discussed. We deal first with the simplest case of secondary effects on $k_a(E^*)$, then primary effects on $k_a(E^*)$ and finally isotope effects on k_{uni} for thermally activated systems.

9.3 SECONDARY KINETIC ISOTOPE EFFECTS ON $k_a(E^*)$ (THEORY AND EXPERIMENT)

The application of (9.7) to calculate secondary isotope effects on $k_a(E^*)$ in monoenergetic systems is fairly straightforward, since it may be assumed that $(E_0)_D = (E_0)_H$ and the values of k_a for the two forms can be compared at the same values of both E^* and E^+. Changes in the moments of inertia of the external (adiabatic) rotations are likely to be small and partially compensating, so the neglect of centrifugal factors will be a reasonable approximation. Deuterium substitution lowers the vibration frequencies and hence the energy level spacing, and both $\sum P(E_{\text{vr}}^+)$ and $N^*(E^*)$ are therefore higher for the deuterio- than for the light molecules.

Since $E^* \gg E^+$, however, $N^*(E^*)$ is always increased more than $\sum P(E_{vr}^+)$, and thus $k_{aH}(E^*) > k_{aD}(E^*)$; a normal statistical-weight secondary isotope effect is predicted.† The figures involved are illustrated by Rabinovitch and Setser's calculations[2] for reactions (9.5) and (9.6); see Table 9.2. The illustrative critical energies of 35 and 85 kcal mol^{-1} were chosen as being typical of C—C fission in an alkyl radical and an alkane molecule respectively. It will be seen that the size of the isotope effect varies markedly with both E_0 and E^+, being largest when E_0 is high and E^+ is low [the difference then being mainly in the $N^*(E^*)$ terms] and smallest when E^+ is large [the maximum compensating effect of the $\sum P(E_{vr}^+)$ terms then being obtained].

The most comprehensive experimental studies of such effects are those of Rabinovitch and co-workers on the decompositions of chemically activated sec-butyl radicals.[6] These radicals were produced by the addition of H or D atoms to butenes or octadeuteriobutenes, e.g. (9.12). By studying

$$H + CH_3CH\!:\!CHCH_3 \longrightarrow$$

$$CH_3CH_2\overset{\cdot}{C}HCH_3^* \begin{array}{c} \xrightarrow{\text{M}} CH_3CH_2\overset{\cdot}{C}HCH_3 \\ \\ \searrow CH_3 + CH_2\!:\!CHCH_3 \end{array} \qquad (9.12)$$

the reactions at two temperatures (195 and 300 K) and at high and low pressures it was possible to produce sec-butyl-d_0, -d_1, -d_8 and -d_9 radicals with a whole range of average energies $\langle E^* \rangle$ and to measure the rates of their unimolecular decomposition to methyl radicals plus propene relative to their rate of collisional stabilization. The technique is discussed in more detail in Chapter 8. As shown there, the excitation is not strictly monoenergetic, but the energy spread in a given experiment is small and the measured rate-constants approximate to the values $k_a(E^*)$ for the appropriate mean energy $\langle E^* \rangle$. The results are presented in Figure 9.2, which shows a large normal isotope effect in favour of the d_0 and d_1 radicals (upper curve) against the d_8 and d_9 radicals (lower curve). The experimental rate-constants are in excellent agreement with those calculated from the RRKM theory over a fiftyfold range of values. The isotopic rate ratio is best obtained from the smooth-curve representations of the results, since the addition of H to C_4H_8 and C_4D_8 gives radicals with slightly different $\langle E^* \rangle$. The difference arises from the different thermal energies of the two olefins at a given temperature; it is typically 0·5–1 kcal mol^{-1} at 300 K, which gives a substantial contribution to the experimental rate

---

† It is interesting that statistical-weight isotope effects are normal in chemically activated systems but inverse in non-equilibrium thermally activated systems (see section 9.5).[5]

Table 9.2[a]   Calculation of secondary kinetic isotope effects for C_2-C_3 fission in butane and butane-d_6 [reactions (9.5) and (9.6)]

| | E^+/kcal mol^{-1} | | | | |
	0	2	10	40	∞
$\sum P_H(E^+_{vr})$	1	$5\cdot1 \times 10^3$	$1\cdot6 \times 10^7$	$6\cdot4 \times 10^{13}$	—
$\sum P_D(E^+_{vr})$	1	$1\cdot3 \times 10^4$	$8\cdot0 \times 10^7$	$1\cdot1 \times 10^{15}$	—
$\sum P_H(E^+_{vr})/\sum P_D(E^+_{vr})$	$1\cdot00$	0.38	$0\cdot20$	$0\cdot060$	$0\cdot0036$
For $E_0 = 35$ $\{N_H^*(E^*)$	$8\cdot7 \times 10^{10}$	$1\cdot9 \times 10^{11}$	$3\cdot3 \times 10^{12}$	$1\cdot9 \times 10^{16}$	—
$(E^* = E^+ + 35)$ $\{N_D^*(E^*)$	$1\cdot2 \times 10^{12}$	$2\cdot7 \times 10^{12}$	$5\cdot7 \times 10^{13}$	$5\cdot9 \times 10^{17}$	—
$N_D^*(E^*)/N_H^*(E^*)$	14	14	18	31	293
For $E_0 = 85$ $\{N_H^*(E^*)$	$2\cdot1 \times 10^{17}$	$3\cdot4 \times 10^{17}$	$2\cdot0 \times 10^{18}$	$7\cdot0 \times 10^{20}$	—
$(E^* = E^+ + 85)$ $\{N_D^*(E^*)$	$7\cdot5 \times 10^{18}$	$1\cdot2 \times 10^{19}$	$7\cdot9 \times 10^{19}$	$3\cdot7 \times 10^{22}$	—
$N_D^*(E^*)/N_H^*(E^*)$	35	36	40	53	293
$k_{aH}(E^*)/k_{aD}(E^*)$ for $\{E_0 = 35$:	14	5·5	3·5	1·8	1·05
$\{E_0 = 85$:	35	14	7·8	3·1	1·05

[a] Energies are in kcal mol^{-1} and state densities in (kcal mol^{-1})$^{-1}$. In the published frequency assignment[2] for the $C_4 \cdot d_6$ molecule the entry 270 (2) should read 207 (2) cm^{-1}. In addition, the symmetry numbers were *included* (as a factor $1/\sigma$ in the rotational partition function) in calculating the entropies of the active degrees of freedom for all the models. With these provisos the above results (recalculated by the present authors) are in good agreement with those of ref. 2. The $N(E)$ values are shown graphically in Figure 9.5.

ratio. The inferred experimental isotopic rate ratio at constant E^* varies from $k_{aH}/k_{aD} \approx 8$ at the lowest energies studied to $\approx 5 \cdot 5$ at the highest energies, and is reproduced very well by RRKM calculations on the basis

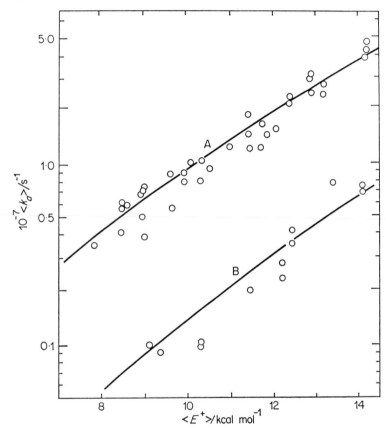

Figure 9.2 Observed rate-constants $\langle k_a \rangle$ for decomposition of chemically activated sec-butyl-d_0 and -d_1 (curve A), and -d_8 and -d_9 (curve B) radicals as a function of their excess energy $\langle E^+ \rangle$

of a pure secondary statistical-weight effect with $(E_0)_H = (E_0)_D = 33$ kcal mol^{-1}.

Similar results have been obtained for the decompositions of chemically activated n-propyl-d_6 radicals[7] and highly deuteriated sec-hexyl and sec-octyl radicals,[8] and for the geometrical and structural isomerizations of cis-1,2-dimethylcyclopropane-d_8.[9] In each case large normal isotope effects were observed and were in reasonable agreement with calculated values.

9.4 PRIMARY KINETIC ISOTOPE EFFECTS ON $k_a(E^*)$ (THEORY AND EXPERIMENT)

For primary isotope effects in monoenergetic systems the appropriate equation is still (9.7), but it is now more difficult to generalize about the nature of the effects because the critical energies $(E_0)_H$ and $(E_0)_D$ are different [usually $(E_0)_H < (E_0)_D$], and different comparisons can arise according to the energy differences involved. Two of the more obvious cases are illustrated in Figures 9.3 and 9.4, which illustrate the energy

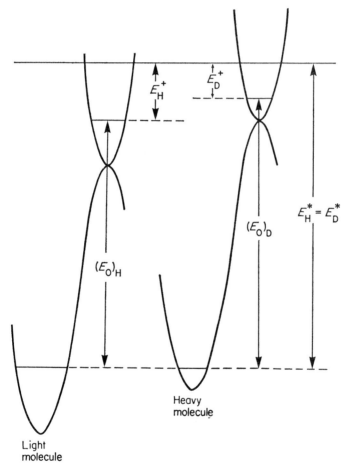

Figure 9.3 Comparative energy diagram for the primary isotope effect $[(E_0)_H < (E_0)_D]$ in monoenergetic systems excited to a common value of E^*

relationships for the cases where $E_H^* = E_D^*$ and $E_H^+ > E_D^+$ (Figure 9.3) or $E_H^+ = E_D^+$ and $E_H^* < E_D^*$ (Figure 9.4).

The first situation is approximately achieved in chemical activation experiments where a fixed amount of energy E^* is released in the molecule

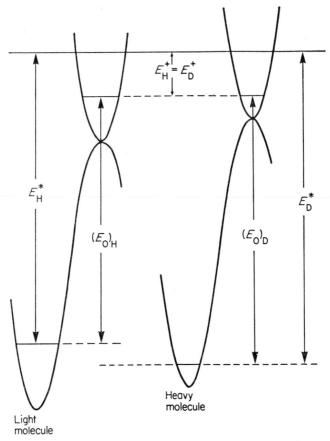

Figure 9.4 Comparative energy diagram for the primary isotope effect in monoenergetic systems excited to a common value of E^+

by a reaction which is energetically unaffected by the isotopic substitution. An example would be reaction (9.13) and the corresponding perdeuterio case, although it will be seen later that even this simple comparison is in practice confused by differences in the thermal energy distributions of the species involved. Similar remarks apply to the photoactivation of isotopic molecules by light of the same wavelength. The comparison is illustrated in Table 9.3 by some calculated data for reactions (9.3) and (9.4);[2] these show the salient features, although it would be difficult to study the reactions experimentally in the forward direction. The tabulated rate ratios can be considered to arise from the combination of three effects.

Table 9.3[a] Calculated primary isotope effect for C_2H_5-H and C_2H_5-D fission [reactions (9.3) and (9.4)] with $E_H^* = E_D^*$

	Data for $(E_0)_D = 35\cdot0$, $(E_0)_H = 33\cdot6$				Data for $(E_0)_D = 85\cdot0$, $(E_0)_H = 83\cdot6$			
E^*	35	40	55	∞	85	90	105	∞
E_D^+	0	5	20	∞	0	5	20	∞
E_H^+	1·4	6·4	21·4	∞	1·4	6·4	21·4	∞
$\dfrac{k_{aH}(E^*)}{k_{aD}(E^*)}$	6·1	3·5	2·0	1·4	7·0	3·9	2·2	1·4

[a] Energies in kcal mol^{-1}; for further details see ref. 2.

Firstly, as noted previously, $N_D^*(E^*)$ is always greater than $N_H^*(E^*)$ (for a common E^*), giving a *normal* statistical-weight contribution. Secondly, for a given E^+ (common to both cases) $\sum P_D(E_{vr}^+)$ would be greater than $\sum P_H(E_{vr}^+)$, tending to give an *inverse* effect. Thirdly, the difference in E_0 makes $E_H^+ > E_D^+$ (by 1·4 kcal mol^{-1} in the present case), and this tends to offset the statistical-weight effect on $\sum P(E_{vr}^+)$. At low E^+ in the present case, the energy difference more than compensates for the statistical-weight effect on $\sum P(E_{vr}^+)$, and the $\sum P(E_{vr}^+)$ terms in (9.7) reinforce the normal isotope effect arising from the $N^*(E^*)$ ratio. At high energies the statistical-weight effect is predominant in the $\sum P(E_{vr}^+)$ ratio, and this offsets the effect of the $N^*(E^*)$ ratio. In all cases, however, a normal isotope effect is predicted.

The second relatively simple case is that in which $E_H^+ = E_D^+$ and therefore $E_D^* > E_H^*$. The isotope effect is still normal, but the rate ratio is generally smaller than in the previous case (see Table 9.4). This is because the effect of a 1·4 kcal mol^{-1} energy difference on $N^*(E^*)$ at relatively high E^* is much less than that of the same energy difference on $\sum P(E_{vr}^+)$ at relatively low E^+.

A practical example of isotope effects in 'monoenergetic' systems is provided by the studies of Setser and co-workers[10] on HCl/DCl elimination from chemically activated C_2H_5Cl/C_2D_5Cl and $CH_2ClCH_2Cl/CD_2ClCD_2Cl$, e.g. (9.13).

$$CH_3 + CH_2Cl \longrightarrow CH_3CH_2Cl^* \begin{array}{c} \xrightarrow{M} C_2H_5Cl \\ \\ \searrow C_2H_4 + HCl \end{array} \qquad (9.13)$$

It might be expected that the combination of CD_3 and CD_2Cl radicals would provide $C_2D_5Cl^*$ at the same E^* as the $C_2H_5Cl^*$ produced in (9.13), so that the situation would be analogous to that in Table 9.3. This would be true at the absolute zero, but at the experimental temperature of 298 K the thermal energies in the active degrees of freedom of the energized molecules (see Chapter 8) differ by an amount which is coincidentally similar to the critical energy difference. The result is a situation in which $\langle E_H^+ \rangle \approx \langle E_D^+ \rangle$, similar to that in Table 9.4. The relevant energies are shown in Table 9.5, together with the observed and calculated isotopic

Table 9.4[a]   Calculated primary isotope effect for C_2H_5—H and C_2H_5—D fission [reactions (9.3) and (9.4)] with $E_H^+ = E_D^+$ and $(E_0)_H = 35 \cdot 0$, $(E_0)_D = 36 \cdot 4$ kcal mol^{-1}

E^+	0	5	20	∞
E_H^*	35	40	55	∞
E_D^*	36·4	41·4	56·4	∞
$\dfrac{k_{aH}(E_H^*)}{k_{aD}(E_D^*)}$	2·2	1·7	1·6	1·4

[a] Energies in kcal mol^{-1}; for further details see ref. 2.

Table 9.5[a]   Primary isotope effects in the decompositions of chemically activated ethyl chloride and 1,2-dichloroethane molecules

	C_2H_5Cl	C_2D_5Cl	CH_2ClCH_2Cl	CD_2ClCD_2Cl
Critical energy E_0	55·0	56·4	55·0	56·3
$\langle E^* \rangle$ (calc.)	91·0	92·5	88·3	89·3
$\langle E^+ \rangle$ (calc.)	36·0	36·1	33·3	33·0
k_{aH}/k_{aD} (calc.)		3·0		2·8
k_{aH}/k_{aD} (obs.)		3·3		3·5

[a] Energies in kcal mol^{-1}; for further details see ref. 10.

rate ratios; normal isotope effects are observed and their magnitudes are in good agreement with the calculated values. The observed effects in this sort of system contain both critical energy and statistical-weight effects, and for this reason are often referred to as 'mixed primary-secondary effects' (cf. section 9.1). Another similar case is the decomposition of the isopropyl-d_6 radical.[7]

An interesting simplification occurs for *intramolecular* isotope effects in monoenergetic systems, e.g. reactions (9.14). For such a pair of reactions

$$C_2H_5D^* \underset{\searrow C_2H_5+D}{\overset{\nearrow C_2H_4D+H}{}} \qquad (9.14)$$

the $N^*(E^*)$ terms in (9.7) are identical and the rate ratio (corrected for the statistical factor of 5) is given simply by

$$k_{a\text{H}}(E_\text{H}^*)/k_{a\text{D}}(E_\text{D}^*) = \sum P_\text{H}(E_{\text{vr}}^+)/\sum P_\text{D}(E_{\text{vr}}^+)$$

The dominant effect here is the critical energy difference, and the calculated rate ratio per bond varies[2] from 4·0 at $E_\text{D}^+ = 0$ to 1·4 at $E_\text{D}^+ \to \infty$; again a normal isotope effect is predicted.

Such effects have been measured experimentally for the decompositions of chemically activated ethyl-d_1, -d_2 and -d_3 radicals produced by the addition of H or D atoms to C_2H_4 or trans-CHD:CHD,[11,12] e.g.:

$$D + CHD:CHD \longrightarrow CHD_2\overset{\bullet}{C}HD^* \begin{array}{l} \xrightarrow{\text{M}} CHD_2\overset{\bullet}{C}HD \\ \longrightarrow D + CHD:CHD \\ \searrow H + CD_2:CHD \end{array}$$

The measured isotopic rate ratios (k_H/k_D per bond) were for ethyl-d_2 10–18 at 195 K and 6–8 at 300 K, and for ethyl-d_3 3·0 at 195 K and 2·0 at 300 K, and were in reasonable agreement with the calculated values. The larger effects for the d_2 case result mainly from the low E^+ values for D-rupture from $CH_2\overset{\bullet}{D}CHD^*$ formed by H-atom addition to CHD:CHD.[11]

9.5 ISOTOPE EFFECTS IN THERMAL UNIMOLECULAR REACTIONS

9.5.1 Theory

It has already been noted that the relatively simple isotope effects for monoenergetic systems become more complex in thermally activated systems because of changes in the energy distribution of the reacting molecules on isotopic substitution. The general equation (9.9) for calculations in the fall-off region is complicated, but considerable insight can be gained from the limiting high-pressure and low-pressure equations (9.10) and (9.11).

Again the discussion is simplest for a purely secondary (statistical-weight) isotope effect. In this case $(E_0)_\text{H} = (E_0)_\text{D}$, and with adiabatic rotation factors ignored, the limiting forms become (9.15) and (9.16):

$$\frac{(k_\infty)_\text{H}}{(k_\infty)_\text{D}} = \frac{Q_{2\text{H}}^+/Q_{2\text{H}}}{Q_{2\text{D}}^+/Q_{2\text{D}}} \tag{9.15}$$

$$\frac{(k_{\text{bim}})_\text{H}}{(k_{\text{bim}})_\text{D}} = \frac{Q_{2\text{H}}^{*\prime}/Q_{2\text{H}}}{Q_{2\text{D}}^{*\prime}/Q_{2\text{D}}} \tag{9.16}$$

Considering first the low-pressure limit (9.16), it is found that both partition functions Q_2 and $Q_2^{*\prime}$ are greater for the deuteriated compound, but that the ratio $Q_{2D}^{*\prime}/Q_{2H}^{*\prime}$ is considerably greater than the ratio Q_{2D}/Q_{2H}. This is because the density of quantum states increases on deuteriation more at the higher energies important in $Q_2^{*\prime}$ than at the lower energies appropriate to Q_2. The effect is clearly seen as a divergence of the $\log[N(E)]$ plots with increasing energy (Figure 9.5), and its possible

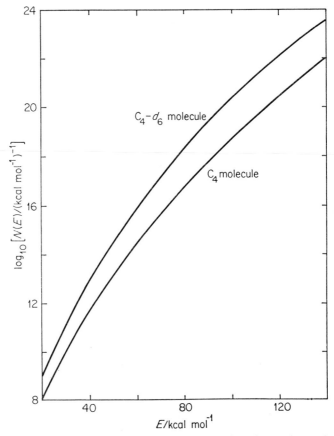

Figure 9.5 State-densities $N(E)$ for the C_4 and C_4-d_6 molecules by the Whitten–Rabinovitch approximation[2]

numerical magnitude is indicated by the figures for reactions (9.5) and (9.6) at 300 °C with $E_0 = 35\ \mathrm{kcal\,mol^{-1}}$:

$$\frac{Q_{2D}}{Q_{2H}} = 4.17, \quad \frac{Q_{2D}^{*\prime}}{Q_{2H}^{*\prime}} = 14.5, \quad \frac{(k_{\mathrm{bim}})_H}{(k_{\mathrm{bim}})_D} \approx 0.29$$

This effect at low pressures is simply a change in the rate of production of energized molecules, all of which react. The rate of energization into a small energy range $E^* \to E^* + \delta E^*$ is given by (4.5):

$$\frac{\delta k_{1(E^* \to E^* + \delta E^*)}}{k_2} = \frac{N^*(E^*) \exp(-E^*/kT)\, \delta E^*}{Q_2}$$

Since $N_D^*(E^*) > N_H^*(E^*)$ and the difference is only partially compensated by the inequality $Q_{2D} > Q_{2H}$, there are more deuteriated molecules than normal molecules excited into any given energy range and the measured rate of reaction is accordingly greater for the deuterio-compound. Thus an *inverse* isotope effect is predicted as a result of statistical-weight effects for unimolecular reactions in the non-equilibrium situation pertaining at low pressures. Such an effect contrasts with the normal statistical-weight iso-tope effect found in chemical activation experiments (section 9.3); it was originally predicted by Rabinovitch and co-workers[13] and has subsequently been found experimentally in several systems (see next section). It has been emphasized[14, 15] that this is purely a statistical-weight effect and can be very much larger than the 'conventional secondary isotope effects', which are usually measured under equilibrium conditions and are due mainly to critical energy differences (cf. section 9.1).

For the high-pressure limit (9.15) may be used. Both Q_2 and Q_2^+ are larger for the deuterio-compound, but this time the effects are roughly compensating. At 300 °C, $Q_{2D}^+/Q_{2H}^+ = 4\cdot 8$ for reactions (9.5) and (9.6), giving $(k_\infty)_H/(k_\infty)_D = 0\cdot 86$, a small inverse effect compared with the ratio $0\cdot 29$ at low pressures. It appears that at high pressures the statistical-weight isotope effect may be either normal or inverse but that the rate ratio will in any event be close to unity; the experimental results discussed in the next section are in line with this conclusion. The situation can alternatively be discussed in terms of the equation (9.17):

$$k_\infty = \int_{E^* = E_0}^{\infty} k_a(E^*)\, \frac{dk_{1(E^* \to E^* + dE^*)}}{k_2} \tag{9.17}$$

For any given energy range $E^* \to E^* + \delta E^*$ it has been seen that $k_{aH}(E^*)/k_{aD}(E^*)$ is substantially greater than unity, while $(\delta k_1/k_2)_H/(\delta k_1/k_2)_D$ is substantially less than unity. The concentration of energized molecules in a given range is higher for the deuterio-compound, but the rate-constant for their reaction is lower, resulting in about the same overall rate of reaction.

Finally, when primary isotope effects are present there is usually a general increase in $(k_{uni})_H/(k_{uni})_D$ at all pressures relative to the curve predicted from the statistical-weight effect alone. Values substantially greater than unity may be found at high pressure, and although the ratio will decline as the pressure is reduced it may or may not invert (i.e. become less than unity) at low pressures. These trends are illustrated by experimental data discussed in the next section.

9.5.2 Experimental studies of isotope effects in thermal unimolecular reactions

The most interesting studies of isotope effects in thermal unimolecular reactions are those which extend over a substantial pressure range, and the results of such studies are summarized in Table 9.6. Prominent among these are the studies on the isomerizations of alkyl isocyanides and their deuteriated forms. These have the useful feature that the measured activation energies are all the same within experimental error; the isotope effects are thus purely secondary statistical-weight effects and comparisons are greatly facilitated. The most extensive studies have been those on the CD_3NC/CH_3NC reactions,[14,16] and Figure 9.6 shows the experimental

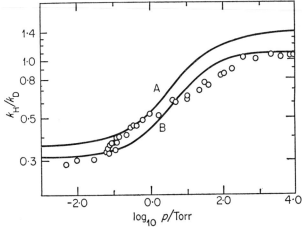

Figure 9.6 Isotope effect for CH_3NC/CD_3NC as a function of pressure.[14] Curves A and B are for the same theoretical model with $(E_0)_H - (E_0)_D = -0.1$ and $+0.1$ kcal mol^{-1} respectively

ratio $k_{uni}(CH_3NC)/k_{uni}(CD_3NC)$ plotted as a function of pressure, together with two of the calculated curves. As predicted in the previous section, there is only a weak isotope effect at high pressures $[(k_\infty)_H/(k_\infty)_D \approx 1.1]$ but a large inverse effect at low pressures $[(k_{bim})_H/(k_{bim})_D \approx 0.3]$. The results are fitted very impressively by several models based on effectively zero difference in critical energy:

$$(E_0)_H - (E_0)_D = \pm 0.1 \text{ kcal mol}^{-1}$$

Table 9.6 Isotope effects in thermally activated unimolecular reactions studied over a range of pressures

Compound	Isotopes involved	Temperature range (°C)	$\Delta E_\infty/$ kcal mol^{-1} (a)	Temperature (°C)	k(light)/k(heavy) at pressures (Torr) in parentheses	Ref.
Secondary intermolecular effects						
MeNC-d_1	H/D	—	(0·0)	245	1·0 (10^4)	0·7 (0·1)	17
MeNC-d_3	H/D	180–250	0·0	230	1·1 (10^4)	0·3 (0·005)	14, 16
EtNC-d_5	H/D	190–260	0·0	231	1·1 (90)	0·2 (0·007)	15
Cyclobutane-d_8	H/D	419–460	0·9	449	1·4 (100)	0·8 (0·005)	18
Cyclobutene-d_6	H/D	140–180	1·1	151	1·3 (10)	0·8 (0·01)	19
Azomethane-d_6	H/D	256–297	(−4·8)^b	—	—	—	b
Primary intermolecular effects						
MeNC-^{13}C	^{12}C/^{13}C	226–243	0·02	226	1·018 (10^3)	1·011 (10)^c	23
Cyclopropane-^{13}C	^{12}C/^{13}C	450–514	0·019	514	1·012 (10^3)	1·003 (1)^{d,e}	24
Cyclopropane-d_6	H/D	407–514	1·3	510	2·0 (760)	1·0 (0·01)	21, 22
Cyclopropane-t_1	H/T	406–492	0·4	490	1·1 (10^3)	1·0 (0·5)^{d,f}	f
C_2D_5Br	H/D	458–692	2·3	553	2·0 (115)	1·8 (6)^g	25
C_2D_5Cl	H/D	439–480	2·1	439	2·0 (6)	2·7 (0·2)^h	26
CD_3CH_2Cl	H/D	439–482	1·9	439	1·9 (10)	1·8 (0·2)^{h,i}	27
Primary intramolecular effects						
CHD_2CD_2Cl	H/D	485–716	1·0	559	2·1 (150)	2·4 (0·7)^j	28
CHD_2CD_2Br	H/D	425–726	0·9	553	2·1 (200)	2·2 (7)^{g,j}	25

^{a} E_{Arr}(heavy)$-E_{\mathrm{Arr}}$(light) at a reasonably high pressure. ^{b} Isotopic comparisons probably invalid because of inaccurate data for azomethane itself; see Do-Ren Chang and O. K. Rice, *Internat. J. Chem. Kinetics*, **1**, 171 (1969). ^{c} Ratios for CH$_3$NC vs the mixture of ^{13}CH$_3$NC and CH$_3$N^{13}C molecules. ^{d} Similar trends were also measured at other temperatures. ^{e} Observed ratios for ^{12}C—^{12}C vs ^{13}C—^{12}C fission derived from the measured rate ratios. ^{f} No apparent temperature dependence. Ratios are those observed for the whole molecules (C$_3$H$_6$ vs C$_3$H$_5$T). R. H. Lindquist and G. K. Rollefson, *J. Chem. Phys.*, **24**, 725 (1956); R. E. Weston, *J. Chem. Phys.*, **26**, 975 (1957). ^{g} Results for inhibited reaction. ^{h} Doubtful results; see text. ^{i} Ratios calculated from experimental points of refs. 26 and 27, not from the 'smoothed' results. ^{j} Isotope effect per β-CH or CD bond.

The corresponding experimental curves for all the isocyanide isomerizations are shown in Figure 9.7 and have been discussed in detail by their originators.[14-17] The fall-off curves for the individual reactions are shown in Figure 9.8; the regular variation from CH_3NC to CH_2DNC and CD_3NC is evidence against the harmonic Slater theory, since on that theory the

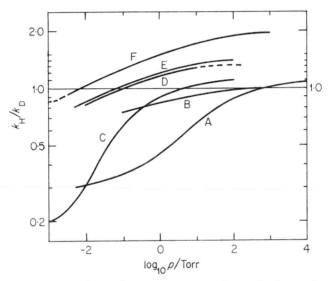

Figure 9.7 Smooth-curve representations of observed isotopic rate ratios for (A) CH_3NC/CD_3NC, (B) CH_3NC/CH_2DNC, (C) C_2H_5NC/C_2D_5NC, (D) cyclobutene/cyclobutene-d_6, (E) cyclobutane/cyclobutane-d_8 and (F) cyclopropane/cyclopropane-d_6

different symmetry classification of CH_2DNC should give rise to a substantially increased value of Slater's n, which is not evident in the experimental results.[17]

Deuterium isotope effects in the decomposition of cyclobutane[18] and the isomerization of cyclobutene[19] are also nominally secondary effects, although there are now significant differences of critical energy which emphasize the somewhat arbitrary nature of a classification into primary and secondary effects according to whether the atoms substituted are 'directly involved in the reaction' or not. The difference in critical energy $[(E_0)_D > (E_0)_H]$ results in a more pronounced normal isotope effect at high pressures (Figure 9.7); however, the ratio $(k_{uni})_H/(k_{uni})_D$ still falls as the pressure is decreased, and inverts at pressures below about 0·01 Torr. Calculated fall-off curves were in good agreement with experiment.[19, 20] It has been found[19] that the inversion pressure and the rate-constants at that pressure depend more critically on the assumed structure of the activated complex than do the individual fall-off curves. These parameters may thus provide a more sensitive probe of activated-complex structure, although it is not clear whether this would still be the case for models which gave a good fit to the experimental k_∞ and fall-off curves.

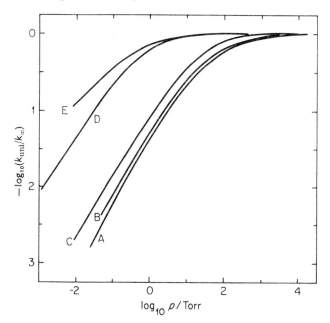

Figure 9.8 Fall-off curves at 230 °C for (A) CH_3NC, (B) CH_2DNC (corrected from experimental results at 245 °C[17]), (C) CD_3NC, (D) C_2H_5NC and (E) C_2D_5NC

Genuine primary isotope effects tend to be even more marked (Table 9.6), and although they fall with decreasing pressure because of the usual non-equilibrium statistical-weight effect, they may not invert at all in the range of pressures which is experimentally accessible. In the cyclopropane-d_6 case, for example, $(k_{uni})_H/(k_{uni})_D$ was measured as 2·0 at high pressures,[21] and the calculated low-pressure limit is about 0·2.[13, 20] The inversion was only just observed (Figure 9.7) because energization at the walls became predominant below 0·01 Torr in the 12 litre reaction vessel used.[22]

The ^{13}C isotope effects in CH_3NC and cyclopropane,[23, 24] although small in magnitude, have been measured with considerable accuracy and display the same sort of pressure dependence as the related deuterium effects.

The primary intermolecular effect in C_2D_5Br behaves in the expected manner,[25] but that for C_2D_5Cl displays the opposite trend with pressure[26] and the results must be considered doubtful (see section 7.2.4). The isotopic rate ratios for CD_3CH_2Cl were based[27] on the same results for C_2H_5Cl and may also be inaccurate.

The primary intramolecular effects in the thermal decompositions of $CHD_2CD_2Cl^{28}$ and $CHD_2CD_2Br^{25}$ (Table 9.6) are similar in their magnitude at high pressure to the intermolecular effects discussed above, but increase as the pressure is decreased. In this case, however, such behaviour is expected. It arises[28] from the reduction in the concentration of energized molecules at the higher energies as the pressure is decreased. This has a proportionately larger effect on the rate of D-elimination, because the complexes for D-elimination have a higher average energy than those for H-elimination. In the case of intramolecular comparison there is no compensating shift in the relative concentration of isotopic energized molecules, and the rate-constant for D-elimination therefore decreases faster with pressure than that for H-elimination.

Finally, for the sake of completeness, some more mechanistically directed studies of isotope effects in unimolecular reactions (usually near the high-pressure limit) are listed below.

Intermolecular effects

$\overline{CH_2CHDCHD}^{40}$ $\overline{CHMeCHDCHD}^{41}$

$CDCl_3{}^{29}$ $C_2D_5OAc^{30}$ $(CD_2)_2O^{31}$

deuteriated pyrazolines $[\overline{CH_2CH_2CH_2N:N}]^{32}$
deuteriated cyclopentenes[33]
$PhCH:CHCMe_2CO_2H$ (H/D and $^{14}C/^{12}C$ effects)[34]
$(CO_2D)_2{}^{35}$ $(Me_2CDCH_2)_3Al^{42}$

Intramolecular effects

$\overline{CH_2CH_2CD_2CD_2}^{36}$
$CHD_2CD_2OAc^{37}$ $CH_2DCH_2Cl^{27}$
$(CO_2D)_2$ ($^{13}C/^{12}C$ effect)[38] $CH_2(CO_2H)_2$ ($^{13}C/^{12}C$ effect)[39]

References

1. J. Bigeleisen and M. Wolfsberg, *Advan. Chem. Phys.*, **1**, 15 (1958); L. Melander, *Isotope Effects on Reaction Rates*, Ronald, New York, 1960; K. J. Laidler, *Chemical Kinetics*, 2nd edn, McGraw–Hill, New York, 1965, pp. 90–98; K. B. Wiberg, *Chem. Rev.*, **55**, 713 (1955).
2. B. S. Rabinovitch and D. W. Setser, *Advan. Photochem.*, **3**, 1 (1964).

3. E. B. Wilson, J. C. Decius and P. C. Cross, *Molecular Vibrations*, McGraw–Hill, New York, 1955, section 8.5.

4. W. E. Buddenbaum and P. E. Yankwich, *J. Phys. Chem.*, **71**, 3136 (1967).

5. B. S. Rabinovitch and J. H. Current, *Can. J. Chem.*, **40**, 557 (1962).

6. J. W. Simons, B. S. Rabinovitch and R. F. Kubin, *J. Chem. Phys.*, **40**, 3343 (1964), and references cited therein.

7. W. E. Falconer, B. S. Rabinovitch and R. J. Cvetanović, *J. Chem. Phys.*, **39**, 40 (1963).

8. M. J. Pearson, B. S. Rabinovitch and G. Z. Whitten, *J. Chem. Phys.*, **42**, 2470 (1965).

9. J. W. Simons and B. S. Rabinovitch, *J. Phys. Chem.*, **68**, 1322 (1964).

10. K. Dees and D. W. Setser, *J. Chem. Phys.*, **49**, 1193 (1968), and references cited therein.

11. J. H. Current and B. S. Rabinovitch, *J. Chem. Phys.*, **38**, 783 (1963), and references cited therein.

12. J. H. Current and B. S. Rabinovitch, *J. Chem. Phys.*, **38**, 1967 (1963).

13. B. S. Rabinovitch, D. W. Setser and F. W. Schneider, *Can. J. Chem.*, **39**, 2609 (1961).

14. F. W. Schneider and B. S. Rabinovitch, *J. Amer. Chem. Soc.*, **85**, 2365 (1963).

15. K. M. Maloney, S. P. Pavlou and B. S. Rabinovitch, *J. Phys. Chem.*, **73**, 2756 (1969).

16. F. J. Fletcher, B. S. Rabinovitch, K. W. Watkins and D. J. Locker, *J. Phys. Chem.*, **70**, 2823 (1966).

17. B. S. Rabinovitch, P. W. Gilderson and F. W. Schneider, *J. Amer. Chem. Soc.*, **87**, 158 (1965).

18. R. W. Carr and W. D. Walters, *J. Amer. Chem. Soc.*, **88**, 884 (1966).

19. H. M. Frey and B. M. Pope, *Trans. Faraday Soc.*, **65**, 441 (1969).

20. M. C. Lin and K. J. Laidler, *Trans. Faraday Soc.*, **64**, 927 (1968).

21. A. T. Blades, *Can. J. Chem.*, **39**, 1401 (1961).

22. B. S. Rabinovitch, P. W. Gilderson and A. T. Blades, *J. Amer. Chem. Soc.*, **86**, 2994 (1964); for a theoretical treatment of wall energization effects see K. M. Maloney and B. S. Rabinovitch, *J. Phys. Chem.*, **72**, 4483 (1968).

23. J. F. Wettaw and L. B. Sims, *J. Phys. Chem.*, **72**, 3440 (1968).

24. L. B. Sims and P. E. Yankwich, *J. Phys. Chem.*, **71**, 3459 (1967).

25. A. T. Blades, P. W. Gilderson and M. G. H. Wallbridge, *Can. J. Chem.*, **40**, 1533 (1962); A. T. Blades, *Can. J. Chem.*, **36**, 1043 (1958).

26. H. Heydtmann and G. W. Völker, *Zeit. Phys. Chem. (Frankfurt)*, **55**, 296 (1967).

27. G. W. Völker and H. Heydtmann, *Zeit. Naturforsch. (B)*, **23**, 1407 (1968).

28. A. T. Blades, P. W. Gilderson and M. G. H. Wallbridge, *Can. J. Chem.*, **40**, 1526 (1962).

29. A. E. Shilov and R. D. Sabirova, *Russ. J. Phys. Chem.*, **34**, 408 (1960).

30. A. T. Blades and P. W. Gilderson, *Can. J. Chem.*, **38**, 1407 (1960).

31. M. L. Neufeld and A. T. Blades, *Can. J. Chem.*, **41**, 2956 (1963).

32. B. H. El-Sader and R. J. Crawford, *Can. J. Chem.*, **46**, 3301 (1968); M. P. Schneider and R. J. Crawford, *Can. J. Chem.*, **48**, 628 (1970).

33. D. A. Knecht, Ph.D. Thesis., University of Rochester, 1968 [*Diss. Abs. (B)*, **29**, 1627 (1968)].

34. D. B. Bigley and J. C. Thurman, *J. Chem. Soc.* (*B*), 1076 (1967).
35. G. Lapidus, D. Barton and P. E. Yankwich, *J. Phys. Chem.*, **70**, 407 (1966).
36. R. Srinivasan and S. M. E. Kellner, *J. Amer. Chem. Soc.*, **81**, 5891 (1959).
37. A. T. Blades and P. W. Gilderson, *Can. J. Chem.*, **38**, 1401 (1960).
38. G. Lapidus, D. Barton and P. E. Yankwich, *J. Phys. Chem.*, **70**, 1575 (1966); T. T.-S. Huang, W. J. Kass, W. E. Buddenbaum and P. E. Yankwich, *J. Phys. Chem.*, **72**, 4431 (1968).
39. P. C. Chang and C. R. Gatz, *J. Phys. Chem.*, **72**, 2602 (1968).
40. E. W. Schlag and B. S. Rabinovitch, *J. Amer. Chem. Soc.*, **82**, 5996 (1960).
41. D. W. Setser and B. S. Rabinovitch, *J. Amer. Chem. Soc.*, **86**, 564 (1964).
42. K. W. Egger, *Internat. J. Chem. Kinetics*, **1**, 459 (1969).

10 Collisional Energy Transfer in Unimolecular Reaction Systems

The basic RRKM theory, like the earlier HRRK and Slater theories, treats the collisional energization and de-energization as essentially single-step processes, i.e. it makes the strong-collision assumption. In thermal unimolecular reactions the reacting molecules have average energies which are typically less than $10 \, kcal \, mol^{-1}$ above the critical energy. It will be seen later that many (but not all) colliding molecules are able to remove this amount of energy on a single collision, so that the strong-collision assumption will often (but not always) be a reasonable approximation for thermal unimolecular reactions. In chemically activated systems, energized molecules may be formed with much higher excess energies (e.g. methylene plus ethylene gives cyclopropane with $E^{+} \approx 40 \, kcal \, mol^{-1}$) and the strong-collision assumption is much more likely to break down. Thus for both thermal and chemical-activation systems it is necessary to develop a treatment of the case where several collisions are needed to energize or de-energize a molecule. This can only be done by considering in detail the unimolecular reaction from each energy level, together with all the collisional processes which transfer molecules from one energy level to another. Calculations of this type are described in the present chapter. The detailed mathematics is difficult, and emphasis is therefore placed on the principles involved and the conclusions which are most important for practical application of the RRKM theory; further details may be sought in the references cited.

10.1 GENERAL EQUATIONS[1-4, 8]

Consider a system in which molecules are distributed among a series of energy levels with a concentration n_i of the molecules in the ith energy level. The processes by which molecules enter or leave this energy level are shown in Figure 10.1. Unimolecular reaction occurs with a rate-constant k_i for molecules with energy $\geqslant E_0$; k_i will usually be calculated from the RRKM treatment [equation (4.16) for k_a or (4.31) for k_{EJ}]. If levels below E_0 are included in a calculation, then $k_i = 0$ for these levels. Next, there

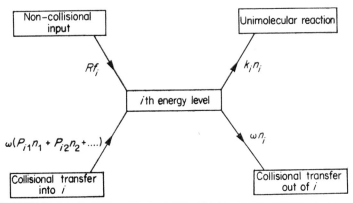

Figure 10.1   Transport of molecules into and out of a given energy level

may be a steady production of reactant molecules in certain energy levels, e.g. by chemical activation or photoactivation. The rate of such non-collisional input of molecules into the ith level is expressed as Rf_i, where R is the total rate into all levels and f_i is the fraction entering the ith level. Finally, there is collisional transfer of molecules into and out of each level. Calculations can often be simplified by considering a system in which the reacting molecules are dispersed in a large excess of inert molecules, so that only collisions with the 'bath' molecules need to be considered. Collisions with the bath molecules occur with a readily calculable first-order rate-constant ω which is equivalent to the $k_2[M]$ or Zp of the earlier chapters [equation (6.21)]. Transition probabilities P_{lm} are defined as the fraction of collisions starting at level m which end at level l; it is helpful to think of '$P_{\text{into }l\text{ from }m}$' or '$P_{l\leftarrow m}$'. Since all collisions from level m must end at some level, the conditions (10.1) are necessary for consistency or

$$\sum_l P_{lm} = 1 \quad \text{for each level } m \tag{10.1}$$

'completeness' of the scheme. Referring again to Figure 10.1, the rate of collisional transfer out of level i is thus $(\omega P_{1i} + \omega P_{2i} + \ldots)n_i = \omega n_i$, while the rate of transfer into i is $(\omega P_{i1}n_1 + \omega P_{i2}n_2 + \ldots + \omega P_{ij}n_j + \ldots)$. When collisions with more than one sort of molecule must be considered, each P_{lm} is taken as a weighted mean of the P_{lm} for the individual bath molecules (see for example ref. 5).

If now the system is assumed to be in the steady state, the processes shown in Figure 10.1 just balance, and (10.2) follows:

$$\frac{dn_i}{dt} = Rf_i - k_i n_i - \omega n_i + \omega \sum_j P_{ij} n_j$$

$$= 0 \quad \text{for steady state} \tag{10.2}$$

A similar equation can be written for each energy level, (10.3), and the complete set of equations is readily expressible in the compact matrix notation shown in (10.4). The meanings of $\mathbf{F}$, $\mathbf{K}$, etc. are indicated in (10.4).

$$\left.\begin{array}{l} Rf_1 - k_1 n_1 - \omega n_1 + \omega(P_{11} n_1 + P_{12} n_2 + P_{13} n_3 + \ldots) = 0 \\ Rf_2 - k_2 n_2 - \omega n_2 + \omega(P_{21} n_1 + P_{22} n_2 + P_{23} n_3 + \ldots) = 0 \\ \quad \cdot \quad \cdot \quad \cdot \quad \cdot \quad \cdot \quad \cdot \quad \cdot \quad \cdot \quad \cdot \quad \cdot \quad \cdot \\ \quad \cdot \quad \cdot \quad \cdot \quad \cdot \quad \cdot \quad \cdot \quad \cdot \quad \cdot \quad \cdot \quad \cdot \quad \cdot \\ Rf_i - k_i n_i - \omega n_i + \omega(P_{i1} n_1 + P_{i2} n_2 + P_{i3} n_3 + \ldots) = 0 \end{array}\right\} \tag{10.3}$$

$$R\begin{bmatrix} f_1 \\ f_2 \\ \vdots \\ f_i \\ \cdot \end{bmatrix} - \begin{bmatrix} k_1 & & & \\ & k_2 & & \\ & & \ddots & \\ & & & k_i \\ & & & & \cdot \end{bmatrix}\begin{bmatrix} n_1 \\ n_2 \\ \vdots \\ n_i \\ \cdot \end{bmatrix} - \omega\begin{bmatrix} 1 & & & \\ & 1 & & \\ & & \ddots & \\ & & & 1 \\ & & & & \cdot \end{bmatrix}\begin{bmatrix} n_1 \\ n_2 \\ \vdots \\ n_i \\ \cdot \end{bmatrix}$$

$$+ \omega\begin{bmatrix} P_{11} & P_{12} & P_{13} & \cdots \\ P_{21} & P_{22} & P_{23} & \cdots \\ \vdots & \vdots & \vdots & \vdots \\ P_{i1} & P_{i2} & P_{i3} & \cdots \\ \cdot & \cdot & \cdot & \cdot \end{bmatrix}\begin{bmatrix} n_1 \\ n_2 \\ \vdots \\ n_i \\ \cdot \end{bmatrix} = \begin{bmatrix} 0 \\ 0 \\ \vdots \\ 0 \\ \cdot \end{bmatrix} \left.\phantom{\begin{bmatrix} 0 \\ 0 \\ \vdots \\ 0 \\ \cdot \end{bmatrix}}\right\} \tag{10.4}$$

or

$$\mathbf{RF} - \mathbf{KN} - \omega\mathbf{IN} + \omega\mathbf{PN} = 0$$

The column vectors $\mathbf{F}$ and $\mathbf{N}$ have as many elements as there are energy levels to be considered; $\mathbf{F}$ gives the fractional rates of input and $\mathbf{N}$ the steady-state populations of the various levels. $\mathbf{K}$, $\mathbf{I}$ and $\mathbf{P}$ are square matrices of the same dimension, $\mathbf{K}$ being a diagonal matrix of the unimolecular rate-constants, $\mathbf{I}$ the corresponding unit matrix and $\mathbf{P}$ the complete transition probability matrix for the system. The matrices $\mathbf{F}$, $\mathbf{K}$,

I and P can be constructed numerically for any given model of the system involved. Equation (10.4) can be written in the form:

$$RF = [K + \omega(I - P)]N$$
$$= JN$$

where

$$J = K + \omega(I - P) \tag{10.5}$$

and the steady-state population distribution $N (\equiv \{n_1 \; n_2 \; ... \; n_i \; ...\})$ can therefore be obtained as (10.6). In (10.6) J^{-1} is the inverse of the matrix J

$$N = RJ^{-1}F \tag{10.6}$$

formed by matrix addition of K to $\omega(I - P)$; this is a matrix such that $J^{-1}J = I$, and can be obtained by standard techniques of matrix algebra. The total rate of unimolecular reaction is given in various equivalent forms in (10.7) and may be calculated from K, J and F by means of the equivalent expressions (10.8). In these equations $(X)_i$ denotes the ith element of the

$$Rate = \sum_i k_i n_i \equiv \sum_i (KN)_i \equiv uKN \tag{10.7}$$

$$= R \sum_i (KJ^{-1}F)_i \equiv RuKJ^{-1}F \tag{10.8}$$

vector X, e.g. $(KN)_i \equiv k_i n_i$, and u is a row vector of unit elements, $u = [1 \; 1 \; 1 \; ... \; 1]$, which on multiplication into a column vector such as KN produces the sum of its elements as the result.

The rate of reaction thus calculated can be related to experimentally observable quantities by calculation of the D/S ratio for chemical activation or the quantum yield for photoactivation (see section 10.3). For thermal reactions a slight extension of the treatment is required in order to calculate the observable quantity k_{uni}, and this will be dealt with in section 10.4.

Equations (10.5)–(10.8) could in principle be applied as such to the whole range of energy levels involved, but the matrices would become very large in most cases. A useful advance can be made by considering separately the energy levels above and below E_0. The matrices and vectors can then be partitioned into blocks some of which are zero, thus simplifying the calculations. The diagonal rate-constant matrix K [cf. (10.4)] now takes the form (10.9), in which the submatrix k contains the k_i for energies above

$$K = \begin{bmatrix} 0 & 0 \\ 0 & k \end{bmatrix} \tag{10.9}$$

E_0 and the leading 0 corresponds to the zero k_i for energies below E_0. The transition probability matrix P is correspondingly factored as in

(10.10) into blocks representing energizing or de-energizing collisions and

$$P = \begin{bmatrix} P_1 & P_2 \\ P_3 & P_4 \end{bmatrix} = \begin{bmatrix} \begin{array}{c} \text{'inactive' transitions} \\ l < E_0, \quad m < E_0 \\ \hline \text{energization} \\ l \geqslant E_0, \quad m < E_0 \end{array} & \begin{array}{c} \text{de-energization} \\ l < E_0, \quad m \geqslant E_0 \\ \hline \text{'active' transitions} \\ l \geqslant E_0, \quad m \geqslant E_0 \end{array} \end{bmatrix} \quad (10.10)$$

transitions above or below the critical energy. J can now be written in the simplified form (10.11). The column vector F representing the fractional

$$J = \begin{bmatrix} J_1 & J_2 \\ J_3 & J_4 \end{bmatrix} = \begin{bmatrix} \omega(I - P_1) & -\omega P_2 \\ -\omega P_3 & k + \omega(I - P_4) \end{bmatrix} \quad (10.11)$$

input into the various levels is expressed as (10.12) in which f_1 (levels below E_0) or f_2 (levels above E_0) may be zero for particular cases (see later).

$$F = \begin{bmatrix} f_1 \\ f_2 \end{bmatrix} \quad (10.12)$$

For numerical calculations with even moderately complex molecules it is still impracticable to consider separately each quantized energy level of the reacting molecules. The energy levels are therefore usually grouped or 'grained' into blocks of finite energy spread, each block being represented as a quasiquantized level with an appropriate degeneracy and average rate-constant k_i. Exact inversion of the J matrix can then be carried out in some cases, but iterative methods are necessary when the graining is reasonably fine, e.g.[1] with step sizes below 0·3 kcal mol^{-1}.

10.2 TRANSITION PROBABILITY MODELS[1,5]

In order to solve the equations developed above, it is necessary to set up a proposed matrix of transition probabilities P_{lm} for the energy levels of interest. The values must satisfy (10.1) for self-consistency, and must also be such as to satisfy the principle of detailed balancing for the system at equilibrium. The rate of transfer to level l from level m is $\omega P_{lm} n_m$ and that to m from l is $\omega P_{ml} n_l$; these are equal at equilibrium and the necessary condition is therefore (10.13), in which g_i is the statistical

$$P_{lm}/P_{ml} = (n_l/n_m)_{\text{eqm}} = (g_l/g_m) \exp\left[-(E_l - E_m)/kT\right] \quad (10.13)$$

weight of level i (see Appendix 2). Under certain conditions it may be reasonable to relax this requirement; for example, calculations for highly excited species are sometimes made with complete neglect of up-transitions.

Apart from these restrictions the calculation can accept any set of P_{lm}, and the usual objective is in fact to discover what models for **P** will reproduce a given set of experimental results. The principal interest lies in the extent of energy transfer at each collision, and the average energy transferred, $\langle \Delta E \rangle$, is thus an important parameter of the calculations. The amount of energy transferred is conveniently expressed for computational purposes as the number of 'quanta' transferred in the quasi-quantized scheme discussed above, i.e.:

$$\Delta E \equiv m \text{ 'quanta'}$$

$$\langle \Delta E \rangle \equiv \bar{m} \text{ 'quanta'}$$

The distribution of the energies removed is also of interest, and various models have been investigated in this respect. In the *stepladder* model, collision is assumed to remove or add one and only one 'quantum' of energy. The relative probabilities of up-and-down transitions from level i are determined by (10.1) and (10.13), so that $P_{i+1,i} + P_{i-1,i} = 1$ and $P_{i+1,i}$ is somewhat less than $P_{i-1,i}$. For the *exponential distribution* the probability of the removal of m 'quanta' is proportional to $\exp(-m/\bar{m})$ and the relative probability of the addition of m quanta follows from (10.13) applied to the whole matrix. Two less-used models involve similarly the *Poisson distribution* $\bar{m}^m/m!$ and the *Gaussian* or *normal distribution*

$$\exp[-(m-\bar{m})^2/2\sigma^2]$$

The last three of these distributions of transition probability allow the occurrence of some large energy transfers which are not possible with the stepladder model. Tardy and Rabinovitch have discussed in detail the construction of the **P** matrix,[1, 5, 6] and have given illustrative numerical values of the P_{lm} for all the models discussed above.[1]

In most of the work to date $\langle \Delta E \rangle$ has been treated as a purely empirical parameter, but some progress is now being made towards the theoretical prediction of this quantity.[7] The collision process can be formulated in terms of a short-lived collision complex. Statistical redistribution of energy is assumed to occur between the internal degrees of freedom of the substrate molecule and the 'transitional modes' of the complex (these being the new vibrational–rotational degrees of freedom produced on formation of the complex from the colliding molecules). Experimental collision efficiencies (see section 10.4) can then be reproduced with reasonable assumptions about the properties of the collision complex.

10.3 CHEMICALLY ACTIVATED AND PHOTOACTIVATED SYSTEMS

For the particular case where energized molecules are created by chemical activation (see Chapter 8) or photoactivation (see later) it can usually be assumed that all of the activation takes place into energy levels above E_0. The input distribution vector $\mathbf{F} = \{\mathbf{f_1}\ \mathbf{f_2}\}$ can thus be written as $\mathbf{F} = \{\mathbf{0}\ \mathbf{f_2}\}$. Under conditions where there is negligible thermal energization it can usually be assumed[4, 8] that $\mathbf{P_3} = \mathbf{0}$, and the resulting equation (10.14) can be obtained.[1, 4] The result for $Rate/R$ will be equated to $D/(D+S)$ for a chemical activation system or to the quantum yield ϕ in a photoactivation

$$Rate/R = \sum_i \{\mathbf{k}[\omega(\mathbf{I}-\mathbf{P_4})+\mathbf{k}]^{-1}\mathbf{f_2}\}_i \qquad (10.14)$$

$$= \mathbf{uk}[\omega(\mathbf{I}-\mathbf{P_4})+\mathbf{k}]^{-1}\mathbf{f_2}$$

system. Under conditions where (10.14) is valid, the calculation thus requires knowledge only of the P_{lm} for levels above E_0, the rate-constants k_i for these levels and the input distribution function $\mathbf{f_2}$. This last will usually be the Boltzmann energy distribution of the original reactants raised by the net energy of the activation process. Details of the computational technique may be found in ref. 5.

There are many published examples of the application of this approach to chemical activation systems. The experimental results are often interpreted in terms of the apparent decomposition rate-constant on a strong collision basis, $\omega D/S$ or $\langle k_a \rangle$ in the nomenclature of Chapter 8. If the excitation is monoenergetic and the strong collision assumption is valid, $\omega D/S$ should be constant with varying ω and equal to the true k_i for the energy level at which the molecules are produced. On a weak collision basis (and often in practice) $\omega D/S$ varies with ω and plots of $\omega D/S$ against ω or against S/D show an increase at low pressures (the 'turn-up'). This is quite different from the often smaller increase of $\omega D/S$ at high pressures, which is due to the finite energy spread of the energized molecules (see section 8.1.1); both effects occur in practice, as is seen for example in Figure 10.2. This figure[8] shows theoretical curves for the decomposition of chemically activated sec-butyl radicals, and illustrates well the effects of collisional inefficiency. The significance of the $\omega D/S$ values in various pressure regions is clarified by approximate treatments[4, 8] of a stepladder model in which the energized molecules cascade down a series of n levels, from the initial level E_{max} at which they are mono-energetically formed, to the first level below E_0, where they are no longer able to react. The high- and low-pressure results are found to be (10.15) and (10.16) respectively. If $n = 1$ (i.e. for a strong collision model) both

$$\left(\frac{\omega D}{S}\right)_{\omega \to \infty} = \sum_{i=1}^{n} k_i \qquad (10.15)$$

$$\left(\frac{\omega D}{S}\right)_{\omega \to 0} = \frac{1}{\omega^{n-1}} \prod_{i=1}^{n} k_i \qquad (10.16)$$

limits give $k_a(E_{\max})$ as expected. Thus in Figure 10.2 the values of $\omega D/S$ for $\langle \Delta E \rangle = 10 \, \text{kcal mol}^{-1}$ are effectively constant since the butyl

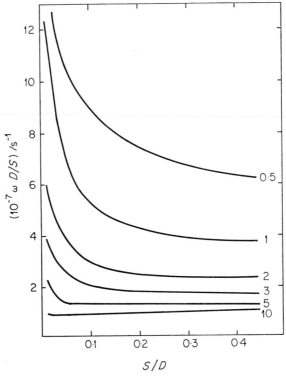

Figure 10.2   Theoretical curves of $\omega D/S$ against S/D for a stepladder model of the sec-butyl radical system, with step sizes ($\langle \Delta E \rangle / \text{kcal mol}^{-1}$) shown on curves[8]

radicals are formed with an average excess energy of about $9 \, \text{kcal mol}^{-1}$ and each collision removes more than this amount of energy. The slight increase to high pressures is due to the energy spread of the radicals (see section 8.1.1). The value $\omega D/S \approx 1 \times 10^7 \, \text{s}^{-1}$ is the true $k_a(E^+)$ for butyl radicals with $E^+ = 9 \, \text{kcal mol}^{-1}$. For lower values of $\langle \Delta E \rangle$, $\omega D/S$ is higher than this value throughout the whole pressure range, reflecting the increased chances for reaction to occur when de-energization occurs by a cascade process. At high pressures $\omega D/S$ is given by (10.15) as the sum of the rate-constants for all the levels the radicals pass through before their eventual de-energization. This follows because the radicals spend equal times in each level [under the approximations involved in (10.15)], and the probability of reaction from each level is thus proportional to the corresponding rate-constant. At low pressures the physical interpretation is less clear, but (10.16) shows that $\omega D/S$ is again higher than $k_a(E_{max})$ and increases indefinitely as the pressure is lowered.

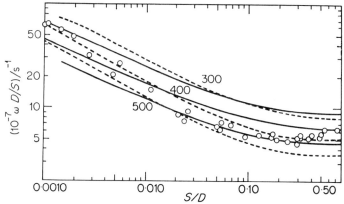

Figure 10.3   Experimental points for chemically activated 2-pentyl radicals compared with theoretical curves of $\omega D/S$ vs S/D for stepladder (broken lines) and exponential (solid lines) models with $\langle \Delta E \rangle \equiv 300$, 400 and 500 cm^{-1} (0·86, 1·14 and 1·43 kcal mol^{-1}).[9] The bath gas is mainly hydrogen

The interpretation of experimental results on the above basis is achieved by superimposing the experimental $\omega D/S$ values on a series of plots such as those shown in Figure 10.2. Both the shape of the curve and its position on the S/D axis are significant. Figure 10.3 shows a set of experimental points[9] for the decomposition of 2-pentyl radicals formed from H atoms and pent-1-ene in a bath gas which is mainly hydrogen. These results cover an exceptionally large range of S/D values, and the turn-up is very marked. Comparison with the theoretical curves also shown does not discriminate unambiguously between the stepladder and exponential models, but for either model the $\langle \Delta E \rangle$ values are fairly closely bracketed at about 400 cm$^{-1} \approx 1$ kcal mol^{-1} for collisions with the hydrogen bath molecules. This is considerably less than the average excitation energy of the initially formed radicals (about 14 kcal mol^{-1}), which accounts for the observation of such pronounced weak-collision characteristics.

Some $\langle \Delta E \rangle$ values determined in chemical activation systems are summarized in Table 10.1. A large number of radical decompositions has

Table 10.1　Some $\langle \Delta E \rangle$ values from chemical activation and photoactivation experiments[a]

Reacting species	Bath molecules	$\langle \Delta E \rangle / \text{kcal mol}^{-1}$	Ref.
Chemical activation:			
sec-butyl radicals	He, Ne, Ar, Kr	1·0–1·7	8
	CO_2	~4	g
	CD_3F, CH_3Cl, SF_6	≥9	g
	cis-but-2-ene	≥9	g, h
alkyl radicals[b]	N_2	2·3	5
	CH_4	3–4	5, 9
alkyl radicals[b, c]	H_2	1·0–1·5	5, 9–11, g
	CF_4	≥4·6[c], ≥9[b]	5, 9–11
cyclopropane	He	6 ($\beta = 0\cdot3$)[d] (0·14[e])	18
	Ar, N_2	6 (0·4[e])	18
	C_2H_4	7–28[f] (0·6[e])	17
1,2-dimethylcyclo-propane	CO	4·6	i
	cis-but-2-ene	11·4	17,
Photoactivation:			
cycloheptatriene	He	0·14	⎫
	CO_2	0·9	⎪
	SF_6	1·4	⎬ 13
	toluene	2·6	⎪
	cycloheptatriene	4·0	⎭

[a] Approximate grouped values. Stepladder (SL) and exponential (E) models usually give similar $\langle \Delta E \rangle$ values, SL being slightly preferable for large deactivating molecules and E for small deactivators. [b] 2-butyl, 2-pentyl, 2-hexyl and 2-octyl radicals. [c] 3-hexyl, 3-methylhexyl-2 and 3,3-dimethylhexyl-2 radicals. [d] Model used had $\langle \Delta E \rangle = 6$ kcal mol^{-1} for only 0·3 of the collisions. [e] Lower values recalculated by Atkinson and Thrush[13] using a more recent value for $\Delta H_f^0(CH_2)$. [f] Result varies with source of methylene radicals. [g] G. H. Kohlmaier and B. S. Rabinovitch, *J. Chem. Phys.*, **38**, 1709 (1963). [h] R. E. Harrington, B. S. Rabinovitch and M. R. Hoare, *J. Chem. Phys.*, **33**, 744 (1960). [i] J. D. Rynbrandt and B. S. Rabinovitch, *J. Phys. Chem.*, **74**, 1679 (1970).

been investigated, including some which decompose by more than one competing process,[10, 11] and the results of these experiments are included only in summary form.

Photoactivation of a unimolecular reaction is a new and promising technique for the study of collisional deactivation, the method being analogous to the more widely studied fluorescence quenching.[12] Atkinson and Thrush have used photochemical excitation to raise cyclohepta-1,3,5-triene molecules to various accurately known energy levels above the critical energy for their isomerization to toluene.[13, 14] This method of energization has the considerable advantage that accurately known and independently variable amounts of energy can be supplied to the molecules. Measurement of the quantum yield ϕ of toluene formation gave effective rate-constants as a function of energy according to (10.17). The rate-constants obtained on this strong-collision basis varied with energy in a

$$[k_a(E)]_{\text{apparent}} = \omega\phi/(1-\phi) \qquad (10.17)$$

manner which was not consistent with RRKM theory. The results could however be interpreted satisfactorily on the basis of a stepladder deactivation model having down-transitions only and $\langle \Delta E \rangle \approx 4 \text{ kcal mol}^{-1}$ for deactivation by the parent molecules. Values for other deactivating molecules are shown in Table 10.1. RRKM theory has also been applied to the calculation of quantum yields for the decomposition of azoethane, but the calculations are at present too arbitrary to provide information on the collision processes since there are no reliable data for the thermal reaction to assist the construction of a model for the activated complex.[15] One disadvantage of the photoactivation technique as so far applied is that (10.17) becomes indeterminate at low pressures.

Inspection of the results in Table 10.1 shows that for a particular reaction system there is always a clear trend to higher $\langle \Delta E \rangle$ values for the larger deactivating molecules. The values for large deactivators are such as to support the idea that the strong-collision assumption is a reasonable approximation for most thermal unimolecular reactions unless the bath molecules are rather small (see section 10.4). Values obtained[13] from the photoactivation of cycloheptatriene were smaller than those obtained for the same deactivators in chemical activation experiments, and were more in line with the results from fluorescence quenching experiments, notably those with β-naphthylamine.[16] The isomerization of chemically activated cyclopropane has also been interpreted[13] in terms of much lower $\langle \Delta E \rangle$ values than previously[17, 18] derived. It has been suggested[13] that energy transfer may be more efficient when it involves open-chain or free-radical species, but further work will clearly be needed to resolve the apparent conflict.†

10.4 THERMAL UNIMOLECULAR REACTION SYSTEMS

For thermal reactions the general equations (10.2) and (10.3) are still appropriate for determining the steady-state population distribution

† A recent re-evaluation of the results of ref. 13 has produced $\langle \Delta E \rangle$ values substantially higher than those originally quoted; see B. S. Rabinovitch, H. F. Carroll, J. D. Rynbrandt, J. H. Georgakakos, B. A. Thrush and R. Atkinson, *J. Phys. Chem.*, in press (1971).

among the various energy levels. There is physically no external input of molecules to replace those removed by reaction, so $R = 0$ unless such an input is invented for computational purposes. If so, the appropriate input distribution vector $\mathbf{F}$ is the Boltzmann distribution of the reservoir of reactant molecules.[1,6] In the absence of such input, however, each energy level is subject to a continuous depletion by the overall loss of reactant molecules to form product. The rate of this loss, $-dn_i/dt$, is very small compared with the rates $\omega P_{ij} n_j$, ωn_i and $k_i n_i$ of the microscopic processes involved, and the solutions of (10.3) with $R = 0$ thus give a close approximation to the steady-state concentration distribution $\mathbf{N} = \{n_1 \; n_2 \; ...\}$. The solution of these equations is best obtained by an iterative technique.[1-3, 6] For energy levels below E_0 there is no reaction to be allowed for $(\mathbf{K} = \mathbf{0})$, and (10.4) leads to (10.18) since $\mathbf{IN} \equiv \mathbf{N}$ by definition. Equation (10.18) is

$$\mathbf{N} \approx \mathbf{PN} \qquad (10.18)$$

in fact a necessary condition for a system at equilibrium, the equality then being exact.[19] Since the vector $\mathbf{PN}$ gives the rates of collisional input into each level, (10.18) implies that in the steady state the fractional population of each level below E_0 is approximately equal to the fractional rate of collisional input into the level. For levels above E_0 a modified matrix $\mathbf{P}'$ is used, in which the ith column (representing transfers out of level i) is normalized to $1 - k_i/(\omega + k_i)$ instead of to unity as required by (10.1) for collisional processes alone. Partitioning of the matrices into different energy regions is again essential for computational purposes, and these aspects, together with the construction of the $\mathbf{P}$ and $\mathbf{P}'$ matrices, have been discussed in detail by Tardy and Rabinovitch.[1, 5, 6] From (10.18) it appears that the steady-state distribution $\mathbf{N}$ may be found by taking *any* initial distribution $\mathbf{N}_0$ and repeatedly operating on it with $\mathbf{P}'$ until no further change occurs, (10.19). The iterations converge, since the operations correspond physically

$$\left.\begin{aligned}
\mathbf{N}_1 &= \mathbf{P}'\mathbf{N}_0 \\
\mathbf{N}_2 &= \mathbf{P}'\mathbf{N}_1 = \mathbf{P}'^2 \mathbf{N}_0 \\
&\cdot \quad \cdot \quad \cdot \quad \cdot \quad \cdot \quad \cdot \\
\mathbf{N}_n &= \mathbf{P}'\mathbf{N}_{n-1} = \mathbf{P}'^n \mathbf{N}_0 \\
&= \mathbf{N}_{n-1} \quad \text{for steady state}
\end{aligned}\right\} \qquad (10.19)$$

to collisional redistribution of the molecules between the energy levels. The number n of iterations needed to reach the steady state corresponds to the number of collisions needed to reach this state in the physical system, i.e. to the induction period.

When the steady-state distribution has thus been calculated, the total rate of reaction is simply obtained (10.20) as the sum of the rates of

$$Rate = \sum_i k_i n_i \equiv \sum_i (\mathbf{KN})_i \equiv \mathbf{uKN} \tag{10.20}$$

reaction from each level. Alternatively (and equivalently) the overall rate may be obtained as the rate of decrease of the number of molecules in all energy levels [cf. (10.2)], the result being (10.21), where $\mathbf{J} = \mathbf{K} + \omega(\mathbf{I} - \mathbf{P})$

$$Rate = \sum_i (- dn_i/dt) = \sum_i (\mathbf{JN})_i \equiv \mathbf{uJN} \tag{10.21}$$

as before. This approach is particularly useful for calculations in the low-pressure region, when it is only necessary to consider the levels below E_0. Valance and Schlag have described a closely related iterative technique which gives the rate-constant directly.[2,3]

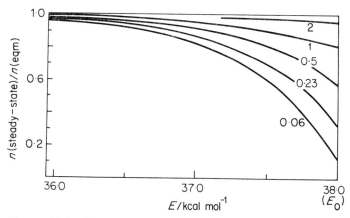

Figure 10.4 Calculated steady-state population distribution[1] of levels below E_0 for thermal methyl isocyanide isomerization at 273 K in the second-order region, using stepladder models with step sizes ($\langle\Delta E\rangle$/kcal mol^{-1}) shown on curves

Tardy and Rabinovitch have given illustrative calculations of weak-collision effects for a number of systems;[1,6] some results for the isomerization of methyl isocyanide are shown in Figures 10.4 and 10.5. Figure 10.4 refers to the second-order region (where non-equilibrium effects are at their maximum) and shows the steady-state concentrations in the energy levels below E_0 as a fraction of the equilibrium (high-pressure) concentrations. The curves represent stepladder models with various $\langle\Delta E\rangle$ values, and illustrate the marked depletion of the levels near E_0 which can result from the irreversible transport of molecules to energies above E_0. Similar curves were obtained with exponential, Poisson or Gaussian distributions of transition probability with the same $\langle \Delta E \rangle$ values. The depletion of the upper levels is negligible for step sizes above 2 kcal mol⁻¹ in Figure 10.4, but becomes quite marked for step sizes below about 0·5 kcal mol⁻¹. Thus the effective collision efficiencies β, defined by (10.22),

$$\beta_{\text{calc}} = \frac{Rate \text{ (weak collision, steady state)}}{Rate \text{ (strong collision, equilibrium distribution)}} \tag{10.22}$$

are about 0·9 and 0·2 for $\langle \Delta E \rangle = 2 \cdot 0$ and $0 \cdot 5$ kcal mol⁻¹ respectively. The disequilibrium increases as the overall reaction rate increases, and is thus more marked at higher temperatures. The effective collision efficiency is

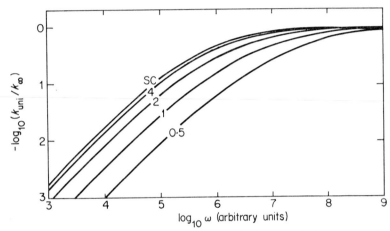

Figure 10.5 Calculated fall-off curves[6] for methyl isocyanide isomerization at 546 K, using stepladder models with step sizes ($\langle \Delta E \rangle$/kcal mol⁻¹) shown on curves; SC = strong-collision curve

roughly a function of $\langle \Delta E \rangle / T$, so that β values of 0·9 and 0·2 are obtained with step sizes of about 4 and about 1 kcal mol⁻¹ respectively at a practical experimental temperature of 546 K. This behaviour is very much as might be expected from the average excitation energy $\langle E^+ \rangle$ of the molecules reacting at this temperature in a strong-collision system, which is 1–3 kcal mol⁻¹ depending on the pressure.[20] Tardy and Rabinovitch[6] have given 'quasiuniversal' plots of collision efficiency against the ratio $\langle \Delta E \rangle / \langle E^+ \rangle$ at various extents of fall-off, where $\langle E^+ \rangle$ is the average excess energy of the reacting molecules under equilibrium (strong-collision) conditions, and can be estimated by empirical correlation.[1]

An important practical aspect of these calculations lies in the derived fall-off curves, some of which are illustrated in Figure 10.5. The illustrated curves refer to stepladder models, but as before an exponential distribution gave similar results. Reduction of the step size is seen to move the fall-off curve to higher pressures with only a minor change of shape. Again, the effect is only significant for step sizes below $4\,\text{kcal}\,\text{mol}^{-1}$ (the temperature being 546 K). The effect can be large, however; a $\langle \Delta E \rangle$ value of $0\cdot5\,\text{kcal}\,\text{mol}^{-1}$ gives a curve which is shifted by $\Delta \log \omega \approx 1\cdot4$ relative to the strong-collision curve. This shift may be interpreted in terms of an effective collision efficiency λ of $10^{-1\cdot4} \approx 0\cdot04$ [cf. section 4.12.2; $\Delta \log \omega$ is similar to, but not in general identical with, the efficiency β defined as (10.22)]. The corresponding λ values for $\langle \Delta E \rangle = 1$ and $2\,\text{kcal}\,\text{mol}^{-1}$ are about $0\cdot1$ and $0\cdot4$ respectively, which are the sort of figures often invoked to obtain agreement between RRKM calculations and experimental fall-off curves (see section 7.2). A closer inspection of Figure 10.5 shows that there is, in fact, a change of shape as the fall-off curve moves to higher pressure. This corresponds to the expected variation of effective collision efficiency with pressure (cf. section 4.12.2); the $0\cdot5\,\text{kcal}\,\text{mol}^{-1}$ stepladder model gives $\lambda = 0\cdot029$, $0\cdot043$ and $0\cdot058$ at $k_{\text{uni}}/k_{\infty} = 0\cdot5$, $0\cdot05$ and $0\cdot005$ respectively. This sort of variation is often insignificant experimentally, and such calculations provide some justification for approximate treatment of weak-collision effects in thermal unimolecular reactions by the simple assumption of a constant collision efficiency λ. The appropriate numerical value of λ for a given reaction with a given $\langle \Delta E \rangle$ can be estimated by empirical correlation.[1, 6]

Experimentally, the effects of limited energy transfer are again most easily assessed for the second-order region. The measured rate of reaction is simply the rate of transport of molecules to levels above E_0, and the variations of rate with the concentrations of reactant and bath gases give relative collision efficiencies β for energy transfer by different molecules. The most extensive series of such measurements are those of Rabinovitch and co-workers for the isomerizations of methyl[21, 22] and ethyl[23] isocyanides. The former are by far the most extensive, and some typical experimental results are shown in Figure 10.6. The relative energization efficiencies per unit pressure are obtained as the relative slopes of these plots, and are converted to relative efficiencies per collision using the kinetic-theory ratio of collision frequencies. The required expression is (10.23), in which A represents the reactant, M and N are bath gases

$$\frac{\beta_{\text{M}}}{\beta_{\text{N}}} = \frac{(dk_{\text{uni}}/dp_{\text{M}})}{(dk_{\text{uni}}/dp_{\text{N}})} \times \left(\frac{\sigma_{\text{AN}}}{\sigma_{\text{AM}}}\right)^{2} \times \left(\frac{\mu_{\text{AM}}}{\mu_{\text{AN}}}\right)^{\frac{1}{2}} \tag{10.23}$$

(one of which is often A in practice), μ represents reduced mass and σ_{ij} is the collision diameter $\frac{1}{2}(\sigma_i + \sigma_j)$ corrected[24, 25] for temperature effects. The efficiencies obtained from (10.23) are differential relative efficiencies, as opposed to the integrated values which could be obtained by application of (10.22) to the mixed reactant–bath system. Various definitions of collision efficiency and their interrelationships have been discussed by Rabinovitch and co-workers.[6, 23] Integral efficiencies have been measured for over 100 bath gases in the second-order region of methyl

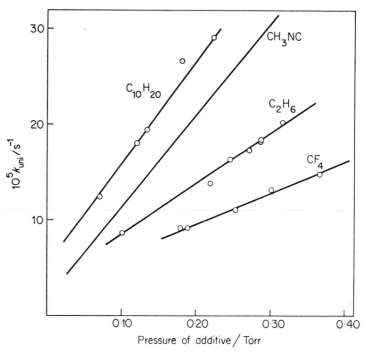

Figure 10.6 Rate data[24, 25] for thermal energization of a constant small pressure of methyl isocyanide in the second-order region at 280·5 °C by various added gases. The line for CH_3NC summarizes many points

isocyanide isomerization. Values relative to $\beta(MeNC) = 1·00$ are plotted in Figure 10.7 against the atomicity of the bath gas for noble gases, alkanes, alkenes, alkynes, nitriles, fluorocarbons and other molecules. The plot shows very clearly that the energization efficiency increases with increasing molecular size and reaches an effectively constant limiting value for molecules containing more than 10 to 12 atoms, i.e. for C_3 or C_4 and higher compounds. It seems very reasonable to assume that the limiting value is unity, and that these sufficiently large molecules behave effectively as strong colliders. This conclusion is entirely in agreement with the theoretical treatment outlined above (e.g. Figure 10.5) together with the $\langle \Delta E \rangle$ values obtained from chemical activation experiments (Table 10.1), and provides a good justification for the use of the strong-collision assumption for many thermal unimolecular reaction systems.

Since the limiting efficiency is also unity *relative to methyl isocyanide itself*, this also implies that methyl isocyanide has unit efficiency despite its relatively small size. For contrast, CD_3CN and C_2H_5CN have appreciably

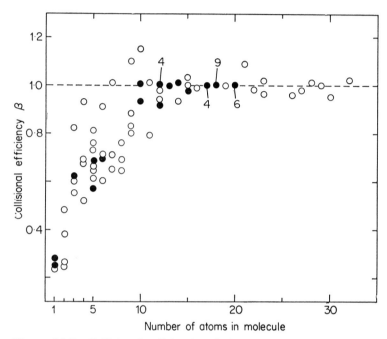

Figure 10.7 Collisional efficiencies β for energization of methyl isocyanide at 280·5 °C by various bath gases relative to β(MeNC) = 1·00, using viscosity-derived cross-sections.[24] Filled points represent coincident points for the stated number of bath gases (two if not stated)

lower efficiencies. Similar remarks may well apply for other reaction systems,[12, 26] although none have been studied as exhaustively as the methyl isocyanide isomerization (note, however, the recent evidence of complications in this reaction; see section 7.1.9). The high efficiency of energy transfer to the substrate itself may possibly be due to the accurate matching of vibrational energy levels.[24]

Extensive study of the reaction in a weak-collider bath gas provides opportunities for a more detailed comparison with the theory described previously, and this is conveniently done by comparing the calculated and experimental values of β over a range of conditions. Efficiencies can be defined and measured at any part of the fall-off region, but the second-order region is most easily dealt with.[6, 23] It emerges from detailed calculations that β is expected to vary with the relative concentrations of reactant and additive. The extreme values in the second-order region (subscript 0) have been denoted by $\beta_0(\infty)$ and $\beta_0(0)$, the parenthetical ∞ and 0 referring respectively to infinite dilution of reactant and to zero dilution (i.e. pure reactant). Moreover, the predicted variation with dilution is such that different transition probability models could be distinguished if the experiments could be made sufficiently accurate, although this has not been unambiguously achieved in practice. An example of such an analysis is the work of Lin and Rabinovitch[27] on methyl isocyanide isomerization with helium, ethane and pent-1-ene as bath gases. The data were best interpreted in terms of an exponential model for helium but stepladder models for ethane and pentene. The $\langle \Delta E \rangle$ values[24] were 1·2 kcal mol^{-1} for helium and > 6 kcal mol^{-1} for pentene, the latter approaching closely to the strong-collision case. The variation of β with temperature has also been measured in the helium system and found to be qualitatively in agreement with theory.[28]

The measurements with homologous series of bath molecules may also be used to obtain information about the collision cross-sections for energy transfer.[29] This is desirable since it cannot be assumed *a priori* that the appropriate figures will be the same as those derived from viscosity or virial-coefficient data. It may be taken as axiomatic, however, that the intrinsic collision efficiency will rise to a constant value for all members of a homologous series above some critical size. If the critical member is M and N is the *i*th member after M, then $\beta_M/\beta_N = 1$. The hypothesis may also be made that the effective collision diameter σ_{AM} increases by a constant increment for each CH$_2$ group added above the critical member; $\sigma_{AN} = \sigma_{AM} + i\Delta\sigma$. Equation (10.24) then follows, and if the assumptions are valid a linear plot of the left-hand side (called R) against i will result.

$$\left[\left(\frac{dk_{\text{uni}}/dp_N}{\mu_{AN}^{\frac{1}{2}}} \right) \middle/ \left(\frac{dk_{\text{uni}}/dp_M}{\mu_{AM}^{\frac{1}{2}}} \right) \right]^{\frac{1}{2}} - 1 = i(\Delta\sigma/\sigma_{AM}) \qquad (10.24)$$

In practice such behaviour has been found for methyl isocyanide isomerization with the homologous series of alkanes, alkenes, alkynes[29] and nitriles and fluorocarbons[25] as bath gases. With the alkanes, for example, the plot is linear through the origin when M is taken as butane, but shows an initial downward curvature if a smaller member of the series is used. The implication is again that the true collision efficiency is constant for butane and higher alkanes and is presumed to be unity for these molecules. The slope of the plot is in fact consistent with the viscosity-derived collision cross-sections, and the use of such data in energy-transfer calculations is thus confirmed, at least for methyl isocyanide isomerization.

References

1. D. C. Tardy and B. S. Rabinovitch, *J. Chem. Phys.*, **45**, 3720 (1966).
2. W. G. Valance and E. W. Schlag, *J. Chem. Phys.*, **45**, 216 (1966).
3. W. G. Valance and E. W. Schlag, *J. Chem. Phys.*, **45**, 4280 (1966).
4. M. Hoare, *J. Chem. Phys.*, **38**, 1630 (1963).
5. D. C. Tardy and B. S. Rabinovitch, *J. Chem. Phys.*, **48**, 5194 (1968).
6. D. C. Tardy and B. S. Rabinovitch, *J. Chem. Phys.*, **48**, 1282 (1968).
7. Y. N. Lin and B. S. Rabinovitch, *J. Phys. Chem.*, **74**, 3151 (1970).
8. G. H. Kohlmaier and B. S. Rabinovitch, *J. Chem. Phys.*, **38**, 1692 (1963).
9. J. H. Georgakakos, B. S. Rabinovitch and E. J. McAlduff, *J. Chem. Phys.*, **52**, 2143 (1970).
10. C. W. Larson and B. S. Rabinovitch, *J. Chem. Phys.*, **51**, 2293 (1969).
11. D. C. Tardy, C. W. Larson and B. S. Rabinovitch, *Can. J. Chem.*, **46**, 341 (1968).
12. B. Stevens, *Collisional Activation in Gases*, Pergamon, Oxford, 1967.
13. R. Atkinson and B. A. Thrush, *Proc. Roy. Soc.* (*A*), **316**, 131 (1970).
14. R. Atkinson and B. A. Thrush, *Proc. Roy. Soc.* (*A*), **316**, 123, 143 (1970); *Chem. Phys. Lett.*, **3**, 684 (1969).
15. P. G. Bowers, *J. Phys. Chem.*, **74**, 952 (1970), and references cited therein.
16. B. S. Neporent, *Zh. Fiz. Khim.*, **21**, 111 (1947); **24**, 1219 (1950); M. Boudart and J. T. Dubois, *J. Chem. Phys.*, **23**, 223 (1955); B. Stevens, *Mol. Phys.*, **3**, 589 (1960).
17. D. W. Setser, B. S. Rabinovitch and J. W. Simons, *J. Chem. Phys.*, **40**, 1751 (1964).
18. J. W. Simons, B. S. Rabinovitch and D. W. Setser, *J. Chem. Phys.*, **41**, 800 (1964).
19. M. Hoare, *J. Chem. Phys.*, **41**, 2356 (1964).
20. F. J. Fletcher, B. S. Rabinovitch, K. W. Watkins and D. J. Locker, *J. Phys. Chem.*, **70**, 2823 (1966).
21. Refs. 24 and 25, and earlier work cited and corrected therein.
22. L. D. Spicer and B. S. Rabinovitch, *J. Phys. Chem.*, **74**, 2445 (1970).
23. S. P. Pavlou and B. S. Rabinovitch, *J. Phys. Chem.*, **75**, 1366, 2171 (1971).
24. S. C. Chan, B. S. Rabinovitch, J. T. Bryant, L. D. Spicer, T. Fujimoto, Y. N. Lin and S. P. Pavlou, *J. Phys. Chem.*, **74**, 3160 (1970).
25. S. C. Chan, J. T. Bryant, L. D. Spicer and B. S. Rabinovitch, *J. Phys. Chem.*, **74**, 2058 (1970).
26. T. L. Cottrell and J. C. McCoubrey, *Molecular Energy Transfer in Gases*, Butterworths, London, 1961.
27. Y. N. Lin and B. S. Rabinovitch, *J. Phys. Chem.*, **72**, 1726 (1968); see also ref. 24.
28. S. C. Chan, J. T. Bryant and B. S. Rabinovitch, *J. Phys. Chem.*, **74**, 2055 (1970).
29. Y. N. Lin, S. C. Chan and B. S. Rabinovitch, *J. Phys. Chem.*, **72**, 1932 (1968); B. S. Rabinovitch, Y. N. Lin, S. C. Chan and K. W. Watkins, *J. Phys. Chem.*, **71**, 3715 (1967); see also refs. 24 and 25.

Appendix 1
Nomenclature

The main object of this appendix is to list and define the principal symbols used in the present work. Terms which are used only locally are mostly omitted, and this omission includes most of the special nomenclature used in the chapters on Slater theory, isotope effects and collisional energy transfer. On the other hand, the terms used in the basic RRKM theory are listed fairly exhaustively. A few symbols are used with more than one meaning, but the appropriate interpretation in these cases will always be clear from the context.

As a secondary object an attempt has been made to correlate the present nomenclature with that of some other prominent authors in the field, especially with regard to RRKM theory. A unique correlation is obviously not possible since authors do not use the same nomenclature in all their papers, but a good general guide is possible. The correlations are in square brackets and the authors' names are abbreviated as follows:

M   R. A. Marcus and co-workers (especially the basic theory papers[1])

R   B. S. Rabinovitch and co-workers (Rabinovitch and Setser review[2] and numerous papers)

B   D. L. Bunker and co-workers (especially Bunker's book[3])

L   K. J. Laidler and co-workers (especially Laidler's book[4])

Thus [M: P; R: Z or Q] would mean that Marcus uses P, Rabinovitch uses sometimes Z, sometimes Q, and Bunker and Laidler either do not use the quantity or use the same symbol as we do.

General nomenclature

$\langle x \rangle$ Average value of a quantity x for molecules of interest (these being defined locally)

$$\sum_{i=1}^{n} x_i = x_1 + x_2 + \ldots + x_n$$

$$\prod_{i=1}^{n} x_i = x_1 \times x_2 \times \ldots \times x_n$$

$\Gamma(n)$ The gamma function of argument n (see Appendix 7)

$n!$ Factorial $n = \Gamma(n-1)$

$(\mathbf{X})_i$ The ith element of a vector $\mathbf{X}$

h, k, N_A, R Planck, Boltzmann, Avogadro and gas constants

p, T Pressure and absolute temperature

Q, Q_v, etc.   Partition functions [M: P; R: Z or Q; L: Q or q]

σ Symmetry number

$S(A)$ Entropy of a species A

$S_v(A), S_r(A)$ Vibrational and rotational contributions to $S(A)$

s Number of vibrational degrees of freedom of a molecule (or Kassel's parameter in Chapter 3)

ν, ν_i Vibration frequencies

v, v_i Vibrational quantum numbers

p Number of rotations of a molecule, i.e. the number of different moments of inertia

d_i Degeneracy of ith rotation

r Number of rotational degrees of freedom, $r = \sum\limits_{i=1}^{p} d_i$

I, I_A, etc.   Moments of inertia

J, J_i Rotational quantum numbers

E, E_v, etc.   Energy per molecule (unless it is stated otherwise or clear from the context that molar energies $N_A E$ are intended) [R: ϵ or E; L: ε or ϵ or E]

E_z Zero-point energy of a molecule [M: E_0; R: ϵ_0 or ϵ_z; B: E_{zp}; L: ε_0 or E_z]

$P(E_n)$ Number of quantum states of a given system at the quantized energy level E_n

$W(E) \equiv \sum\limits_{E_n=0}^{E} P(E_n)$ Number of quantum states of a system at all energies up to and including E (see section 4.11 and Chapter 5)

$[\equiv \sum P(E_n)]$ [R: $\sum P(\epsilon_{vR})$ or $G_{vR}(\epsilon)$; see also $\sum P(E_{vr}^+)$ below]

General rate parameters

k Rate-constants in general

k_{uni} First-order rate-constant, $-(1/[A]) \, d[A]/dt$ [M and R: k or k_{uni}; B: k; L: k^1]

k_∞ Limiting high-pressure value of k_{uni} [M, B, and L: k^∞]

k_{bim} Second-order rate-constant, $-(1/[A]^2) \, d[A]/dt = k_{uni}/[A]$ [M and R: k_0]

k_1, k_2 Rate-constants (second order) for energization and de-energization respectively [L: k_1, k_{-1}]

k_3 Rate-constant for conversion of energized molecules to products when this is independent of energy (HL theories)

$k_a(E)$ Same when energy-dependent as in RRK and RRKM theories [M and B: k_a; R: k_ϵ or k_a; L: k_2 or k_E]

$L(E_1, E_2, ..., E_n)$ Slater's specific dissociation probability for molecules having energies $E_1, ..., E_n$ in normal modes 1 to n

Z, σ_d Second-order collision rate-constant and collision diameter

ω $= Zp =$ first-order rate-constant for collisions

λ, β Collisional energy transfer efficiencies (sections 4.12.2 and 10.4)

E_0 Critical energy for reaction [In RRKM theory, M: E_a; R: ϵ_0 or E_0; L: ε_a or E_0]

E_{Arr}, A Arrhenius activation energy [equation (1.6)] and A-factor [$A = k \exp(E_{\text{Arr}}/kT)$]

$A_\infty, E_\infty, E_{\text{bim}}$ Limiting high-pressure values of A and E_{Arr}, and low-pressure value of E_{Arr}

$p_{\frac{1}{2}}$ Pressure at which $k_{\text{uni}} = \frac{1}{2}k_\infty$

L^+ Statistical factor or reaction path degeneracy

ΔS^+ Entropy of activation; $\Delta S^+ = S(A^+) - S(A)$

D Rate of production of decomposition or isomerization products in a chemical activation experiment

S Rate of production of stabilized molecules in a chemical activation experiment

$f(E)\,\delta E$ Fractional rate of energization into the energy range E to $E + \delta E$ in a chemical activation system (Chapter 8)

$\phi(E)\,\delta E$ Fractional rate of reaction from the energy range E to $E + \delta E$ in a chemical activation system (Chapter 8)

$f'(E)\,\delta E$ Fractional steady-state concentration of energized molecules in the range E to $E + \delta E$ in a thermal reaction (Chapter 9)

$K(E)\,\delta E$ $= \delta k_{1(E \to E + \delta E)}/k_2 =$ fractional concentration of molecules in the energy range E to $E + \delta E$ at thermal equilibrium (the Boltzmann distribution) [equation (8.11)]

$\langle k_a \rangle$ Average value of $k_a(E)$ for the reacting molecules in a chemical activation experiment [R: k_a]

$\langle k_a \rangle_0, \langle k_a \rangle_\infty$ Low- and high-pressure values of $\langle k_a \rangle$ [R: $k_{a0}, k_{a\infty}$]

Nomenclature used particularly in connection with RRKM theory

(Other authors' symbols differing in minor respects, e.g. replacement of E by ε, are not all listed)

A^*, A^+ Energized molecule and activated complex [R: Λ^*, ($A\dagger$); L: A^*, A^+]

$E^*(\equiv E_{vr}^*)$ Total non-fixed energy in the active degrees of freedom of a given energized molecule A^* ($E^* = E_v^* + E_r^*$; called E_{active}^* in section 4.10)

E_v^*, E_r^* Vibrational and rotational parts of E^*

E^+ Total non-fixed energy in the active degrees of freedom of a given activated complex A^+ ($E^+ = E_v^+ + E_r^+ + x$)

E_v^+, E_r^+ Vibrational and rotational contributions to E^+

E_{vr}^+ ($E_v^+ + E_r^+$) [M: E_n^+; L: $\varepsilon_n^{\ddagger}$]

x Translational energy of A^+ in the reaction coordinate [M: x or E_t^+; B: E_t^+; L: $\varepsilon_t^{\ddagger}$]

E_J, E_J^+ Energy of adiabatic rotations in their Jth energy level in A^* and A^+ respectively; $\Delta E_J = E_J^+ - E_J$

$N^*(E^*)$ Density of quantum states of A^* at energy E^* (see section 4.11) [M: $N^*(E^*)$ or $\Omega^*(E^*)$; B: $G(E^*)$]

$N^+(E_{vr}^+, x)$ Density of quantum states of A^+ having energy E_{vr}^+ in the active degrees of freedom and energy x in the reaction coordinate [M: $N_2(E^+ - x)$; B: $G(E^+ - E_t^+)$; L: $N_2(\varepsilon^{\ddagger} - \varepsilon_t^{\ddagger})$]

$N_{rc}^+(x)$ Density of quantum states for the translational motion of A^+ in the reaction coordinate with energy x [M: $N_1(x)$; B: $G(E_t^+)$; L: $N_1(\varepsilon_t^{\ddagger})$]

$P(E_{vr}^+)$ Number of vibrational–rotational quantum states of A^+ at the quantized energy level E_{vr}^+ [M: $P(E_n^+)$ or $\Omega^+(E_n^+)$ or $W^+(E_n^+)/\sigma$]

$\sum\limits_{E_{vr}^+=0}^{E^+} P(E_{vr}^+)$ or $\sum P(E_{vr}^+)$ Number of quantum states of A^+ at all energies up to and including E^+ (see section 4.11)

δ Arbitrary length of region at top of potential energy barrier which is taken to define the activated complex [M: b]

μ Characteristic mass for motion in the reaction coordinate [M: m; L: $m^{\ddagger}$]

$k^+(x)$ Rate constant with which complexes of energy x cross the barrier [M: k_3; B: ν_r; L: $\dot{x}$]

Q_1, Q_1^+ Partition functions for adiabatic rotations in A and A^+ respectively [M: P_1, P_1^+; R: $Z_1^*, Z_1\dagger$ or $Q_1, Q_1\dagger$; B: Q_r, Q_r^+; L: $Q_R, Q_R^{\ddagger}$]

Q_2, Q_2^+ Partition functions for active degrees of freedom of A and A^+ respectively (A^+ has one fewer degree of freedom) [M: P_2, P_2^+; R: Z_{vr}, etc; L: $Q_a, Q_a^{\pm}$]

Q, Q^+ Complete vibrational–rotational partition functions for A and A^+ respectively; $Q = Q_1 Q_2$ and $Q^+ = Q_1^+ Q_2^+$ [M: P, P^+; B: Q, Q^+ or $Q^{\pm}$; L: Q_i, Q_+]

σ, σ^+ Symmetry numbers of A and A^+

$Q_2^*, Q_2^{*\prime}$ See section 4.8

$k_{E,J}(E)$ Symbol for $k_a(E)$ when it depends on the rotational quantum number J as well as the total energy E

E' Reduced energy E/E_z

$a, \beta, w(E')$ Whitten–Rabinovitch functions; see sections 5.4.3, 5.4.4

β Laplace transform parameter in section 5.5

References

1. R. A. Marcus, *J. Chem. Phys.*, **20**, 359 (1952); G. M. Wieder and R. A. Marcus, *J. Chem. Phys.*, 37, 1835 (1962); R. A. Marcus, *J. Chem. Phys.*, **43**, 2658 (1965); **52**, 1018 (1970).
2. B. S. Rabinovitch and D. W. Setser, *Advan. Photochem.*, **3**, 1 (1964).
3. D. L. Bunker, *Theory of Elementary Gas Reaction Rates*, Pergamon, Oxford, 1966.
4. K. J. Laidler, *Theories of Chemical Reaction Rates*, McGraw-Hill, New York, 1969.

Appendix 2
Statistical Mechanics

In this appendix we outline some parts of Statistical Mechanics which are important in rate theories. Particular emphasis is placed on some features which are important in the RRKM theory but which are not dealt with prominently in the books on Statistical Mechanics. The treatment is mainly in terms of quantum statistical systems, but section A2.7 refers briefly to the classical treatment with its related concept of phase-space.

Fuller details and more of the background can be found in the standard works on Statistical Mechanics.[1]

A2.1 PARTITION FUNCTIONS: DEFINITION

If a molecular system is capable of existing in a series of quantized energy levels with total energy $E_0, E_1, E_2, \ldots$, then the *partition function Q* for the molecule is defined as

$$Q = g_0 \exp(-E_0/kT) + g_1 \exp(-E_1/kT) + g_2 \exp(-E_2/kT) + \ldots$$
$$= \sum_{i=0}^{\infty} g_i \exp(-E_i/kT) \tag{A2.1}$$

where g_i is the *degeneracy* or *statistical weight* of the energy level E_i, defined as the number of distinct (i.e. physically distinguishable) quantum states of that energy. The degeneracy is alternatively the number of different independent wave-functions of the system with total energy E_i, or the number of physically distinct ways this energy can be distributed in the molecule. These factors are introduced into (A2.1) so that the sum can be taken over all energy levels rather than evaluating $\sum \exp(-E_i/kT)$ over all quantum states (some of the same energy), which is mathematically more complicated. For a given system (i.e. for a given set of energy levels E_i) the numerical value of Q depends only on the temperature and on the zero chosen for the energy scale; if the energy zero is shifted down by an amount ΔE all the energies are increased by ΔE so that each exponential term in (A2.1) is multiplied by $\exp(-\Delta E/kT)$ and Q is simply multiplied by the same factor.

It is also possible to consider the partition function not for a whole molecule but for certain degrees of freedom of the molecule, the definition being as before but with E_i being the energy specifically in these degrees of freedom, and g_i the number of different ways this energy can be distributed in these degrees of freedom. For example, it is common to consider separately the electronic, vibrational, rotational and translational partition functions, Q_e, Q_v, Q_r and Q_t, and it is not hard to show that if the energies are simply additive, i.e. if the total energy can be written as $E_{tot} = E_e + E_v + E_r + E_t$, then the molecular partition function is the product of the individual partition functions:

$$Q = Q_e \cdot Q_v \cdot Q_r \cdot Q_t$$

In the same way it is not essential to consider together all the degrees of freedom of one type; for example Q_v can be factorized into contributions for different vibrations, provided that these are independent so that their energies are additive. Such an approach may be useful if an approximate treatment of Q_v is valid for some vibrations but not for others. Expressions for the partition functions of some specific systems are given in section A2.6.

A2.2 SIGNIFICANCE OF Q: POPULATION OF STATES

The basic usefulness of a partition function derives from the fact that each term of the series (A2.1) is proportional to the number of molecules existing in the corresponding energy level at thermal equilibrium. Since the sum of the terms (Q) therefore represents the total number of molecules (N) the fraction of molecules in the ith energy level is given by

$$\frac{N_i}{N} = \frac{g_i \exp(-E_i/kT)}{Q} \tag{A2.2}$$

It is interesting to note that for molecules of the same (or nearly the same) energy E_i the population is governed merely by the statistical weights g_i (the number of quantum states corresponding to each energy level). This is an expression of the fundamental postulate of Statistical Mechanics, *that all quantum states of a system are* a priori *equally probable* and will be equally populated in the absence of constraints imposed (for example) by energetic considerations. This point will be further illustrated in section A2.5.

It may be useful by way of background information to discuss briefly the significance of numerical values of Q in terms of the distribution of molecules between the energy levels. The result obtained by summing (A2.1) depends very much on the relative values of the energies E_i and the quantity kT which is the yardstick of thermal energy—the amount of energy which is readily available in a system. If E_0 is taken as zero, as is common, then for $E_i \gg kT$ there will be few molecules in the excited state since energy E_i will seldom be available to activate the molecules. If, on the other hand, $E_i \ll kT$, there will be almost as many molecules with energy E_i as with energy E_0. Thus only at high energies (relative to kT) will there be a significant decrease in the population of the energy levels.

These ideas are well illustrated by the cases of a typical vibration and a translation at room temperature. Consider, for example, the vibration of a bromine molecule Br_2; this has the fairly low vibration frequency of $\nu = 9 \cdot 7 \times 10^{12}$ s^{-1} and hence energy levels (relative to the ground state) of $\nu h \nu = 0$, $6 \cdot 4 \times 10^{-21}$, $12 \cdot 8 \times 10^{-21}$, etc. J (per molecule). At 25 °C kT is $4 \cdot 1 \times 10^{-21}$ J so the series for the partition function is (to 3 decimal places):

$$Q_V(Br_2) = 1 \cdot 000 + 0 \cdot 210 + 0 \cdot 044 + 0 \cdot 009 + 0 \cdot 002 + 0 \cdot 000 + \ldots$$

$$= 1 \cdot 265$$

$$(\equiv [1 - \exp(-h\nu/kT)]^{-1} \text{ by algebraic summation})$$

It will be seen that this situation, in which kT is only of the same order of magnitude as the spacing of the energy levels, gives a partition function not much greater than unity. This in turn shows that the population of the ground state is high; from (A2.2),

$$N_0/N = \exp(-E_0/kT)/Q = 1 \cdot 000/1 \cdot 265 = 0 \cdot 79 \quad \text{or} \quad 79\%$$

Correspondingly, the population falls off rapidly in the excited energy levels, the figures being 17%, 3·5%, 0·7%, etc. in successive levels.

In contrast, the energy levels for translation are very closely spaced relative to kT for most molecules even at very low temperatures. For example, for the one-dimensional translation of a bromine molecule (mass $m = 2 \cdot 7 \times 10^{-22}$ g) in a container of length $b = 10$ cm, the energy levels are given (relative to the ground state) by $(h^2/8mb^2)(n^2-1) = 0$, $6\cdot2 \times 10^{-41}$, $1\cdot6 \times 10^{-40}$, $3\cdot1 \times 10^{-40}$, ..., (for $n = 10^9$) $2\cdot1 \times 10^{-23}$, ..., (for $n = 10^{10}$) $2\cdot1 \times 10^{-21}$, ..., (for $n = 10^{11}$) $2\cdot1 \times 10^{-19}$ J, ... compared with $kT = 4\cdot1 \times 10^{-21}$ J at 25 °C. There are clearly a very large number of energy levels easily accessible to molecules with the typical energy kT, and this is reflected in the series for the partition function, again given to 3 decimal places:

$$Q_t(Br_2) = 1\cdot000 + 1\cdot000 + 1\cdot000 + 1\cdot000 + \ldots$$

$$+ (\text{for } n = 10^9)\ 0\cdot995 + \ldots + (\text{for } n = 10^{10})\ 0\cdot603 + \ldots$$

$$+ (\text{for } n = 10^{11})\ 0\cdot007 + \ldots$$

$$\approx 6\cdot2 \times 10^{10}$$

$$[\equiv (2\pi mb^2 kT/h^2)^{\frac{1}{2}} \text{ by integration}]$$

Thus when the thermal energy kT is large compared with the spacing of the energy levels, the partition function has a very large value and there is a correspondingly low fractional population of all energy levels. In the above example all the levels up to about the 10^9th have $N_i/N = 1/(6\cdot2 \times 10^{10})$ to within $\frac{1}{2}\%$, and only at around the 10^{10}th level is the number of molecules decreasing significantly. As before, all quantum states are equally probable when the availability of energy imposes no restraint, i.e. at levels for which $E_i \ll kT$. It is under these circumstances, where the appropriate degrees of freedom are highly excited, that the effects of the quantization of the energy disappear and a classical treatment of the motion in terms of a continuously variable energy becomes adequate.

A2.3 Q FOR MOLECULES RESTRICTED TO A SPECIFIED ENERGY RANGE

This aspect does not figure very prominently in the books on Statistical Mechanics but is fundamental to the RRKM theory and is therefore dealt with separately here. It is quite in order to consider the partition function for molecules restricted to a certain range of energies in the relevant degrees of freedom, but in this case only the relevant terms in (A2.1) must be included. The partition function for a molecule A in the energy range E to E' is therefore given by (A2.3). For molecules with

$$Q(A_{(E \to E')}) = \sum_{E < E_i < E'} g_i \exp(-E_i/kT) \tag{A2.3}$$

energy in a *small* range E to $E + \delta E$, the exponential terms in (A2.3) will be nearly the same for all the states of interest and can be replaced throughout the range by $\exp(-E/kT)$. The partition function then becomes (A2.4), in which the summation is now simply the total number

$$Q(A_{(E \to E + \delta E)}) = \left(\sum_{E < E_i < E + \delta E} g_i \right) \exp(-E/kT) \qquad (A2.4)$$

of quantum states with energies in the relevant range $E \to E + \delta E$. If further the energy levels are very closely spaced so that there are many levels in this small range, $\sum g_i$ can well be replaced by $N(E)\,\delta E$ where $N(E)$ is a continuous distribution function, *the density of quantum states,* or the number of quantum states per unit energy range at energy E (see also sections 4.4 and 4.11). In this case:

$$Q(A_{(E \to E + \delta E)}) = N(E) \exp(-E/kT)\,\delta E \qquad (A2.5)$$

A2.4 ENERGY AND ENTROPY

If the partition function for a system is known as a function of temperature, the standard statistical mechanical results (A2.6) and (A2.7) can be used to calculate the average thermal energy per molecule, $\langle E \rangle$,

$$\langle E \rangle = kT^2\, \mathrm{d} \ln Q / \mathrm{d}T \qquad (A2.6)$$

$$\left. \begin{aligned} S = R\frac{\mathrm{d}\,T \ln Q}{\mathrm{d}T} &= R \ln Q + RT \frac{\mathrm{d} \ln Q}{\mathrm{d}T} \\ &= R \ln Q + N_A \langle E \rangle / T \end{aligned} \right\} \qquad (A2.7)$$

and the molar entropy S of the system. The average energy $\langle E \rangle$ is given relative to the chosen ground state and is identical with the internal energy U of the system relative to the same energy zero. Equations (A2.6) and (A2.7) can also be applied to individual degrees of freedom of a system if the partition function is factorized into individual contributions as discussed in section A2.1.

A2.5 EQUILIBRIUM CONSIDERATIONS

One of the main uses of partition functions, and the use which is especially important in rate theories, is their use to calculate equilibrium constants. It is shown in the works on statistical mechanics that the equilibrium constant K for a reaction

$$A \rightleftharpoons B$$

is given by

$$K = \frac{Q(B)}{Q(A)} \exp(-\Delta E^\circ / kT)$$

where $Q(A)$ and $Q(B)$ are the partition functions for A and B respectively and ΔE° is the difference between the energy zeros of A and B used for the evaluation of the partition functions; ΔE° is closely related to the heat of the reaction.

It happens in the RRKM theory that the equilibria under consideration involve molecules for which the partition functions are calculated using the same energy zero, and hence $\Delta E^\circ = 0$ and the equilibrium constant becomes

$$K = Q(B)/Q(A)$$

In one case both partition functions refer to excited molecules with energies in the same small range E to $E + \delta E$, in which case the partition functions can be expressed as in (A2.4) and we obtain

$$K = \sum g_i(B) / \sum g_i(A)$$

where the $\sum g_i$ are the number of quantum states of A and B in the small energy range. This gives the simple result, already mentioned in section A2.2, that if there are two conditions X and Y of a system, corresponding to n_X and n_Y quantum states respectively, the molecules will distribute themselves between these conditions in the ratio n_X/n_Y provided there are no constraints on the system.

Finally, if the energy levels of A and B in the range E to $E + \delta E$ are close enough together for a continuous distribution function to be used, (A2.5) gives

$$K = N_B(E)/N_A(E)$$

where $N_B(E)$ and $N_A(E)$ are the densities of quantum states of B and A respectively at energy E.

A2.6 EXPRESSIONS FOR SPECIFIC SYSTEMS

In the numerical application of reaction rate theories, specific expressions are required for the partition functions Q, average energies $\langle E \rangle$ and entropies S of the various degrees of freedom of the species involved. Translational contributions are of no concern in unimolecular reactions, and electronic contributions, which are only occasionally significant, may be obtained by direct summation of (A2.1). The equations required most are those for vibrational and rotational degrees of freedom, and some of the commonly used expressions for Q and $\langle E \rangle/kT$ are given in Table A2.1. The corresponding results for S are obtained simply from (A2.7). The most widely used expressions are those for the simple harmonic oscillator, and the results here are conveniently expressed in terms of the dimensionless quantity $x = h\nu/kT = 1\cdot4388(\bar{\nu}/\mathrm{cm}^{-1})/(T/\mathrm{K})$. The oscillator is sometimes alternatively characterized by the quantum energy $E_q = N_A hc\bar{\nu}$,

Table A2.1 Commonly used expressions for Q and $\langle E \rangle/kT$ for vibrational and rotational degrees of freedom; the corresponding molar entropies are given by
$$S = R \ln Q + R(\langle E \rangle/kT)$$

Type of motion	Degrees of freedom	Q	$\langle E \rangle/kT$
simple harmonic oscillator	1	$[1 - \exp(-h\nu/kT)]^{-1}$	$\dfrac{(h\nu/kT)}{[\exp(h\nu/kT) - 1]}$
free internal rotation	1	$\pi^{\frac{1}{2}}(8\pi^2 kT/h^2)^{\frac{1}{2}} I^{\frac{1}{2}}$	$\tfrac{1}{2}$
overall rotation of linear molecule	2	$(8\pi^2 kT/h^2) I$	1
overall rotation of non-linear molecule[a]	3	$\pi^{\frac{1}{2}}(8\pi^2 kT/h^2)^{\frac{3}{2}} (I_A I_B I_C)^{\frac{1}{2}}$	$\tfrac{3}{2}$
independent rotations (Marcus expression)[b]	$r = \sum\limits_{i=1}^{p} d_i$	$\left\{ \prod\limits_{i=1}^{p} \Gamma(\tfrac{1}{2}d_i) \right\} (8\pi^2 kT/h^2)^{\frac{1}{2}r} \prod\limits_{i=1}^{p} I_i^{\frac{1}{2}d_i}$	$\tfrac{1}{2}r$

[a] Expression for Q applies strictly to spherical tops ($I_A = I_B = I_C$) and symmetric tops ($I_A \neq I_B = I_C$), and approximately to asymmetric tops ($I_A \neq I_B \neq I_C$).
[b] See Appendix 5 and section 5.2.

where $N_A hc = 0\cdot0028591$ kcal(thermochem) mol^{-1} cm, in which case $x = (E_q/\mathrm{kcal\,mol}^{-1})/(0\cdot0019872T/\mathrm{K})$. In many calculations (see, for example, sections 6.2.3 and 6.4) a coarse tabulation of S as a function of $\bar{\nu}$ and T is useful, and Table A2.2 presents such a tabulation. For more precise calculations this may be supplemented by a finer tabulation of S as a function of $x = h\nu/kT$, and this is given in Table A2.3. If the effects of anharmonicity or vibrational–rotational coupling are to be considered, more complicated specific formulae are required and reference must be made to other works.[2]

The expressions for molecular rotations given in Table A2.1 are all based on the classical treatment, as is clear from the $\langle E \rangle$ values which are all $\frac{1}{2}kT$ per rotational degree of freedom. Some care may be needed in the selection of the appropriate expressions for Q for a given calculation, and this problem is discussed in section 5.2. If moments of inertia are given in $g\,cm^2$ the appropriate value of $(8\pi^2 k/h^2)$ is $2\cdot4831 \times 10^{38}\,g^{-1}cm^{-2}\,K^{-1}$.

Table A2.2 Harmonic oscillator entropies (cal $K^{-1}\,mol^{-1}$) as a function of frequency and temperature

$\bar{v}/cm^{-1}$				Temperature/K				
	300	400	500	600	800	1000	1200	1500
50	4·83	5·40	5·84	6·20	6·77	7·22	7·58	8·02
75	4·03	4·60	5·04	5·40	5·97	6·41	6·77	7·22
100	3·47	4·03	4·47	4·83	5·40	5·84	6·20	6·65
125	3·03	3·59	4·03	4·39	4·96	5·40	5·76	6·20
150	2·68	3·24	3·67	4·03	4·60	5·04	5·40	5·84
200	2·14	2·68	3·11	3·47	4·03	4·47	4·83	5·27
250	1·74	2·26	2·68	3·03	3·59	4·03	4·39	4·83
300	1·43	1·93	2·34	2·68	3·24	3·67	4·03	4·47
350	1·18	1·66	2·05	2·39	2·94	3·37	3·73	4·17
400	0·97	1·43	1·81	2·14	2·68	3·11	3·47	3·90
500	0·67	1·07	1·43	1·74	2·26	2·68	3·03	3·47
600	0·46	0·80	1·13	1·43	1·93	2·34	2·68	3·11
700	0·31	0·61	0·90	1·18	1·66	2·05	2·39	2·82
800	0·21	0·46	0·72	0·97	1·43	1·81	2·14	2·56
900	0·14	0·34	0·57	0·80	1·23	1·61	1·93	2·34
1000	0·10	0·26	0·46	0·67	1·07	1·43	1·74	2·14
1200	0·04	0·14	0·29	0·46	0·80	1·13	1·43	1·81
1500	0·01	0·06	0·14	0·26	0·53	0·80	1·07	1·43
2000	0·00	0·01	0·04	0·10	0·26	0·46	0·67	0·97
2500	0·00	0·00	0·01	0·03	0·12	0·26	0·41	0·67
3000	0·00	0·00	0·00	0·01	0·06	0·14	0·26	0·46
3500	0·00	0·00	0·00	0·00	0·03	0·08	0·16	0·31

Moments of inertia are sometimes alternatively given in amu $\mathring{A}^2 = 1\cdot6604 \times 10^{-40}\,g\,cm^2$. The case of hindered internal rotation is again more complicated,[2,3] but the thermodynamic functions can be readily evaluated from tables given by Pitzer and Brewer.[2] State-counting for such systems is complicated and is generally avoided by substituting in the model a harmonic vibration or a free rotation which has the same entropy at the temperature of interest as the actual hindered rotation.

Table A2.3 Entropy (cal K^{-1} mol^{-1}) of a simple harmonic oscillator as a function of the dimensionless parameter $x = h\nu/kT$

	0·000	0·002	0·004	0·006	0·008	0·010	0·012	0·014	0·016	0·018
0·00	∞	14·337	12·959	12·154	11·582	11·139	10·776	10·470	10·205	9·971
0·02	9·761	9·572	9·399	9·240	9·093	8·956	8·827	8·707	8·593	8·486
0·04	8·384	8·287	8·195	8·106	8·022	7·941	7·863	7·788	7·715	7·646
0·06	7·578	7·513	7·450	7·389	7·330	7·272	7·216	7·162	7·109	7·057
0·08	7·007	6·958	6·910	6·863	6·818	6·773	6·729	6·687	6·645	6·604

	0·000	0·010	0·020	0·030	0·040	0·050	0·060	0·070	0·080	0·090
0·10	6·564	6·374	6·202	6·043	5·896	5·759	5·631	5·511	5·398	5·290
0·20	5·189	5·092	5·000	4·912	4·828	4·747	4·670	4·595	4·523	4·454
0·30	4·387	4·323	4·260	4·199	4·141	4·084	4·028	3·974	3·922	3·871
0·40	3·821	3·773	3·726	3·680	3·635	3·591	3·548	3·506	3·465	3·425
0·50	3·385	3·347	3·309	3·272	3·236	3·200	3·165	3·131	3·097	3·064
0·60	3·032	3·000	2·969	2·938	2·908	2·878	2·849	2·820	2·791	2·764
0·70	2·736	2·709	2·682	2·656	2·630	2·605	2·580	2·555	2·531	2·507
0·80	2·483	2·459	2·436	2·414	2·391	2·369	2·347	2·325	2·304	2·283
0·90	2·262	2·242	2·222	2·202	2·182	2·162	2·143	2·124	2·105	2·086
1·00	2·068	2·050	2·032	2·014	1·996	1·979	1·962	1·945	1·928	1·911
1·10	1·895	1·879	1·863	1·847	1·831	1·815	1·800	1·785	1·770	1·755
1·20	1·740	1·725	1·711	1·697	1·682	1·668	1·654	1·641	1·627	1·613
1·30	1·600	1·587	1·574	1·561	1·548	1·535	1·523	1·510	1·498	1·485
1·40	1·473	1·461	1·449	1·438	1·426	1·414	1·403	1·391	1·380	1·369
1·50	1·358	1·347	1·336	1·325	1·315	1·304	1·294	1·283	1·273	1·263
1·60	1·252	1·242	1·232	1·223	1·213	1·203	1·194	1·184	1·175	1·165
1·70	1·156	1·147	1·138	1·129	1·120	1·111	1·102	1·093	1·085	1·076
1·80	1·067	1·059	1·051	1·042	1·034	1·026	1·018	1·010	1·002	0·994
1·90	0·986	0·978	0·971	0·963	0·955	0·948	0·940	0·933	0·926	0·918

(cont.)

Table A2.3 (*cont.*)

	0·000	0·020	0·040	0·060	0·080	0·100	0·120	0·140	0·160	0·180
2·00	0·911	0·897	0·883	0·869	0·855	0·842	0·829	0·816	0·803	0·790
2·20	0·778	0·766	0·754	0·742	0·731	0·719	0·708	0·697	0·686	0·675
2·40	0·665	0·654	0·644	0·634	0·624	0·614	0·605	0·595	0·586	0·577
2·60	0·568	0·559	0·550	0·542	0·533	0·525	0·517	0·508	0·501	0·493
2·80	0·485	0·477	0·470	0·462	0·455	0·448	0·441	0·434	0·427	0·420
3·00	0·414	0·407	0·401	0·395	0·388	0·382	0·376	0·370	0·364	0·359
3·20	0·353	0·347	0·342	0·336	0·331	0·326	0·321	0·316	0·311	0·306
3·40	0·301	0·296	0·291	0·287	0·282	0·278	0·273	0·269	0·264	0·260
3·60	0·256	0·252	0·248	0·244	0·240	0·236	0·232	0·229	0·225	0·221
3·80	0·218	0·214	0·211	0·207	0·204	0·201	0·198	0·194	0·191	0·188
4·00	0·185	0·182	0·179	0·176	0·173	0·171	0·168	0·165	0·162	0·160
4·20	0·157	0·155	0·152	0·150	0·147	0·145	0·142	0·140	0·138	0·135
4·40	0·133	0·131	0·129	0·127	0·125	0·123	0·121	0·119	0·117	0·115
4·60	0·113	0·111	0·109	0·107	0·106	0·104	0·102	0·100	0·099	0·097

	0·000	0·050	0·100	0·150	0·200	0·250	0·300	0·350	0·400	0·450
5·00	0·081	0·078	0·074	0·071	0·068	0·065	0·063	0·060	0·058	0·055
5·50	0·053	0·051	0·049	0·047	0·045	0·043	0·041	0·039	0·038	0·036
6·00	0·035	0·033	0·032	0·030	0·029	0·028	0·027	0·026	0·024	0·023
6·50	0·022	0·021	0·021	0·020	0·019	0·018	0·017	0·017	0·016	0·015
7·00	0·015	0·014	0·013	0·013	0·012	0·012	0·011	0·011	0·010	0·010
7·50	0·009	0·009	0·009	0·008	0·008	0·007	0·007	0·007	0·007	0·006
8·00	0·006	0·006	0·005	0·005	0·005	0·005	0·005	0·004	0·004	0·004
8·50	0·004	0·004	0·004	0·003	0·003	0·003	0·003	0·003	0·003	0·003
9·00	0·002	0·002	0·002	0·002	0·002	0·002	0·002	0·002	0·002	0·002
9·50	0·002	0·001	0·001	0·001	0·001	0·001	0·001	0·001	0·001	0·001
10·00	0·001	0·001	0·001	0·001	0·001	0·001	0·001	0·001	0·001	0·001
10·50	0·001	0·001	0·001	0·001	0·001	0·001	0·000	0·000	0·000	0·000

A2.7 PHASE-SPACE AND CLASSICAL STATISTICAL MECHANICS

The concept of phase-space is not used extensively in this book, and only a brief indication of some salient features will be given. A system of N particles behaving classically can be completely described by $3N$ position coordinates together with $3N$ momenta, i.e. by the positions and momenta of each particle in three independent directions. The state of the whole system can thus be specified by the position of a point (the *representative point*) in a $6N$-dimensional space known as the *phase-space* of the system. Any change in the position or velocity of one or more of the particles is described by a movement of the representative point in the phase-space. Since changes in the system are subject to certain requirements such as the conservation of total momentum, the representative point can only move on certain defined lines in phase-space, and these paths are known as *trajectories*. The concept is somewhat similar to the well-known mechanical analogy[4] of a ball rolling on a potential energy surface; once the ball is given a certain initial position and velocity its subsequent path is pre-determined. For a system capable of chemical reaction some regions of the phase-space will be recognizable as reactants and some as products; these regions may be separated by a $(6N-1)$-dimensional hypersurface, and reaction corresponds to the crossing of this hypersurface by the representative point.[5] Crossings will occur most frequently at the lowest energy regions of the hypersurface, and these critical molecular configurations correspond loosely to the activated complexes of Absolute Rate Theory.

The application of Statistical Mechanics to classical systems depends on the replacement of summations over discrete energy levels by integrations over regions of phase-space. For example, the classical partition function is given by (A2.8) instead of (A2.1). In (A2.8) $x_1, ..., x_{3N}$ are the $3N$ position

$$Q_{\text{class}} = \frac{1}{h^{3N}} \int\int \cdots \int\int \exp\left(-E/kT\right) dx_1 dx_2 \ldots dx_{3N} dp_1 \ldots dp_{3N} \quad \text{(A2.8)}$$

coordinates and $p_1, ..., p_{3N}$ the $3N$ momenta. The energy E is expressed as a function of these $6N$ variables, and the integration limits are such that the integration covers all relevant regions of phase-space. The factor $(1/h^{3N})$ has been introduced so that for comparable systems (A2.8) gives the same result as (A2.1), although the multiplication of Q by a constant factor makes no difference for many purposes, including rate theory applications. The classical treatment can obviously be derived as a limiting case of the quantum version for cases where energy differences become small compared with kT, and treatments of this sort have been used in Appendices 4 and 6.

13

References

1. See, for example, R. C. Tolman, *Foundations of Statistical Mechanics*, Oxford University Press, 1938; R. H. Fowler and E. A. Guggenheim, *Statistical Thermodynamics*, Cambridge University Press, 1939; N. Davidson, *Statistical Mechanics*, McGraw–Hill, New York, 1962; G. S. Rushbrooke, *Introduction to Statistical Mechanics*, Oxford University Press, 1949.
2. See, for example, G. N. Lewis and M. Randall, revised by K. S. Pitzer and L. Brewer, *Thermodynamics*, 2nd edn, McGraw–Hill, New York, 1961, Chap. 27.
3. G. Herzberg, *Molecular Spectra and Molecular Structure*, Vol. 2, *Infrared and Raman Spectra of Polyatomic Molecules*, van Nostrand, Princeton, 1945, sections IV, 5 and V, 1.
4. H. Eyring and M. Polanyi, *Zeit. Phys. Chem.* (*B*), **12**, 279, (1931).
5. D. L. Bunker, *Theory of Elementary Gas Reaction Rates*, Pergamon, Oxford, 1966, p. 52.

Appendix 3
Computer Programs for the Direct Count of Vibrational Quantum States

It is not profitable to list a complete program for the direct count procedure, since the programming is on the whole straightforward and programs can be obtained on request from various reliable sources (see Chapter 5, ref. 16). The small section which actually performs the counting is somewhat intricate, however, and is exemplified in Figures A3.1 and A3.2 by the relevant part of a program written in 'Atlas Autocode', together with the corresponding flow diagram. The programming language is largely self-explanatory, and conversion to other languages is not difficult. The use of zeroth array elements [e.g. $e(0)$] must be avoided in a FORTRAN translation, e.g. by increasing all indices by 1.

LINE NO.

101	$i = 0;$ $e(0) = 0;$ $d'(0) = 1$
102	$10: \quad i = i+1;$ $n(i) = -1;$ $e(i) = e(i-1) - v(i)$
103	$11: \quad n(i) = n(i)+1;$ $ensum = e(i) + v(i)$
104	$level = \text{intpt}(ensum + 1)$
105	$- > 12 \text{ IF } level > maxe$
106	$e(i) = ensum;$ $d'(i) = d'(i-1) * d(i, n(i))$
107	$- > 10 \text{ IF } i < nv$
108	$states(level) = states(level) + d'(nv);$ $- > 11$
109	$12: \quad - > 14 \text{ IF } i = 1$
110	$i = i - 1;$ $- > 11$
111	$14:$ $\begin{bmatrix} \text{Instructions to print out the array } states(i) \\ \text{and the sum of states up to each energy} \\ \text{required} \end{bmatrix}$

nv = No. of oscillator groups
i = serial No. of oscillator group under consideration ($i = 1$ to nv)
$v(i)$ = No. of energy units corresponding to 1 quantum in the ith group
$d(i, j)$ = degeneracy of ith group when containing j quanta (section 5.3.1)
$states(level)$ = No. of quantum states counted in the $level$th energy range ($level = 0$ to $maxe$); set to zero before counting starts
$maxe$ = serial No. of highest energy range of interest
$n(i)$ = No. of quanta in ith group
$e(i)$ = total No. of energy units in groups 1 to i with the current set of $n(i)$
$ensum$ = temporary storage of $e(i)$
$d'(i)$ = combined degeneracy in groups 1 to i with the current set of $n(i)$
$level$ = serial No. of the energy range in which energy $ensum$ falls

Figure A3.1 Part of an Atlas Autocode program for the direct count of vibrational states. The quantities i, nv, $level$, $maxe$ and $n(i)$ are integers and the remainder are real variables

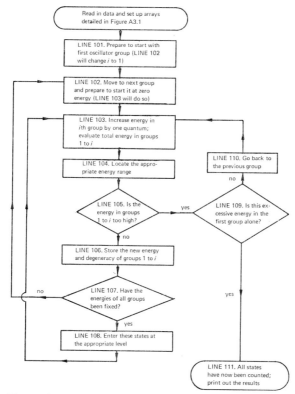

Figure A3.2 Flow diagram for the direct count of vibrational states (only the actual counting is described in detail)

Appendix 4
Classical Approximation to $W(E_r)$, the Sum of Rotational States

The classical sum of rotational states is given (section 5.2.2) by the multiple integral

$$W(E_r) \equiv \sum_{E_r=0}^{E_r} P(E_r)$$

$$= \int_{J_p=0}^{\cdot J_p'} \int \cdots \int \int_{J_1=0}^{\cdot J_1'} \left(\prod_{i=1}^{p} 2 J_i^{d_i-1} \right) dJ_1 \ldots dJ_p$$

where

$$J_n' = \left(\frac{8\pi^2 I_n}{h^2} \right)^{\frac{1}{2}} \left(E_r - \sum_{i=n+1}^{p} \frac{h^2 J_i^2}{8\pi^2 I_i} \right)^{\frac{1}{2}} \quad (n \neq p) \tag{A4.1}$$

and

$$J_p' = \left(\frac{8\pi^2 I_p}{h^2} \right)^{\frac{1}{2}} E_r^{\frac{1}{2}}$$

We give three methods for performing this integration, and these vary considerably in the amount of mathematical background required. The first and shortest is a simple application of a general result known as Dirichlet's integral. The second is a more explicit method involving use of the beta and gamma functions, while the third and longest avoids the beta function and should be within the grasp of anyone who is reasonably familiar with integral calculus. Another approach, not elaborated here, regards an integral of this type as the volume of a multidimensional ellipsoid in phase-space.[1] Some general observations on the gamma function $\Gamma(n)$ are made in Appendix 7.

A4.1   DERIVATION FROM DIRICHLET'S INTEGRAL

Dirichlet's result, cast in the notation of the present application, refers to the multiple integral

$$\int\int \cdots \int\int \left(\prod_{i=1}^{p} J_i^{d_i-1}\right) dJ_1\, dJ_2 \ldots dJ_p$$

in which all the J_i and d_i are positive and the limits of the integrations are determined by the condition

$$\sum_{i=1}^{p} (J_i/a_i)^{b_i} \leqslant E_r$$

It can be shown[2] that the result of this integration is

$$\frac{E_r^{t}}{\Gamma(1+t)} \prod_{i=1}^{p} \left[\frac{a_i^{d_i}\,\Gamma(d_i/b_i)}{b_i}\right]$$

where

$$t = \sum_{i=1}^{p} (d_i/b_i)$$

The integral under discussion in this appendix is clearly 2^p times a special case of Dirichlet's integral in which $a_i = (8\pi^2 I_i/h^2)^{\frac{1}{2}}$ and $b_i = 2$ for all i, and is therefore given by (A4.2).

$$W(E_r) = \frac{E_r^{\frac{1}{2}r}}{\Gamma(1+\frac{1}{2}r)} \prod_{i=1}^{p} \left[\left(\frac{8\pi^2 I_i}{h^2}\right)^{\frac{1}{2}d_i} \Gamma(\tfrac{1}{2}d_i)\right] \tag{A4.2}$$

where

$$r = \sum_{i=1}^{p} d_i$$

A more compact form of this result may be obtained by using the equation derived in Appendix 5 for the rotational partition function Q_r of the system of classical rotors under discussion; it readily follows that

$$W(E_r) = \frac{Q_r}{\Gamma(1+\frac{1}{2}r)} \left(\frac{E_r}{kT}\right)^{\frac{1}{2}r}$$

A4.2   AN ALTERNATIVE PROOF

The above derivation suffers from the disadvantage that Dirichlet's result must be taken on trust, and this is avoided at the expense of considerable complexity in the following methods. We proceed essentially in three stages. Firstly, the innermost integral $I_1 = \int 2J_1{}^{d_1-1} \, dJ_1$ is evaluated and cast as a particular case of the general expression. It is then shown that if this form is valid for the nth integral I_n it must also be valid for the next integral I_{n+1}. Finally a slight extension is made to give the final result of the last (outside) integral.

A4.2.1 Evaluation of I_1

$$I_1 = \int_{J_1=0}^{J_1'} 2J_1{}^{d_1-1} \, dJ_1 \quad \text{where} \quad J_1' = \left(\frac{8\pi^2 I_1}{h^2}\right)^{\frac{1}{2}} \left(E_r - \sum_{i=2}^{p} \frac{h^2 J_i{}^2}{8\pi^2 I_i}\right)^{\frac{1}{2}}$$

$$= \frac{2}{d_1}\left[J_1{}^{d_1}\right]_{J_1=0}^{J_1'} = \frac{2}{d_1}(J_1')^{d_1}$$

$$= \left(E_r - \sum_{i=2}^{p} \frac{h^2 J_i{}^2}{8\pi^2 I_i}\right)^{\frac{1}{2}d_1}\left(\frac{8\pi^2 I_1}{h^2}\right)^{\frac{1}{2}d_1}\left(\frac{2}{d_1}\right)$$

This expression is now cast as a particular case ($n = 1$) of the general result to be derived, bearing in mind some properties of the gamma function (Appendix 7).

$$I_n = \left(E_r - \sum_{i=n+1}^{p} \frac{h^2 J_i{}^2}{8\pi^2 I_i}\right)^{\frac{1}{2}r_n} \prod_{i=1}^{n}\left(\frac{8\pi^2 I_i}{h^2}\right)^{\frac{1}{2}d_i} \prod_{i=1}^{n} \Gamma(\tfrac{1}{2}d_i) \Big/ \Gamma(1+\tfrac{1}{2}r_n) \quad \text{(A4.3)}$$

where

$$r_n = \sum_{i=1}^{n} d_i \quad \text{(A4.4)}$$

and therefore $r_1 = d_1$ and $\Gamma(1+\tfrac{1}{2}r_1) = \tfrac{1}{2}d_1 \Gamma(\tfrac{1}{2}d_1)$.

A4.2.2 Evaluation of I_{n+1} from I_n

If the form (A4.3) is valid for I_n, then I_{n+1} is given by

$$I_{n+1} = \int_{J_{n+1}=0}^{J_{n+1}'} 2(J_{n+1})^{d_{n+1}-1} I_n \, dJ_{n+1}$$

where

$$J_{n+1}' = \left(\frac{8\pi^2 I_{n+1}}{h^2}\right)^{\frac{1}{2}} \left(E_r - \sum_{i=n+2}^{p} \frac{h^2 J_i{}^2}{8\pi^2 I_i}\right)^{\frac{1}{2}}$$

Now for present purposes I_n is of the following form, with the constant as shown in (A4.3):

$$\mathbf{I}_n = \text{constant}\left(E_r - \sum_{i=n+2}^{p} \frac{h^2 J_i^2}{8\pi^2 I_i} - \frac{h^2 J_{n+1}^2}{8\pi^2 I_{n+1}}\right)^{\frac{1}{2}r_n}$$

$$= \text{constant}\left[\frac{h^2}{8\pi^2 I_{n+1}}(J_{n+1}'^2 - J_{n+1}^2)\right]^{\frac{1}{2}r_n}$$

[using (A4.1)], and therefore $\mathbf{I}_{n+1}$ is given by

$$\mathbf{I}_{n+1} = \prod_{i=1}^{n}\left(\frac{8\pi^2 I_i}{h^2}\right)^{\frac{1}{2}d_i} \frac{\prod_{i=1}^{n}\Gamma(\tfrac{1}{2}d_i)}{\Gamma(1+\tfrac{1}{2}r_n)}\left(\frac{h^2}{8\pi^2 I_{n+1}}\right)^{\frac{1}{2}r_n}\mathbf{I}' \qquad (A4.5)$$

where

$$\mathbf{I}' = \int_{J_{n+1}=0}^{J_{n+1}'}(J_{n+1}'^2 - J_{n+1}^2)^{\frac{1}{2}r_n}\,2(J_{n+1})^{d_{n+1}-1}\,\mathrm{d}J_{n+1} \qquad (A4.6)$$

The integral $\mathbf{I}'$ can be evaluated by a change of variable. Let

$$Y = (J_{n+1}/J_{n+1}')^2 \quad \text{and} \quad \delta Y = 2J_{n+1}\,\delta J_{n+1}/J_{n+1}'^2$$

Then

$$\mathbf{I}' = (J_{n+1}')^{r_n+d_{n+1}}\int_{Y=0}^{1}(1-Y)^{\frac{1}{2}r_n}\,Y^{\frac{1}{2}d_{n+1}-1}\,\mathrm{d}Y$$

The integral is now of the type which defines the beta function

$$B(m,n) = \int_{x=0}^{1} x^{m-1}(1-x)^{n-1}\,\mathrm{d}x$$

and it is a well-known theorem[3] that

$$B(m,n) = \Gamma(m)\,\Gamma(n)/\Gamma(m+n)$$

Since $r_n + d_{n+1} = r_{n+1}$ [from (A4.4)], we therefore have

$$\mathbf{I}' = (J_{n+1}')^{r_{n+1}}\,\Gamma(\tfrac{1}{2}d_{n+1})\,\Gamma(1+\tfrac{1}{2}r_n)/\Gamma(1+\tfrac{1}{2}r_{n+1}) \qquad (A4.7)$$

Insertion of this expression into (A4.5), followed by some rearrangement and insertion of the expression analogous to (A4.1) for J_{n+1}', leads to the result

$$\mathbf{I}_{n+1} = \left(E_r - \sum_{i=n+2}^{p}\frac{h^2 J_i^2}{8\pi^2 I_i}\right)^{\frac{1}{2}r_{n+1}}\prod_{i=1}^{n+1}\left(\frac{8\pi^2 I_i}{h^2}\right)^{\frac{1}{2}d_i}\prod_{i=1}^{n+1}\Gamma(\tfrac{1}{2}d_i)\bigg/\Gamma(1+\tfrac{1}{2}r_{n+1})$$

This is of the same form as was assumed for $\mathbf{I}_n$, and since it was shown in the previous section to be valid for $\mathbf{I}_1$ it must also be correct for $\mathbf{I}_2$, and therefore for $\mathbf{I}_3, \mathbf{I}_4$, etc. It only remains to formulate correctly the last integration from $\mathbf{I}_{p-1}$ to $\mathbf{I}_p$.

A4.2.3 Final result

The last integral is very similar to the others, but the upper limit J'_p is simply $(8\pi^2 I_p/h^2)^{\frac{1}{2}} E_r^{\frac{1}{2}}$, without the meaningless term

$$\sum_{i=p+1}^{p} h^2 J_i^2/8\pi^2 I_i$$

The integration follows, and the final result, slightly rearranged for clarity, is

$$W(E_r) = \frac{E_r^{\frac{1}{2}r}}{\Gamma(1+\frac{1}{2}r)} \prod_{i=1}^{p} \left[\left(\frac{8\pi^2 I_i}{h^2}\right)^{\frac{1}{2}d_i} \Gamma(\tfrac{1}{2}d_i) \right] \quad \text{where} \quad r = \sum_{i=1}^{p} d_i$$

in agreement with (A4.2).

A4.3 DERIVATION AVOIDING USE OF THE BETA FUNCTION

As a final stage in the avoidance of esoteric mathematics we evaluate the integral $\mathbf{I}'$ (A4.6) by a method which does not use the beta function. We consider separately the cases where d_{n+1} is 2 or 1.

A4.3.1 Case where $d_{n+1} = 2$

In this case the integration (A4.6) is fairly simple:

$$\mathbf{I}' = -\int_{J_{n+1}=0}^{J'_{n+1}} (J'^2_{n+1} - J^2_{n+1})^{\frac{1}{2}r_n} \, d(J'^2_{n+1} - J^2_{n+1})$$

$$= -\int_{x=J'^2_{n+1}}^{0} x^{\frac{1}{2}r_n} \, dx = \left[\frac{x^{(1+\frac{1}{2}r_n)}}{1+\frac{1}{2}r_n}\right]_{x=0}^{J'^2_{n+1}}$$

$$= (J'_{n+1})^{(r_n+2)} \cdot \frac{2}{r_n+2} \tag{A4.8}$$

This may be cast in the desired form by noting that

$$\frac{\Gamma(1+\frac{1}{2}r_n)}{\Gamma(1+\frac{1}{2}r_{n+1})} = \frac{\Gamma(1+\frac{1}{2}r_n)}{\frac{1}{2}r_{n+1}\Gamma(\frac{1}{2}r_{n+1})} = \frac{2}{r_n+2}$$

since $d_{n+1} = 2$ and therefore $r_{n+1} = r_n + 2$. In addition $\Gamma(\tfrac{1}{2}d_{n+1}) = 1$, and (A4.8) is therefore equivalent to (A4.9), identical with the form derived in A4.2.2:

$$\mathbf{I}' = (J'_{n+1})^{r_{n+1}} \frac{\Gamma(\tfrac{1}{2}d_{n+1})\,\Gamma(1+\frac{1}{2}r_n)}{\Gamma(1+\frac{1}{2}r_{n+1})} \tag{A4.9}$$

A4.3.2   Case where $d_{n+1} = 1$

In this case the integral can be evaluated by means of the substitution $\theta = \sin^{-1}(J_{n+1}/J'_{n+1})$, whereupon $\mathbf{I}'$ becomes

$$\mathbf{I}' = 2(J'_{n+1})^{(1+r_n)} \int_{\theta=0}^{\pi/2} (\cos\theta)^{1+r_n}\,d\theta$$

$$= 2(J'_{n+1})^{r_{n+1}} \int_{\theta=0}^{\pi/2} (\cos\theta)^{r_{n+1}}\,d\theta \qquad (A4.10)$$

Since r_{n+1} is integral and positive we can apply the standard reduction formula:

$$\int_0^{\pi/2} \cos^m\theta\,d\theta = \frac{m-1}{m}\int_0^{\pi/2}\cos^{m-2}\theta\,d\theta \quad (m\geq 2)$$

and this reduces (A4.10) to a form involving one of the readily evaluated integrals

$$\int_0^{\pi/2}\cos\theta\,d\theta = 1 \quad \text{or} \quad \int_0^{\pi/2}d\theta = \frac{\pi}{2}$$

In this way it is easily shown that when r_{n+1} is odd:

$$\mathbf{I}' = 2(J'_{n+1})^{r_{n+1}}\left(\frac{r_{n+1}-1}{r_{n+1}}\cdot\frac{r_{n+1}-3}{r_{n+1}-2}\cdots\cdots\frac{4}{5}\cdot\frac{2}{3}\right).1$$

$$= 2(J'_{n+1})^{r_{n+1}}\left(\frac{\frac12 r_{n+1}-\frac12}{\frac12 r_{n+1}}\cdot\frac{\frac12 r_{n+1}-\frac32}{\frac12 r_{n+1}-1}\cdots\cdots\frac{2}{5/2}\cdot\frac{1}{3/2}\right)\left(\frac{1}{\frac12\Gamma(\frac12)}\cdot\frac{\Gamma(\frac12)}{2}\right)$$

(where the last expression has replaced 1 in order to complete the series in the denominator, cf. Appendix 7)

$$= 2(J'_{n+1})^{r_{n+1}}\frac{\Gamma(\frac12 r_{n+1}+\frac12)}{\Gamma(\frac12 r_{n+1}+1)}\cdot\frac{\Gamma(\frac12)}{2}$$

or

$$\mathbf{I}' = (J'_{n+1})^{r_{n+1}}\frac{\Gamma(1+\frac12 r_n)\,\Gamma(\frac12 d_{n+1})}{\Gamma(1+\frac12 r_{n+1})} \qquad (A4.11)$$

Similarly, when r_{n+1} is even:

$$\mathbf{I}' = 2(J'_{n+1})^{r_{n+1}}\left(\frac{r_{n+1}-1}{r_{n+1}}\cdot\frac{r_{n+1}-3}{r_{n+1}-2}\cdots\cdots\frac{5}{6}\cdot\frac{3}{4}\cdot\frac{1}{2}\right)\frac{\pi}{2}$$

$$= 2(J'_{n+1})^{r_{n+1}}\left(\frac{\frac12 r_{n+1}-\frac12}{\frac12 r_{n+1}}\cdot\frac{\frac12 r_{n+1}-\frac32}{\frac12 r_{n+1}-1}\cdots\cdots\frac{5/2}{3}\cdot\frac{3/2}{2}\cdot\frac{1/2}{1}\right)\left(\Gamma(\tfrac12)\cdot\frac{\Gamma(\frac12)}{2}\right)$$

$$= 2(J'_{n+1})^{r_{n+1}}\frac{\Gamma(\frac12 r_{n+1}+\frac12)}{\Gamma(1+\frac12 r_{n+1})}\cdot\frac{\Gamma(\frac12)}{2}$$

or

$$\mathbf{I}' = (J'_{n+1})^{r_{n+1}} \frac{\Gamma(1 + \frac{1}{2}r_n)\,\Gamma(\frac{1}{2}d_{n+1})}{\Gamma(1 + \frac{1}{2}r_{n+1})} \tag{A4.12}$$

The two results (A4.11) and (A4.12) are identical with each other and with the equation (A4.7) derived by using the beta function.

References

1. See, for example, O. K. Rice, *Statistical Mechanics Thermodynamics and Kinetics*, Freeman, San Francisco, 1967, p. 570; R. H. Fowler, *Statistical Mechanics*, Cambridge University Press, 2nd edn, 1936, Chap. 2.
2. H. Jeffreys and B. S. Jeffreys, *Methods of Mathematical Physics*, Cambridge University Press, 3rd edn, 1962, p. 468.
3. Ref. 2, p. 463.

Appendix 5
Partition Function of a System of Classical Rotors

For a system of classical rotors the partition function may be written

$$Q_{\mathrm{r}} = \int_{E_{\mathrm{r}}=0}^{\infty} N(E_{\mathrm{r}}) \exp(-E_{\mathrm{r}}/kT) \, dE_{\mathrm{r}}$$

This is equivalent to (A2.1) with the summation replaced by an integration over a continuous range of energy and the degeneracy g_i replaced by $N(E_{\mathrm{r}}) \, dE_{\mathrm{r}}$. The state density $N(E_{\mathrm{r}})$ is obtained from the sum of states $W(E_{\mathrm{r}})$ (A4.2) by differentiation:

$$N(E_{\mathrm{r}}) = \frac{d}{dE_{\mathrm{r}}} W(E_{\mathrm{r}})$$

whence, with some rearrangement,

$$Q_{\mathrm{r}} = \frac{(\tfrac{1}{2}r)}{\Gamma(1+\tfrac{1}{2}r)} \prod_{i=1}^{p} \left[\left(\frac{8\pi^2 I_i kT}{h^2}\right)^{\tfrac{1}{2}d_i} \Gamma(\tfrac{1}{2}d_i) \right] \int_{x=0}^{\infty} x^{\tfrac{1}{2}r-1} \exp(-x) \, dx$$

where $x = E_{\mathrm{r}}/kT$. The integral in this expression is the defining function for $\Gamma(\tfrac{1}{2}r)$ (see Appendix 7), and since $\tfrac{1}{2}r \, \Gamma(\tfrac{1}{2}r) = \Gamma(1+\tfrac{1}{2}r)$ it follows that

$$Q_{\mathrm{r}} = \prod_{i=1}^{p} \left[\left(\frac{8\pi^2 I_i kT}{h^2}\right)^{\tfrac{1}{2}d_i} \Gamma(\tfrac{1}{2}d_i) \right]$$

Appendix 6
Classical Approximation
to $W(E_v)$, the Sum of
Vibrational States

The classical sum of vibrational states is given (section 5.4.1) by the multiple integral

$$W(E_v) \equiv \sum_{E_v=0}^{E_v} P(E_v)$$

$$= \int_{v_s=0}^{v_s'} \int \cdots \int \int_{v_1=0}^{v_1'} dv_1 \, dv_2 \ldots dv_s$$

where

$$v_n' = \frac{1}{hv_n} \left(E_v - \sum_{i=n+1}^{s} v_i \, hv_i \right) \quad (n \neq s) \tag{A6.1}$$

and

$$v_s' = E_v / hv_s$$

This integral is similar to that evaluated in Appendix 4 and may also be evaluated by the use of Dirichlet's integral (see section A4.1), in which we replace J_i and p by v_i and s respectively, and note that $a_i = 1/hv_i$, all $d_i = 1$ and all $b_i = 1$. It follows immediately that

$$W(E_v) = \frac{E_v^s}{\Gamma(1+s)} \prod_{i=1}^{s} \frac{(1/hv_i) \, \Gamma(1)}{1} = \frac{E_v^s}{s! \prod_{i=1}^{s} hv_i} \tag{A6.2}$$

Alternatively a stepwise derivation similar to that of section A4.2 may be used. The inner integral $\mathbf{I}_1$ is given by

$$\mathbf{I}_1 = \int_{v_1=0}^{v_1'} dv_1 = v_1'$$

which may be cast in the required general form (A6.3) with $n = 1$.

$$\mathbf{I}_n = \frac{1}{n! \prod\limits_{i=1}^{n} h\nu_i} \left(E_\mathrm{v} - \sum_{i=n+1}^{s} v_i\, h\nu_i \right)^n \qquad (A6.3)$$

We next show that if this form is valid for $\mathbf{I}_n$ it is also valid for $\mathbf{I}_{n+1}$:

$$\mathbf{I}_{n+1} = \int_{v_{n+1}=0}^{v'_{n+1}} \mathbf{I}_n\, dv_{n+1}$$

$$= \frac{1}{n! \prod\limits_{i=1}^{n} h\nu_i} \int_{v_{n+1}=0}^{v'_{n+1}} \left(E_\mathrm{v} - \sum_{i=n+2}^{s} v_i\, h\nu_i - v_{n+1} h\nu_{n+1} \right)^n dv_{n+1}$$

$$= \frac{1}{n! \prod\limits_{i=1}^{n} h\nu_i} (h\nu_{n+1})^n \int_{v_{n+1}=0}^{v'_{n+1}} (v'_{n+1} - v_{n+1})^n\, dv_{n+1}$$

[using (A6.1)]. The integral here is of the form

$$\int_{x=0}^{a} (a-x)^n\, dx = a^{n+1}/(n+1),$$

whence [again using (A6.1)]

$$\mathbf{I}_{n+1} = \frac{1}{n! \prod\limits_{i=1}^{n} h\nu_i} (h\nu_{n+1})^n \frac{1}{h\nu_{n+1}} \left(E_\mathrm{v} - \sum_{i=n+2}^{s} v_i\, h\nu_i \right)^{n+1} \bigg/ (n+1)$$

$$= \frac{1}{(n+1)! \prod\limits_{i=1}^{n+1} h\nu_i} \left(E_\mathrm{v} - \sum_{i=n+2}^{s} v_i\, h\nu_i \right)^{n+1}$$

which is the same as (A6.3) with n replaced by $n+1$. Thus if (A6.3) is valid for $n = 1$, which has been shown, it is also valid for $n = 2$ and therefore for $n = 3, 4$, etc. In the last (outside) integration the upper limit v'_s is simply $E_\mathrm{v}/h\nu_s$, so that the final result is

$$W(E_\mathrm{v}) = \frac{E_\mathrm{v}^s}{s! \prod\limits_{i=1}^{s} h\nu_i}$$

in agreement with (A6.2).

Appendix 7
The Gamma Function $\Gamma(n)$

The gamma function $\Gamma(n)$ is defined, for $n > 0$, by the integral:

$$\Gamma(n) = \int_{x=0}^{\infty} x^{n-1} \exp(-x)\,dx \quad (n > 0)$$

The function has many interesting and useful properties, but in the present work its argument n is restricted to positive integral or half-integral values, and only three points need to be noted:

$$\Gamma(n+1) = n\Gamma(n)$$
$$\Gamma(1) = 1$$
$$\Gamma(\tfrac{1}{2}) = \sqrt{\pi}$$

Thus, if n is integral we have:

$$\Gamma(n) = (n-1)\,\Gamma(n-1) = (n-1)(n-2)\,\Gamma(n-2) = \ldots\ldots$$
$$= (n-1)(n-2)\ldots3.2.1.\Gamma(1) = (n-1)!$$

If n is half-integral:

$$\Gamma(n) = (n-1)(n-2)\ldots \tfrac{5}{2}.\tfrac{3}{2}.\tfrac{1}{2}.\Gamma(\tfrac{1}{2})$$

e.g.

$$\Gamma(\tfrac{7}{2}) = \tfrac{5}{2}.\tfrac{3}{2}.\tfrac{1}{2}.\sqrt{\pi}$$

Table A7.1 gives values of $\Gamma(n)$ for small values of n. Values for larger n will generally be evaluated as required in computer programs, or may occur as ratios of gamma functions which are better simplified by cancellation of corresponding terms.

Table A7.1 Gamma functions $\Gamma(n)$ for integral and half-integral n

n	$\Gamma(n)$	n	$\Gamma(n)$
0·5	$\sqrt{\pi} = 1\cdot7725$	3·5	$\tfrac{15}{8}\sqrt{\pi} = 3\cdot3234$
1·0	$0! = 1\cdot0000$	4·0	$3! = 6\cdot0000$
1·5	$\tfrac{1}{2}\sqrt{\pi} = 0\cdot8862$	4·5	$\tfrac{105}{16}\sqrt{\pi} = 11\cdot6317$
2·0	$1! = 1\cdot0000$	5·0	$4! = 24\cdot0000$
2·5	$\tfrac{3}{4}\sqrt{\pi} = 1\cdot3293$	5·5	$\tfrac{945}{32}\sqrt{\pi} = 52\cdot3428$
3·0	$2! = 2\cdot0000$	6·0	$5! = 120\cdot0000$

Index